T0281302

Introductory Medical Statistics
3rd edition

Other books in the series

The Physics and Radiobiology of Fast Neutron Beams
D K Bewley

Biomedical Magnetic Resonance Technology
C-N Chen and D I Hoult

Rehabilitation Engineering Applied to Mobility and Manipulation
R A Cooper

Linear Accelerators for Radiation Therapy, second edition
D Greene and P C Williams

Health Effects of Exposure to Low-Level Ionizing Radiation
W R Hendee and F M Edwards

Radiation Protection in Hospitals
R F Mould

RPL Dosimetry – Radiophotoluminescence in Health Physics
J A Perry

The Physics of Medical Imaging
S Webb

The Physics of Three-Dimensional Radiation Therapy:
Conformal Radiotherapy, Radiosurgery and Treatment Planning
S Webb

The Physics of Conformal Radiotherapy: Advances in Technology
S Webb

The Design of Pulse Oximeters
S Webb

Other titles by the same author

Mould's Medical Anecdotes: Omnibus Edition

A Century of X-Rays and Radioactivity in Medicine

Introductory Medical Statistics

3rd edition

Richard F Mould

CRC Press
Taylor & Francis Group
Boca Raton London New York

CRC Press is an imprint of the
Taylor & Francis Group, an **informa** business

First edition 1976
Second edition 1989

First published 1998 by IOP Publishing Ltd

Published 2019 by CRC Press
Taylor & Francis Group
6000 Broken Sound Parkway NW, Suite 300
Boca Raton, FL 33487-2742

ISBN 13: 978-0-367-45580-4 (pbk)
ISBN 13: 978-0-7503-0513-6 (hbk)

Visit the Taylor & Francis Web site at
http://www.taylorandfrancis.com

and the CRC Press Web site at
http://www.crcpress.com

British Library Cataloguing-in-Publication Data

A catalogue record for this book is available from the British Library.

Library of Congress Cataloging-in-Publication Data are available

Series Editors:
  R F Mould, Croydon, UK
  C G Orton, Karamanos Cancer Institute, Detroit, USA
  J A E Spaan, University of Amsterdam, The Netherlands
  J G Webster, University of Wisconsin-Madison, USA

Typeset using the IOP Bookmaker macros, from the author's Microsoft Word 6 files

The Medical Science Series is the official book series of the International Federation for Medical and Biological Engineering (IFMBE) and the International Organization for Medical Physics (IOMP).

IFMBE

The IFMBE was established in 1959 to provide medical and biological engineering with an international presence. The Federation has a long history of encouraging and promoting international cooperation and collaboration in the use of technology for improving the health and life quality of man.

The IFMBE is an organization that is mostly an affiliation of national societies. Transnational organizations can also obtain membership. At present there are 42 national members, and one transnational member with a total membership in excess of 15 000. An observer category is provided to give personal status to groups or organizations considering formal affiliation.

Objectives

• To reflect the interests and initiatives of the affiliated organizations.

• To generate and disseminate information of interest to the medical and biological engineering community and international organizations.

• To provide an international forum for the exchange of ideas and concepts.

• To encourage and foster research and application of medical and biological engineering knowledge and techniques in support of life quality and cost-effective health care.

• To stimulate international cooperation and collaboration on medical and biological engineering matters.

• To encourage educational programmes which develop scientific and technical expertise in medical and biological engineering.

Activities

The IFMBE has published the journal *Medical and Biological Engineering and Computing* for over 34 years. A new journal *Cellular Engineering* was established in 1996 in order to stimulate this emerging field in biomedical engineering. In *IFMBE News* members are kept informed of the developments in the Federation. *Clinical Engineering Update* is a publication of our division of Clinical Engineering. The Federation also has a division for Technology Assessment in Health Care.

Every three years, the IFMBE holds a World Congress on Medical Physics and Biomedical Engineering, organized in cooperation with the IOMP and the IUPESM. In addition, annual, milestone, regional conferences are organized in different regions of the world, such as the Asia Pacific, Baltic, Mediterranean, African and South American regions.

The administrative council of the IFMBE meets once or twice a year and is the steering body for the IFMBE. The council is subject to the rulings of the General Assembly which meets every three years.

For further information on the activities of the IFMBE, please contact Jos A E Spaan, Professor of Medical Physics, Academic Medical Centre, University of Amsterdam, PO Box 22660, Meibergdreef 9, 1105 AZ, Amsterdam, The Netherlands. Tel: 31 (0) 20 566 5200. Fax: 31 (0) 20 6917233. E-mail: IFMBE@amc.uva.nl. WWW: http://vub.vub.ac.be/ ifmbe.

IOMP

The IOMP was founded in 1963. The membership includes 64 national societies, two international organizations and 12 000 individuals. Membership of IOMP consists of individual members of the Adhering National Organizations. Two other forms of membership are available, namely Affiliated Regional Organization and Corporate Members. The IOMP is administered by a council, which consists of delegates from each of the Adhering National Organizations; regular meetings of Council are held every three years at the International Conference on Medical Physics (ICMP). The Officers of the Council are the President, the Vice-President and the Secretary-General. IOMP committees include: developing countries; education and training; nominating; and publications.

Objectives

• To organize international cooperation in medical physics in all its aspects, especially in developing countries.

• To encourage and advise on the formation of national organizations of medical physics in those countries which lack such organizations.

Activities

Official publications of the IOMP are *Physiological Measurement, Physics in Medicine and Biology* and the *Medical Science Series*, all published by Institute of Physics Publishing. The IOMP publishes a bulletin *Medical Physics World* twice a year.

Two Council meetings and one General Assembly are held every three years at the ICMP. The most recent ICMPs were held in Kyoto, Japan (1991), Rio de Janeiro, Brazil (1994) and Nice, France (1997). A future conference is scheduled for Chicago, USA (2000). These conferences are normally held in collaboration with the IFMBE to form the World Congress on Medical Physics and Biomedical Engineering. The IOMP also sponsors occasional international conferences, workshops and courses.

For further information contact: Gary D Fullerton, Professor, University of Texas HSC–San Antonio, Department of Radiology, 7703 Floyd Curl Drive, San Antonio, TX 78284-7800, USA, e-mail: fullerton@uthscsa.edu, telephone: (210) 567-5550, and fax: (210) 567-5549.

With love to Imogen, who kept her Grandad entertained while this book was being written

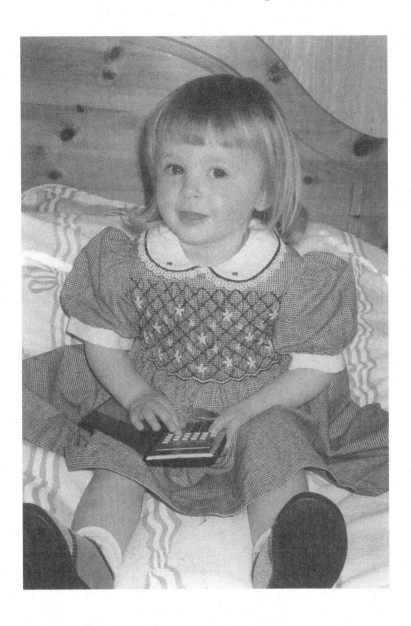

Contents

Preface

The first edition of *Introductory Medical Statistics* was published in 1976 and was based on a course given to medical students at the Westminster Hospital, University of London. The second edition, which incorporated additional examples and some new material on life table/actuarial survival rate calculations, was published in 1989. In the 22 years between the first edition and this greatly expanded third edition, much has changed in the world of medical statistics. For example, desktop personal computers are now the standard and several commercial statistical software packages are available (at great expense!).

This improved access to computing power has meant that some statistical techniques which were previously seldom applied because they were too labour intensive are now in common use, for example, modelling techniques. This advance in computers has also meant that users of commercial statistical software do not have to think too deeply before they number crunch and problems can arise as there is no built-in software guarantee that the correct statistical significance test has been used. Basic training in statistics is essential for software package users without any real previous statistical experience.

Not only has medical statistics changed direction but also the work of the author. With the closure of the Westminster Hospital early retirement beckoned from my former life as a Director of Medical Physics and a Hospital Cancer Registry in the British National Health Service, and since 1990 I have been giving *Basic Medical Statistics Courses with Special Reference to Cancer* in hospitals worldwide: Austria, Australia, Belgium, Canada, France, Germany, Japan, the Netherlands, Saudi Arabia, Switzerland, the UK and the USA.

The planning of this third edition has been helped enormously by comments and material for examples from those (radiation oncologists, radiologists, gynaecologists, physiologists, physicians, surgeons, biochemists, physicists, opthalmologists, dermatologists, anaesthesiologists, to mention some of the specialties) who have attended these courses.

I am also pleased to learn that my cartoons and anecdotes (see also *Mould's Medical Anecdotes: Omnibus Edition* published in 1996 by Institute of Physics Publishing) have stood the test of time in that they continue to keep course participants awake in the 1990s. In the 1970s the *British Medical Journal* in the review of the first edition stated 'The book is enlivened by the author's engaging sense of humour. He is to be congratulated on his highly original and elegant contribution to professional education in a subject which would be much more widely understood and applied if his book achieves the circulation it deserves'. The first edition was translated into Spanish and the second edition into Japanese.

This is still an *Introductory* textbook although by popular demand I have also included a chapter on multivariate analysis and the Cox proportional hazards model but this should be considered as optional (advanced) reading. New chapters which are in this edition are chapters 17–23 and include much expanded material on clinical trials, t-tests, epidemiology and analysis of variance. In addition, new topics are included such as risk specification (using Hiroshima, Nagasaki and Chernobyl as examples), sensitivity and specificity, McNemar's test, Bayesian statistics, decision theory, meta-analysis, and a full discussion on treatment success, cure from cancer and quality of life assessment.

The volume is intended to have widespread appeal, not only for radiation oncologists, diagnostic radiologists and medical physicists, but also for those medical specialties mentioned above, since it is no longer practical for any doctor to ignore statistics as they form part of the syllabus of many professional examinations and journal editors often insist on the inclusion of statistics (usually at least the *famous P* value!) in papers.

Finally I would like to express my thanks to all the participants over almost a quarter of a century who have attended my Statistics courses and given me much constructive advice. I am also most grateful to my Institute of Physics Publishing editor, Kathryn Cantley, for her support and her allowance of *elastic* deadlines for the manuscript, to Sharon Toop of IOPP production and Adrian Corrigan for his production expertise, and to Pamela Whichard of IOPP marketing for ensuring that this third edition reached the bookshops in Olympic record time!

Dick Mould
January 1998
Croydon

Chapter 1

Data Presentation

1.1 INTRODUCTION

Data presentation is an essential but sometimes neglected topic for any introduction to statistics. Lack of a proper understanding of the various possibilities of data presentation can lead to confusion for those who are expected to interpret statistical results. Even if the results are correct, they will not be of much use if nobody can understand them. A useful adage is that simplicity of presentation should be a priority.

The following diagrams, charts and graphs are only a few examples but they clearly show the range of possible presentations. Real data has been used, including some examples referring to the problems of cancer of the cervix in developing countries[1].

With the graphics capabilities of computer software, three-dimensional data presentations, such as isometric charts, are also readily available. These are charts or graphs that portray three dimensions on a plane surface.

1.2 BAR CHART

A bar chart, bar diagram or bar graph, is a series of horizontal or vertical bars of equal width for a two-dimensional chart, or equal cross-section for a three-dimensional chart as in Figures 1.1 and 1.2. The width or cross-section does not have any significance for a bar chart; only the height presents the data of interest.

1.3 PIE CHART

Figure 1.3 is a pie chart or pie diagram and is a circle which is divided into segmental areas representing proportions. Since a circle consists of 360°, the segments are calculated by dividing these 360° in the relevant proportions. Thus

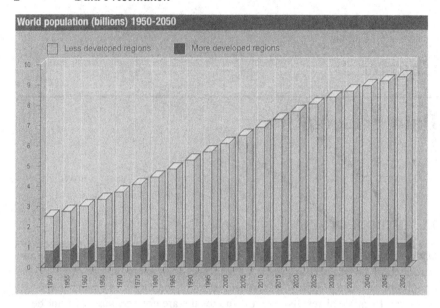

Figure 1.1. Bar chart of the world population[2]. Three-dimensional vertical blocks are used in this chart and they are an example of how to present data for two populations (less and more developed regions of the world) in a single diagram. The world's largest countries in 1996, ranked in order 1–10, are as follows. (Courtesy: United Nations.)

Rank	Country	Population (millions)	Percentage of world population
1	China	1232	21.4
2	India	945	16.4
3	USA	269	4.7
4	Indonesia	200	3.5
5	Brazil	161	2.8
6	Russia	148	2.6
7	Pakistan	140	2.4
8	Japan	125	2.2
9	Bangladesh	120	2.1
10	Nigeria	115	2.0

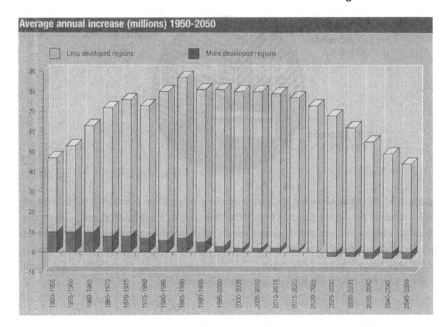

Figure 1.2. Bar chart of the average annual changes in the world population[2]. This is an example where the vertical scale can be negative to show a decrease. (Courtesy: United Nations.)

for three segments of 10, 30 and 60%, the segmental angles are 36°, 108° and 216°, respectively.

1.4 HISTOGRAM

In a histogram, the height of each vertical block does not always represent the value of the variable of interest (unless the width of the block is unity), as is the case of a bar in a bar chart. Also, in a histogram, the horizontal scale is continuous and not, like the bar charts, discrete. Also, unlike a bar chart width, a histogram block width *does have a meaning*. Histogram blocks are usually of a constant width, indicating equal intervals on the horizontal scale, although this is not absolutely necessary, since it is the *area of each histogram block* which is important, in that it is this which represents the value of the variable of interest. Figure 1.4(*a*) is a histogram of the distribution of ages of 667 cancer of the cervix patients treated in Algeria, and the constant intervals on the horizontal scale are 5 year age ranges. The vertical scale indicates the observed frequency, which is the number of patients in a given 5 year age group. The dimensions of these particular histogram blocks are 'patient numbers ×5 years of age' and thus the first two blocks have values of 3×5 and 10×5 in these dimensions,

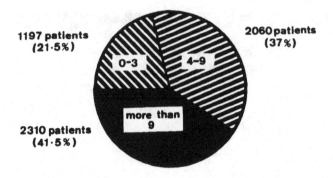

Figure 1.3. Pie chart showing the parity (number of children born) of a population of 5567 cancer of the cervix patients treated at Chittaranjan Hospital, Calcutta[1]. Also, the association between early marriage and the incidence of cancer of the cervix has been well documented. For this cancer patient population it was found that the age at marriage for 7% was below the age of 12 years, 27% in the range 12–15 years, 50% in the range 15–20 years, and only 16% with an age above 20 years.

totalling 65. If, alternatively, the first histogram block is to be made with a width of 10 years, from 20–29 years, then the height to signify 13 patients must be $\frac{1}{2} \times 13 = 6.5$ since the area of this single histogram block must equal that of the previous two. The remaining blocks in the histogram, for 30–34, 35–39 years etc, can then remain the same. This is a trivial example, but the importance of the area of a histogram block must be remembered. The vertical scale of a histogram or bar chart may either be in absolute numbers or in percentages.

Figure 1.4(*b*) looks at first glance like a histogram, but in reality it is a vertical bar chart. The horizontal scale which represents the number of pregnancies is not continuous and it can only take on integer values, including zero. However, for zero pregnancies no histogram block can be drawn because its area would always be zero no matter how high the number of patients. The height of each histogram block can be indicated on the top of the block, as well as on a vertical scale, as in Figure 1.4(*a*), but this is optional.

1.5 PICTOGRAM

Pictograms can be very useful means of presenting some types of data and may be maps, as in Figure 1.5, or a series of symbols arranged in an appropriate manner as in Figure 1.6. In particular, pictograms are valuable for medical education programs for the general public.

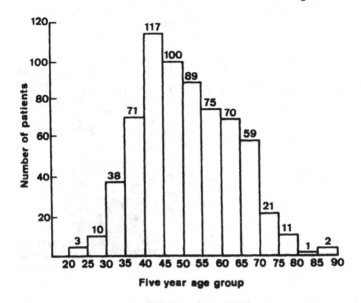

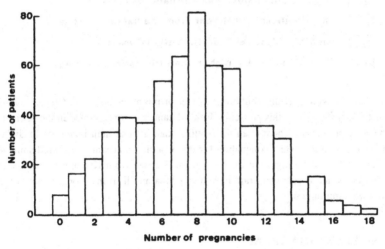

Figure 1.4. (*a*) Histogram of the ages of 667 cancer patients in Algeria. (*b*) Bar chart showing the number of pregnancies of 667 Algerian cancer of the cervix patients. The association between the number of pregnancies and the incidence of cancer of the cervix has been well documented; see also Figures 1.3 and 1.6.

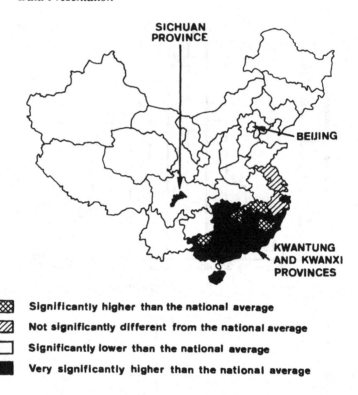

SICHUAN
PROVINCE

BEIJING

KWANTUNG
AND KWANXI
PROVINCES

Significantly higher than the national average

Not significantly different from the national average

Significantly lower than the national average

Very significantly higher than the national average

Figure 1.5. Geographical distribution of cancer of the nasopharynx in China, 1973–79. The pictogram shows that this particular form of cancer is concentrated in the Kwantung and Kwanxi provinces. There is an interesting small area of high incidence in Sichuan province and an unexpected explanation for this is a civil war which occurred some 400 years ago and caused emigration to this area of Sichuan from Kwantung and Kwanxi. This cancer, which is relatively rare except in China and in Hong Kong, is also known as the Kwantung tumour.

1.6 SCATTER DIAGRAM

A scatter diagram is a form of pictogram and Figure 1.7(a) is an example in which the annual lung cancer incidence rate per 100,000 population is shown for selected countries. The horizontal axis does not have a scale since it is not appropriate, but some scatter diagrams will have numerical scales for both vertical and horizontal axes. In this case, the observed patterns can then be studied by the statistical techniques of correlation and regression. Such an example is given in Figure 1.7(b) which indicates a complicated relationship between chromosome aberrations and radiation dose for the Hiroshima population who survived the atomic bomb in 1945.

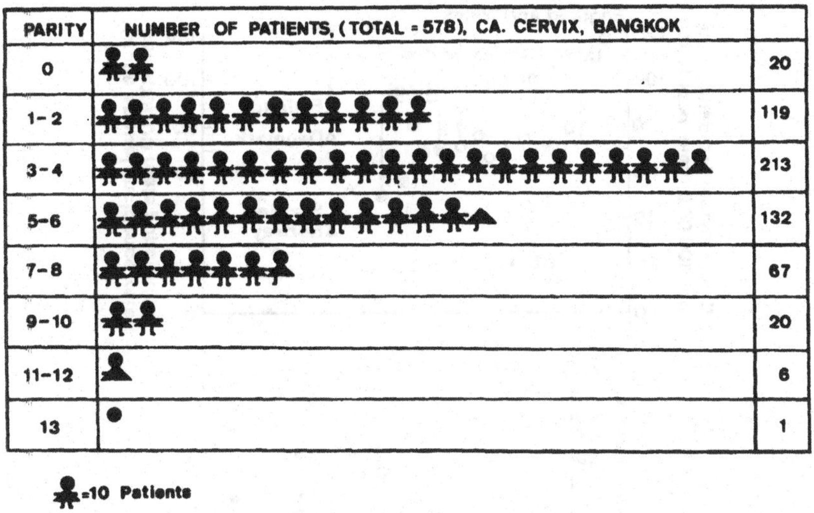

Table 7

PARITY	NUMBER OF PATIENTS, (TOTAL = 578), CA. CERVIX, BANGKOK	
0	👥👥	20
1-2	👥👥👥👥👥👥👥👥👥👥👥	119
3-4	👥👥👥👥👥👥👥👥👥👥👥👥👥👥👥👥👥👥👥👥👥	213
5-6	👥👥👥👥👥👥👥👥👥👥👥👥👥	132
7-8	👥👥👥👥👥👥👥	67
9-10	👥👥	20
11-12	👤	6
13	•	1

👤=10 Patients

Figure 1.6. Each symbol represents 10 cancer of the cervix patients out of a total of 578 treated in Bangkok, as a function of parity.

1.7 TABLE

It should be recognised that some forms of data defy most methods of illustration and there is no better alternative than a table. An example of such data is given in Table 1.1 and is a list of the reasons why patients with cancer of the cervix in rural Thailand do not attend for treatment. These are the results of a questionnaire in which the patient was allowed to give more than one answer. A pie chart would thus be inappropriate and, although a bar chart would technically be possible, a table is better.

1.8 GRAPHS: LINEAR AND LOGARITHMIC AXES

In graphs on rectangular coordinate graph paper, the magnitude of one variable is plotted along the horizontal axis, called the X-axis, and all the values along this axis are known as abscissae. The magnitude of the other variable is plotted along the vertical axis, called the Y-axis, and the values along this axis are known as ordinates. The position of a point on a graph is defined by its abscissa and ordinate, which together form its coordinates. The point of intersection of the two axes at $X = 0$, $Y = 0$ is called the origin but is not automatically included in all graph plots and for a logarithmic axis the value 0 can never be included. The mathematical equation for a straight line, such as in Figure 1.8, is

$$Y = a + bX$$

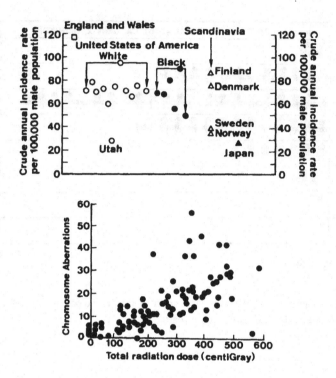

Figure 1.7. (*a*) Male lung cancer in selected populations. England and Wales has the highest incidence, and the state of Utah in the USA which has a 70% Mormon population, has a much lower lung cancer incidence than other American states. This is to be expected because smoking is against the Mormon religion. A variety of symbols may be used in scatter diagrams or in pictograms. (*b*) Raw data on chromosome aberrations and radiation dose. The points are so widely scattered that you can deduce from it almost any relationship.

where X and Y are the variables and a and b are constants. The points (X_1, Y_1) and (X_2, Y_2) are shown and the constant b defines the slope of the straight line. This is the ratio of the increase in Y corresponding to an increase in X:

$$b = (Y_2 - Y_1)/(X_2 - X_1).$$

The constant a is known as the intercept, and is the distance between the origin ($X = 0$, $Y = 0$) and the point on the Y-axis which intersects the straight line. Straight-line equations will be met again later in the book when correlation and regression are discussed.

There are many different types of graph paper but the simplest and the one most often used is linear graph paper in which equal numerical intervals are equally spaced on a linear graph axis. Figure 1.9 is an example of a graph (not

Table 1.1. Reasons for failure to complete treatment for cancer of the cervix in rural Thailand. Data for 167 patients.

Patient's problem	Percentage of responders
No money for transport	61.1
No accommodation near hospital	57.5
Too weak, poor general condition	25.2
Too advanced disease—given up	10.2
Influenced by relatives to quit	47.9
Reluctance to attend	14.4
Seeks herb and magic treatment	51.5
Did not understand doctor's advice	50.3
Delay due to ignorance of cancer	35.9
Delay due to old fashioned beliefs	55.1
No motivation for early detection of cancer	100

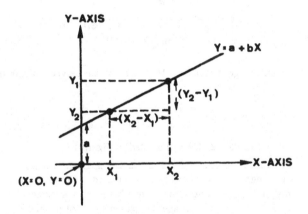

Figure 1.8. The straight line. Three points are shown: •, the origin at $X = 0$, $Y = 0$ and the two points (X_1, Y_1) and (X_2, Y_2). Graph points are usually written in this manner, inside brackets with the X value preceding the Y value.

a straight line in this instance) on linear graph paper. There are equal intervals between each 10% spread of values on the Y-axis and equal intervals between each 10 year age spread of values on the X-axis. The data refer to the age of cancer of the cervix patients in Bombay, see Table 1.2. The meaning of the adjective *cumulative*, as in cumulative number or cumulative percentage, can clearly be seen from the table. The age corresponding to the 50% cumulative percentage value is the median age (see Chapter 2 for *median*) of the patients and is 45 years for these data. The cumulative curve in Figure 1.9 is S-shaped and this is sometimes referred to as a sigmoid curve.

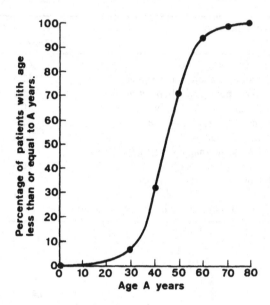

Figure 1.9. Cumulative age graph for 416 cancer of the cervix patients in Bombay, drawn on linear graph format.

Table 1.2. Data illustrated in Figure 1.9.

Age range in years	Number of patients in a given age range	Cumulative number of patients	Cumulative percentage of patients
Less than 30	30	30	7.2
31–40	102	132	31.7
41–50	162	294	70.7
51–60	96	390	93.8
61–70	22	412	99.0
71–80	4	416	100

Total number of patients = 416

Figure 1.10 is an illustration of logarithmic–linear graph paper, sometimes known as semi-log graph paper, in which the X-axis is a linear scale, such as in Figure 1.9, but the Y-axis has a scale with unequal intervals. These intervals are in proportion to the values of $\log_e X$ which are termed natural logarithms (as distinct from $\log_{10} X$ which are logarithms to the base 10). Table 1.3 lists the values of $\log_e X$ and, in the last column, differences between successive $\log_e X$ values. These differences reduce as X increases from 2 to 10, just as the intervals on the Y-axis reduce in Figure 1.10. The logarithmic scale 1–10 is repeated for each power of 10, such as 10–100, 100–1000 and 1000–10 000. Each power of 10 is called a cycle and, for example, a logarithmic scale from 1–1000 is called a 3-cycle scale. An advantage of using a log scale is that a wider range of values can be plotted than when using linear graph paper. For example, if the Y values extended from, say, 1.5 to 800, linear graph paper would either have to make the Y-scale very compressed or the graph paper very tall. On the other hand, log 3-cycles could extend from 1 to 10 for the first cycle, and then through 20, 30, ..., to 100 for the second cycle, and finally through 200, 300, ..., to 1000 for the third cycle.

Figure 1.11 is an example of a log–linear graph plot using four cycles and illustrates the typical shape of age-specific lung cancer incidence. That for males is higher than for females because there is a much greater lung cancer

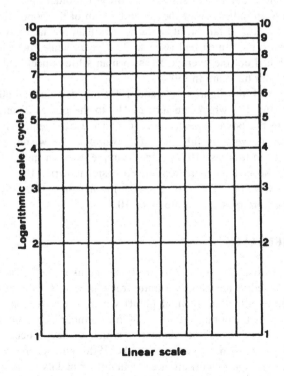

Figure 1.10. Log–linear graph paper.

Table 1.3. Variation of $\log_e X_i$ with X_i where i extends from 1 to 10.

X_i	$\log_e X_i$	Difference between $\log_e X_i$ and $\log_e X_{i-1}$
1	0	
2	0.693	0.693
3	1.098	0.405
4	1.386	0.288
5	1.609	0.223
6	1.792	0.183
7	1 946	0 154
8	2.079	0.133
9	2.197	0.118
10	2.303	0.106

incidence in males. The curves show a fall-off at the older ages and where this fall-off begins is an indication of the general health of the population. Thus for example, data for Russia fall-off at an earlier age than for the USA. In 1996 the average lifespan for US men had risen to 72.4 years while more heart disease and other risks had cut the average Russian man's life expectancy to about 57 years in 1995 from 62 years in 1984.

In Figure 1.11 the figures 100 and 1000 could equally well have been written as 10^2 and 10^3 which are termed '10 to the power 2' and '10 to the power 3' where the power value is also called the exponent. This shorthand terminology can be very useful when, for example, one wishes to talk about millions (10^6) or millionths (10^{-6}). One example is given in Table 1.4 for the electromagnetic spectrum where it is much easier to write 10^6 (i.e. 1×10^6) rather than 1 000 000 or 10^{-6} (i.e. 1×10^{-6}) rather than 0.000 001. Table 1.5 lists the standard prefixes for multiples of 10.

1.9 EXPONENTIALS

The symbol e in $\log_e Y$ refers to an irrational number (i.e. a number which cannot be expressed as an integer, a finite fraction or as a decimal with a finite number of figures after the decimal point) which to seven decimal places is 2.7182818. Another example of an irrational number is π, the ratio of the circumference of a circle to its diameter, which to seven decimal places is 3.1415927. A German mathematician in the 19th century, in the age before computers, spent his entire working life calculating π and its value was inscribed on his gravestone!

Table 1.4. The electromagnetic spectrum. In external beam radiation therapy using X-ray beams from linear accelerators there are three classifications. Superficial therapy is 10–150 kilovolts (kV), i.e. 10–150 $\times 10^3$ volts. Deep (or orthovoltage) therapy is 200–300 kV, i.e. 200–300 $\times 10^3$ volts. Megavoltage (or supervoltage) therapy is above 1 million volts, i.e. $> 10^6$ volts.

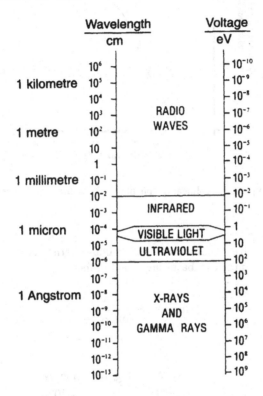

Table 1.5. Prefixes and symbols for multiples of 10.

Multiple	Prefix	Symbol	Multiple	Prefix	Symbol
10^2	hecto	h	10^{-2}	centi	c
10^3	kilo	k	10^{-3}	milli	m
10^6	mega	M	10^{-6}	micro	μ
10^9	giga	G	10^{-9}	nano	n
10^{12}	tera	T	10^{-12}	pico	p
10^{15}	peta	P	10^{-15}	femto	f
10^{18}	exa	E	10^{-18}	atto	a

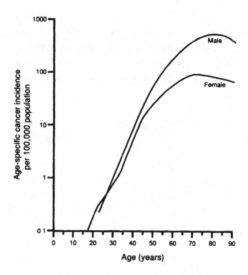

Figure 1.11. Example of log–linear graph plots. Data for lung cancer incidence in Canada.

Logarithms to base e are known as natural logarithms, but logs exist to bases other than e and if two bases are a and b, then

$$\log_a N = \frac{\log_b N}{\log_b a}$$

Thus

$$\log_{10} N = \frac{\log_e N}{\log_e 10} = \frac{\log_e N}{e} = \frac{\log_e N}{2.3026}$$

also

$$\log_a MN = \log_a M + \log_a N$$
$$\log_a M/N = \log_a M - \log_a N$$
$$\log_a (N)^p = p \log_a N$$

e^{bx} is called an exponential series and is given by the formula

$$e^{bx} = 1 + bx + (bx)^2/2! + (bx)^3/3! + (bx)^4/4! + \ldots$$

where the symbol ! means factorial, and

$$2! = 1 \times 2$$
$$3! = 1 \times 2 \times 3$$
$$4! = 1 \times 2 \times 3 \times 4$$
$$n! = 1 \times 2 \times 3 \times 4 \times \ldots \times n$$

The constant b may be positive or negative and thus e^{bx} are known as positive exponentials and e^{-bx} as negative exponentials. An example of a phenomenon which can be described by a negative exponential law is radioactive decay. Figure 1.12 illustrates the decay of three different radioactive isotopes, iodine-131 with a half-life of 8 days, gold-198 with a half-life of 2.7 days and technetium-99m with a half-life of 6 hours. The decay data are plotted on semi-logarithmic graph paper, with a 3-cycle logarithmic axis. The decay data are shown as straight lines, and this always occurs for an exponential law when the data are plotted on semi-logarithmic paper. The data would appear as curves if plotted on linear graph paper. Hence by changing the type of graph paper, we have 'transformed' a curve into a straight line.

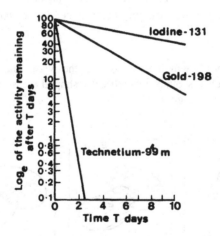

Figure 1.12. Exponential decay of three different radioactive isotopes. Iodine-131 concentrates in the thyroid gland and is used to treat and diagnose thyroid disorders. It is also one of the radioactive isotopes which was a constituent of the Chernobyl radioactive fallout and caused much concern about contaminated milk in the early days following the accident. Gold-198 is used in the form of small 'seeds' for implantation into tumours as a method of radiotherapy. Technetium-99m, when bound to certain pharmaceuticals is used in nuclear medicine organ imaging.

1.10   VENN AND EULER DIAGRAMS

An example of a Venn diagram is given in Figure 1.13 and as seen it consists of three overlapping circles: Venn circles. In theory, there could be any number of overlapping circles but with more than three or four circles the diagram becomes rather cluttered.

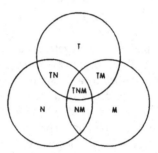

Figure 1.13. Most patients with cancer fail for three reasons: inability to control the primary tumour (T failure), nodal involvement (N failure), dissemination to distant sites (M failure). Thus failure may occur from any one of these categories of from a combination TN, TM, NM, TNM, as shown in this Venn diagram. The number of failures in each category can be equated to the areas in the diagram or, alternatively the actual number of failures in each category can be stated in the appropriate part of the Venn diagram.

A form of diagram which is similar to the Venn diagram is the Euler diagram in which circles or other geometrically shaped areas are within a square or rectangular boundary, Figure 1.14.

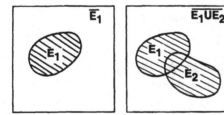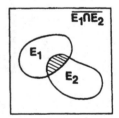

Figure 1.14. Euler diagrams. The mathematical symbol $\cup$ is called a *union* and is the shaded area (centre) $E_1 \cup E_2$ in that this is the combination of E_1 and E_2. The symbol $\cap$ is called an *intersection* and (right) is the shaded area which represents the parts common to both E_1 and E_2. When a horizontal bar is placed over, say, $\overline{E_1 \cup E_2}$ it signifies that the area in the square excluding $E_1 \cup E_2$ represents what is not $E_1 \cup E_2$.

1.11 BOX AND WHISKERS PLOT

A box and whiskers plot, sometimes referred to only as a box plot, is a graphical method of presenting the distribution of a variable measured on a numerical scale. An example is seen later in Figure 21.5. The midpoint of the distribution is represented by a horizontal line and the values above and below the midpoint line are divided into quartiles by horizontal lines. The *hinges* (or ends) of the box are at the two quartiles nearest to the midpoint and the quartiles extending to the extreme values are represented by vertical lines (the *whiskers*) which end in a horizontal line.

REDRAWING THE DOW

This is an example from the *New York Times* of 2 March 1997 of when it is not appropriate to use a logarithmic scale. Under the title of *The Bulls and Bears and the Little Pigs: Redrawing the Dow*, it was stated that 'This chart, plotted on a logarithmic scale, gives the same visual weight to comparable percentage changes in the Dow Jones industrial average. A 100 point rise when the Dow is at 1000 looks the same here as a 700 point rise at 7000. Seen this way, it's easy to see why investors have ignored recent declines'.

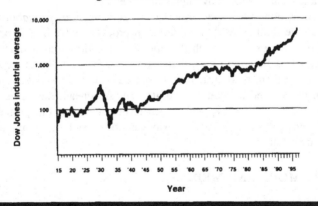

Chapter 2

Describing Curves and Distributions

2.1 INTRODUCTION

To describe a curve so that it can be reproduced, a quantitative description is needed. The terms *curve* and *distribution* used in this book are synonymous, in the sense that a particular curve depicts the distribution of a quantity such as age, death due to a particular cause, blood pressure, drug response, etc. It is also quite common to see a series of histogram blocks representing an observed distribution of a quantity, with a good fit curve superimposed. This curve can be termed a *distribution curve*. Any curve must be drawn relative to two axes, and the description of the curve must include information about both its position and its shape. Measures of central tendancy of location (i.e. of position) are arithmetic mean, median and mode. Measures of dispersion (i.e. shape) are variance, standard deviation, covariance and skewness.

The final section in this chapter defines what is meant by *probability density function* since we shall be working in terms of probability from the next chapter onwards. Thus, for example, with the normal curve drawn as a function of x (that is, y equals a function of x), if the total area beneath the curve equals 1 the entire area beneath this particular normal curve (which is the the *standard* normal) can be equated to total probability. Areas beneath the curve between given values of x, such as x_1 and x_2, therefore represent the probability of values of y lying between y_1 and y_2 always assuming (for this example) that y is normally distributed.

2.2 MEAN, MODE AND MEDIAN

The three most important and commonly used measures of position are the arithmetic mean, mode and median. For the following ten numbers

$$7, 6, 3, 11, 5, 7, 7, 9, 6, 5$$

which we will assume are observed values of a quantity, the arithmetic mean or average is the sum of all the observations divided by their total number. Thus the mean is

$$\frac{(7+6+3+11+5+7+7+8+6+5)}{10} = 66/10 = 6.6.$$

The mode of a distribution is the observation that occurs most frequently and in the above group of ten observations the mode is 7, which occurs three times. In figure 1.4(*b*), which is data for number of pregnancies, the mode is 8.

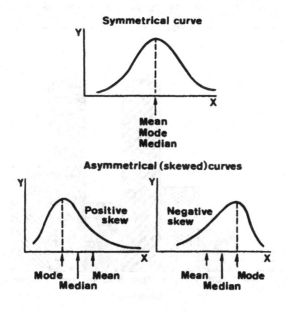

Figure 2.1. Curve shapes.

The median is the middle-valued observation when all the observations are ranked in order of value;

$$3, 5, 5, 6, 6, 7, 7, 7, 9, 11.$$

If the total number of observations is even, as above, the median is the value between the two middle observations, that is between 6 and 7. For grouped distribution data as in table 1.2, the median can be found from the cumulative curve in figure 1.9, as already noted in Chapter 1.

For the special case when the distribution curve is symmetrical,

$$\text{mean} = \text{mode} = \text{median}$$

as in the normal curve of Chapter 3. However, most distributions are at least slightly asymmetrical and in this instance

$$\text{mean} - \text{mode} = 3 \times (\text{mean} - \text{median}).$$

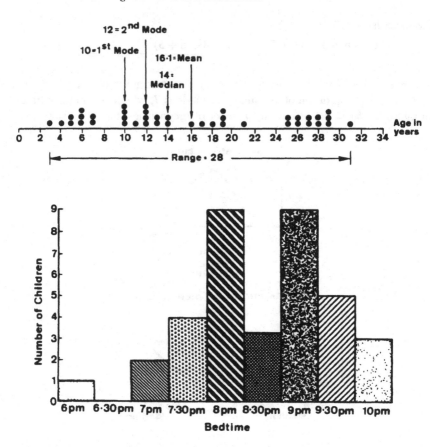

Figure 2.2. (*a*) Scatter diagram of the ages of 41 Russian czars. (*b*) Histogram of the bedtimes of the 36 girls in class IK at Old Palace School, Croydon. 25% of the class went to bed at 8 pm, 25% also went to bed at 9 pm, which accounts for the bimodal shape of the distribution. (Courtesy of Fiona Mould (Imogen's mother!), ex-member of class IK.)

Not all observed data can be fitted by a curve but a mean, mode and median of the observations can still be computed. Such an example is the ages of 41 Russian czars who lived between the years 1341 and 1696 and are now buried in one of the cathedrals inside the Kremlin walls in Moscow. Figure 2.2(*a*) shows the distribution of ages in a scatter diagram. This particular distribution has two modes§, but it is a good example of when a mode is not really useful. A better

§ Figure 2.2(b) is also a distribution of observed data with two modes, in this case presented in the form of a histogram. When one observation is far removed from the remainder, such as the 6 pm bedtime of one child, this observation is known as an *outlier*.

description of these data would be to state only that the mean is 16.1 years, the median 14 years, and the range 28 years, extending from 3 to 31 years. More data than the 41 observations in this example would be necessary before any recognisable distribution shape might appear. However, only 41 observations are available.

To look at another example, consider the ages of male doctors given an obituary in *The Lancet* for the first six months of 1985. Of the 30 obituaries, three were female and 27 male and two of the males did not have their ages stated. Of the 25 males for analysis, the ranked ages are

49
56
57
60
67
70
72
73, 73, 73   (Mode = 73)
74
75, 75   (Median = 75)
76, 76
77
80, 80
81, 81
82
84
87
88
90

with a mean age of 1856/25 = 74.2 years.

As with the data of figure 2.2, a statement of the modal value is not very informative. However, fortunately with this *Lancet* obituary data more observations are available and when the 3-year period 1983–1985 is reviewed the number of observations increases to 194 for male doctors. The range is large, from 27 to 99 years and although there are 11 deaths at age 82 (the modal value), there are 10 deaths at age 76, and 9 deaths each at ages 67, 78 and 81 years. The best method of illustrating these data is as grouped data (see table 2.1) using a histogram, figure 2.3, rather than a ranking list as for the 25 observations in 1985, since this is visually more informative. The data are negatively skewed, see figure 2.1, with a mode between 76 and 84 and a median between 75 and 76: since the 97th ranked observation is 75 and the 98th ranked observation is 76 (the total number of observations is an even number and equal to 194). From table 2.1 it is seen that the mean is 73.4. This is a more accurate estimate of the mean age than the previous value of 74.2 years since more data are available.

Table 2.1. Grouped data for the ages of 194 male doctors whose obituaries appeared in *The Lancet* during the years 1983–1985. The mean calculated using the 194 ungrouped data values is 14, 219/194 = 73.4 and therefore very little accuracy is lost in this example in estimating the mean from the grouped data.

Age interval in years	Frequency F	Mid-interval year A	F × A
26–30	1	28	28
31–35	0	33	0
36–40	0	38	0
41–45	2	43	86
46–50	5	48	240
51–55	9	53	477
56–60	14	58	812
61–65	14	63	882
66–70	23	68	1564
71–75	29	73	2117
76–80	37	78	2886
81–85	37	83	3071
86–90	16	88	1408
91–95	5	93	465
96–100	2	98	196
Totals	194		14,232

Mean = 14, 232/194 = 73.4 years

Figure 2.3 is an example of a unimodal distribution and a bimodal distribution for population numbers far larger than those in figure 2.2. Age distributions for most cancer sites are unimodal, such as that for newly diagnosed corpus uteri (endometrium) cancer and lung cancer. In such a case, specification of a mean age is appropriate. However, for cancer of the cervix when the population includes both *in situ* cancer and invasive cancer the distribution is bimodal and the age distributions should be demonstrated using two separate distributions to maximise the available information in the diagram.

2.3 SKEWNESS

The terminology for asymmetry in a distribution curve is skewness. When the mode of the curve is 'pushed to the right' the curve is called negatively skewed and when 'pushed to the left' is called positively skewed, see figure 2.1. A

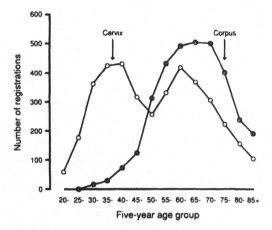

Figure 2.3. Age distribution in England and Wales for registered new cases of cancer of the cervix uteri and corpus uteri, demonstrating both a unimodal and a bimodal distribution.

symmetrical curve has no skewness. Two measures of skewness which are useful for some grouped frequency distributions when the mode cannot easily be found are the *first coefficient of skewness* which is defined as

$$(\text{Mean} - \text{Mode})/\text{Standard deviation}$$

and the *second coefficient of skewness* which is defined as

$$3 \times (\text{Mean} - \text{Median})/\text{Standard deviation}$$

An example of a negatively skewed distribution is shown in figure 2.4 and a second example could be constructed from data for mortality in England and Wales in 1984 from chronic renal failure, where for the 882 male deaths, the mode is in the 5 year age group 80–84. An example of a positively skewed distribution is mortality in England and Wales in 1984 from motor vehicle traffic accidents, where for 3547 male deaths, the mode is in the 5 year age group 15–19.

A further example of a positively skewed distribution is seen in figure 9.2 which is a lognormal distribution curve for the distribution of survival times of patients with cancer of the cervix who died with their cancer present. This distribution is the basis of the lognormal model, which with properly chosen values for its mean and standard deviation can be used to predict the proportion of long-term survivors following treatment for cancer. It has, however, only been verified for certain specific cancer sites.

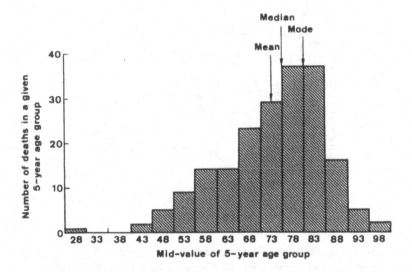

Figure 2.4. Histogram of the ages of 194 male doctors whose obituaries appeared in *The Lancet* during the years 1983–1985.

2.4 STANDARD DEVIATION AND VARIANCE

Mean, mode and median are measures of location of a curve or series of observations, but in addition to measures of location, measures of shape are also necessary for a full description of a curve or distribution of observations. Skewness is a measure of symmetry or asymmetry but it will not completely describe the shape in terms of dispersion. Range is a measure of dispersion, for example, two groups of eight values of x_i

$$N = 8$$

x_i values are 12, 6, 7, 3, 15, 10, 18, 5

range is 3–18 and mean $= \bar{x} = 9.5$

$$N = 8$$

x_i values are 9, 3, 8, 8, 9, 8, 9, 18

range is 3–18 and mean $= \bar{x} = 9.0$

may have the same range and not very different means but, even so, their shape if ranked from 3–18 and presented in the form of a vertical bar chart is very different. Range is therefore not the best measure of dispersion of a distribution of observations. An improvement on range is the semi-interquartile range, which

for the first group is 6–12 and for the second is 8–9

$$N = 8$$

ranked x_i values are 3, 5, 6, 7, 10, 12, 15, 18

$$N = 8$$

ranked x_i values are 3, 8, 8, 8, 9, 9, 9, 18.

This indicates that for the second group, the extreme ends of the range are widely different from the central 50% of the values.

However, the best measure of dispersion is the standard deviation, which is the square root of the *variance*:

$$(\text{Standard deviation})^2 = \text{Variance}.$$

To illustrate the calculation of standard deviation a series of four observations is used

$$N = 4$$

x_i values are 4, 5, 6, 9

$$\text{mean} = \bar{x} = 6$$

and two methods which may be termed the *direct method* and the *quick* method are illustrated. Both give a value for the standard deviation of 2.2.

Using the notation $\bar{x}$ = mean of all x_i values and the total number of x_i values = N, the formula for standard deviation is

$$\text{Standard deviation} = \sqrt{\frac{\text{Sum of all } (x_i - \bar{x})^2 \text{ values}}{N - 1}}$$

However, a quicker method than using the direct formula above is to use the following formula:

$$\sqrt{\frac{\text{Sum of all } (x_i^2) \text{ values} - [(\text{Sum of all } (x_i) \text{ values}^2)/N]}{N - 1}}.$$

Calculation of a standard deviation (*direct method*).

x_i	$(x_i - \bar{x})$	$(x_i - \bar{x})^2$
4	−2	4
5	−1	1
6	0	0
9	+3	9
Sum = 24		Sum = 14

$N = 4$

$$SD = \sqrt{\frac{Sum(x_i - \bar{x})^2}{N - 1}}$$

$\bar{x} = (24/4) = 6$

$SD = \sqrt{14/(4 - 1)} = 2.2$

Calculation of a standard deviation (*quick method*).

x_i	x_i^2
4	16
5	25
6	36
9	81
Sum = 24	Sum = 158

$N = 4$

$$SD = \sqrt{\frac{Sum(x_i^2) - \frac{(Sum(x_i))^2}{N}}{N - 1}}$$

$$SD = \sqrt{\frac{158 - (24)^2/4}{4 - 1}}$$

$SD = 2.2$

For the two groups of eight observations with identical ranges 3–18 and means of 9.5 and 9.0, the standard deviations are 4.9 and 3.9. This indicates, as it should, that the x_i values 3, 8, 8, 8, 9, 9, 9, 18 are more closely grouped together than are 3, 5, 6, 7, 10, 12, 15, 18.

The $N = 4$ and $N = 8$ groups are trivial and used only to illustrate computational methods. In practice, the number of observations will be larger. For the data in figure 2.2(*a*), the computation is given in table 2.2, using the *quick method*. Some observations will have the same value: there are 41 observations but only 21 age values, termed data group i. This is why the frequency column of F_i-values is required.

Table 2.2. Computation of the standard deviation of the data in figure 2.2(a).

Data group i	Age in years x_i	Frequency F_i	$F_i x_i$	$(x_i)^2$	$F_i(x_i)^2$
1	3	1	3	9	9
2	4	1	4	16	16
3	5	2	10	25	50
4	6	3	18	36	108
5	7	2	14	49	98
6	10	4	40	100	400
7	11	1	11	121	121
8	12	4	48	144	576
9	13	2	26	169	338
10	14	2	28	196	392
11	16	1	16	256	256
12	17	1	17	289	289
13	18	1	18	324	324
14	19	3	57	361	1083
15	21	1	21	441	441
16	25	2	50	525	1050
17	26	2	52	676	1352
18	27	2	54	729	1458
19	28	2	56	784	1568
20	29	3	87	841	2523
21	31	1	31	961	961

$$N = \text{Sum}(F_i) = 41$$

$$\text{Sum}(F_i x_i) = 661$$

$$\text{Sum}[F_i(x_i)^2] = 13,413$$

$$[\text{Sum}(F_i x_i)]^2 = 436,920$$

$$\text{SD} = \sqrt{\frac{13,413 - (436,920)/41}{40}} = \sqrt{68.9} = 8.3$$

2.5 COEFFICIENT OF VARIATION

The coefficient of variation compares the spread of the observations with their magnitude and is

$$100 \times \text{Standard deviation/Mean.}$$

As with standard deviation, a low value of the coefficient of variation corresponds to high precision, while a high value corresponds to low precision. From the data in figure 2.2, with mean= 16.1, standard deviation = 8.3, range = 28 years, the coefficient of variation is 51.6 whereas for the 25 age values on page 21 the mean = 74.2, standard deviation = 10.2, range = 41 and the coefficient of variation is 13.7. Coefficients of variation are particularly useful when observations with different dimensions are being compared, such as £sterling and US $. A dimensionless measure of dispersion is then very convenient.

2.6   PROBABILITY DENSITY FUNCTION

The probability density function of x which is also sometimes called the *distribution function* of x or, loosely, the distribution of x, is shown schematically as a graph of $y = f(x)$ *versus* x in figure 2.5. The probability that a random observation x (for which $f(x)$ is its probability density function) will fall in any interval x_1 to x_2 is the area under the curve of $f(x)$ from x_1 to x_2. Expressed as an integral this probability is

$$\text{Prob}(x_1 < x \leqslant x_2) = \int_{x_1}^{x_2} f(x)\mathrm{d}x$$

and since a random observation x is certain to have some value, the total area under the curve $f(x)$ is equal to unity, i.e.

$$\int_{-\infty}^{\infty} f(x)\mathrm{d}x = 1.$$

In the next chapter we shall be discussing various features of the normal probability density function, which we will abbreviate to normal distribution, and we will again encounter *areas under the curve* which can be equated to probabilities.

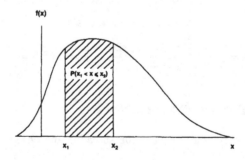

Figure 2.5. Schematic diagram of a probability density function. If this were the normal, then it would be symmetrical and bell shaped with mean=mode=median, see figure 2.1, and x would be called a normal variable.

MEAN, MODE AND MEDIAN

This rather novel diagram to describe the mean, median and mode is taken from the book *How to Lie with Statistics* by Darrell Huff (1954) when the author is discussing salary announcements by a corporation executive or a business proprietor about the average wage of the workers. 'The boss might like to express the situation as 'average wage $5700' to be deceptive. The mode is more revealing: the most common rate of pay is $2000. The median tells even more about the situation: half the people get more than $3000 and half get less'.

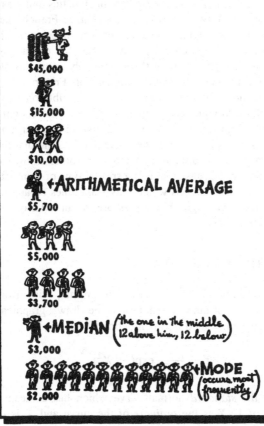

Chapter 3

The Normal Distribution Curve

3.1 INTRODUCTION

The normal distribution curve is very important in statistics and was discovered in 1733 by Abraham de Moivre (1667–1754), a refugee French mathematician living in London. He was solving problems for wealthy gamblers! However, the curve was apparently forgotten until later in the 18th century when it was rediscovered by those investigating the theory of probability and the theory of errors. They included the German mathematician Carl Friedrich Gauss (1777–1855) whose name is now often associated with the distribution curve, so much so that it is also known as Gaussian. When illustrated, it is recognisable by its distinctive bell shape, Figure 3.1. Thus, when a distribution curve is described as normal the term is not being used in the sense that it is the distribution curve which represents more observational data than any other known distribution curve. It is being used in the sense that it is a special type of symmetrical curve. Examples of frequency distributions which may sometimes be approximated by a normal curve are those of height, blood pressure, mean red blood cell volume and certain age distributions.

3.2 MATHEMATICAL FORMULA

Any well defined distribution curve such as the normal, will always have an associated mathematical formula to enable it to be drawn graphically and the area beneath the curve calculated between defined limits.

$$Y = \frac{1}{\sqrt{2\pi}} \exp(-\tfrac{1}{2}X^2)$$

is the formula for the *standard* normal curve which has a mean of 0 and a standard deviation of 1. Y is the *ordinate* of the curve and X is the abscissa and is called the *unit normal deviate*, Figure 3.1.

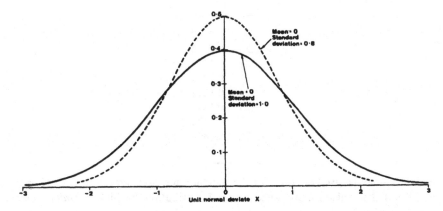

Figure 3.1. Two examples of normal curves. The standard normal curve is the curve with a mean of 0 and a standard deviation of 1. The values of the X-axis are in unit normal deviates, which can take positive or negative values. The value of Y is tending to 0 as X approaches $+3$ or -3 unit normal deviates.

3.3 MATHEMATICAL TABLES

Most books of log tables also contain tables of exponentials and therefore Y, in the formula just quoted can be calculated. For example, for the normal curve with $M = 0$ and $S = 0.8$ (Figure 3.1) we have

$$(X/S) = 1.25 \text{ and } (X/S)^2 = 1.563$$

$$S\sqrt{2\pi} = 2.00$$

and from tables of exponentials

$$\exp[-\tfrac{1}{2}(X/S)^2] = e^{-0.781} = 0.458$$

and thus

$$Y = 0.458/2.00 = 0.288.$$

The point $X = 1.0$, $Y = 0.228$ is on the dotted normal curve in Figure 3.1 and since the normal curve is symmetrical it also contains the point $X = -1.0$, $Y = 0.288$. Although X may be positive or negative, Y is always positive.

Use of the formula to calculate Y-values for constructing a normal curve will be laborious if the above procedure has to be repeated many times. However, this is not necessary in practice because there are available tables of ordinate (Y) values of the standard normal distribution for various values of the unit normal deviate (X). For example,

X	Y
1.20	0.1942
1.21	0.1919
1.22	0.1895
1.23	0.1872
1.24	0.1849
1.25	0.1826
1.26	0.1804
1.27	0.1781

and these may be used not only for the standard normal distribution with $M = 0$ and $S = 1.0$, but also for the general normal distribution in which the *unit normal deviate* which is

$$\frac{X - M}{S}$$

can be calculated for any value of M or S. The formula for the *general normal curve* is

$$Y = \frac{1}{S\sqrt{2\pi}} \exp\left[\frac{-(X - M)^2}{2S^2}\right]$$

This formula is also to be found on the current German 10 deutschmark bank note (see below) together with a picture of Carl Freidrich Gauss. On this 10 DM note the mean is symbolised by μ rather than by M, the standard deviation by σ rather than by S and Y by $f(x)$. There are no internationally recommended symbols for the mean and standard deviation and different authors make their own choices. This is also the case for the presentation of tables (see Figure 3.2) giving areas beneath the standard normal curve as a function of the unit normal deviate (X in Figure 3.1) as will be seen later. This spectrum of tabular presentations (although in the end they can all be used to find a given areas which is to be equated to a probability, see Figure 2.5) can be most confusing to those encountering such tables for the first time.

The unit normal deviate (X) axis for the standard normal curve with $M = 0$ and $S = 1$, Figure 3.1, is often symbolised by z rather than by X and in effect this horizontal $(X$ or $z)$ axis gives multiples of standard deviations of the standard normal curve. When the z-test is used (see later) it is multiples of standard deviations which are described by z.

Using, as an example, the curve in Figure 3.1 with $M = 0$ and $S = 0.8$

$$(X - M)/S = 1.0/0.8 = 1.25$$

$Y = 0.1826$ when $X = 1.25$ for the standard normal curve. However, for the *general normal curve* $1/S\sqrt{2\pi}$ is the first factor in the formula and this has an additional $1/S$ when compared with the first factor in the formula for the *standard normal curve*. The Y-value from the table must therefore be divided by the S-value. Thus

$$Y(\text{for curve } M = 0, S = 0.8) = Y(\text{for curve } M = 0, S = 1)/S$$
$$= 0.1826/0.8 = 0.228$$

When an event is guaranteed to occur the probability equals 1, when there is a 50% possibility of the event occurring the probability equals 0.5 and when there is no possibility of the event occurring the probability equals 0. Thus for a probability distribution, in this case the normal distribution, when the area beneath the curve equals 1 this represents total probability. If it is required to calculate the probability of occurrence of an event between defined limits of A and B then this probability can be obtained from the integral of the standard normal curve from $X = A$ to $X = B$.

$$\int_{X=A}^{X=B} Y \, dx$$

The total probability of 1 is the integral from minus infinity to plus infinity and if the area beneath the standard normal curve is denoted by P between the limits minus infinity and ζ then this refers to the shaded area in Figure 3.2(a). The unshaded area in this figure will be $1 - P$.

$$1 - P = \int_{X=-\infty}^{X=\zeta} \frac{1}{\sqrt{2\pi}} \exp(-\tfrac{1}{2}X^2) dx.$$

The white area in Figure 3.2(a), symbolised by P will therefore be the integral between the lower limit of $X = \zeta$ and the upper limit of $X = +\infty$ and it is this probability, P, which is given in Table 3.1(a).

Figure 3.3 is a repetition of the standard normal curve, $M = 0$, $S = 1$, of Figure 3.1, but in this instance the X-values are stated in terms of standard deviations in order to demonstrate the meaning of *standard deviation for a normal curve* in terms of areas beneath the curve between defined limits. It also introduces the term *tail area* which refers to the small areas under the

Table 3.1. (*a*) Area P beneath the standard normal curve, between the limits $X = \zeta$ and $X = +\infty$, shown as the white area in the schematic diagram Figure 3.2(*a*): the ᵤaded area is therefore $1 - P$. The areas P are read from the table by the following method. For $\zeta = 0.54$ first look down the first column of the table to 0.5 and then look along the line of figures level with 0.5 until you reach the number below the column heading 4. This gives $P = 0.29460$. The notation 0.0^2 means that there are two zeros before the 5-digit number in the table. Thus for $\zeta = 2.54$, the required area is $P = 0.0055426$. (From Fisher and Yates, *Statistical Tables for Biological, Agricultural and Medical Research* (6th edn, 1974, table lii, p 45). Courtesy Longman Group UK Limited.)

ζ		0	1	2	3	4	5	6	7	8	9
0.0	0.	50000	49601	49202	48803	48405	48006	47608	47210	46812	46414
0.1		46017	45620	45224	44828	44433	44038	43644	43251	42858	42465
0.2		42074	41683	41294	40905	40517	40129	39743	39358	38974	38591
0.3		38209	37828	37448	37070	36693	36317	35942	35569	35197	34827
0.4		34458	34090	33724	33360	32997	32636	32276	31918	31561	31207
0.5		30854	30503	30153	29806	29460	29116	28774	28434	28096	27760
0.6		27425	27093	26763	26435	26109	25785	25463	25143	24825	24510
0.7		24196	23885	23576	23270	22965	22663	22363	22065	21770	21476
0.8		21186	20897	20611	20327	20045	19766	19489	19215	18943	18673
0.9		18406	18141	17879	17619	17361	17106	16853	16602	16354	16109
1.0		15866	15625	15386	15151	14917	14686	14457	14231	14007	13786
1.1		13567	13350	13136	12924	12714	12507	12302	12100	11900	11702
1.2		11507	11314	11123	10935	10749	10565	10383	10204	10027	98525
1.3	0.0	96800	95098	93418	91759	90123	88508	86915	85343	83793	82264
1.4		80757	79270	77804	76359	74934	73529	72145	70781	69437	68112
1.5		66807	65522	64255	63008	61780	60571	59380	58208	57053	55917
1.6		54799	53699	52616	51551	50503	49471	48457	47460	46479	45514
1.7		44565	43633	42716	41815	40930	40059	39204	38364	37538	36727
1.8		35930	35148	34380	33625	32884	32157	31443	30742	30054	29379
1.9		28717	28067	27429	26803	26190	25588	24998	24419	23852	23295
2.0		22750	22216	21692	21178	20675	20182	19699	19226	18763	18309
2.1		17864	17429	17003	16586	16177	15778	15386	15003	14629	14262
2.2		13903	13553	13209	12874	12545	12224	11911	11604	11304	11011
2.3		10724	10444	10170	99031	96419	93867	91375	88940	86563	84242
2.4	0.0^2	81975	79763	77603	75494	73436	71428	69469	67557	65691	63872
2.5		62097	60366	58677	57031	55426	53861	52336	50849	49400	47988
2.6		46612	45271	43965	42692	41453	40246	39070	37926	36811	35726
2.7		34670	33642	32641	31667	30720	29798	28901	28028	27179	26354
2.8		25551	24771	24012	23274	22557	21860	21182	20524	19884	19262
2.9		18658	18071	17502	16948	16411	15889	15382	14890	14412	13949
3.0		13499	13062	12639	12228	11829	11442	11067	10703	10350	10008

Table 3.1. (*a*) (continued).

ζ		0	1	2	3	4	5	6	7	8	9
3.1	0.0^3	96760	93544	90426	87403	84474	81635	78885	76219	73638	71136
3.2		68714	66367	64095	61895	59765	57703	55706	53774	51904	50094
3.3		48342	46648	45009	43423	41889	40406	38971	37584	36243	34946
3.4		33693	32481	31311	30179	29086	28029	27009	26023	25071	24151
3.5		23263	22405	21577	20778	20006	19262	18543	17849	17180	16534
3.6		15911	15310	14730	14171	13632	13112	12611	12128	11662	11213
3.7		10780	10363	99611	95740	92010	88417	84957	81624	78414	75324
3.8	0.0^4	72348	69483	66726	64072	61517	59059	56694	54418	52228	50122
3.9		48096	46148	44274	42473	40741	39076	37475	35936	34458	33037
4.0		31671	30359	29099	27888	26726	25609	24536	23507	22518	21569

extremities of the curve. The total tail area defined in Figure 3.3 is 5%, but there can equally well be a defined tail area of 1% in which each tail is 0.5%.

In Figure 3.3 the standard deviation is $S = 1$ and therefore once again we have the unit normal deviate X (which we can also call z) and Table 3.1(c) relates probability to multiples of standard deviations (or *standard errors* which we will meet in the next chapter) for a normal distribution.

The probabilities in Table 3.1(c) can be calculated from the areas given in Table 3.1(a). Thus considering $\pm 2S$ first look up the area (i.e. probability) in this table which corresponds to $\zeta = 2.0$. This is seen to be 0.022750. Taking into account the symmetry of the normal distribution, the probability that an observation showing *at least as large a deviation* from the population mean as $\pm 2S$ is 2×0.022750 which equals 0.046.

As another easy example of the use of Table 3.1(a), in this case where for the basic data we do not have $M = 0$ and $S = 1$, we therefore have to first calculate the unit normal deviate $(X - M)/S$ before we can use Table 3.1(a). Let us assume that the height of males is normally distributed with $M = 67.5''$ and $S = 2.5''$ and we wish to know the probability of men having a height above 70.5".

The unit normal deviate for $X = 70.5''$ is $\{(70.5 - 67.5)/2.5\} = +1.2$. The probability we are interested in corresponds to area P in Figure 3.2(a) and from Table 3.1(a) this is 0.11507. If, alternatively, we were interested in the probability of men having a height in the range 65.0"–68.0", the two unit normal deviates we would require are $\{(65.0 - 67.5)/2.5\} = -1.0$ and $\{(68.0 - 67.5)/2.5\} = +0.20$.

We now use the fact that the *normal distribution is symmetrical* because we have a negative value of X. From Table 3.1(a) we find that area P for $X = +1.0$ is 0.15866 and therefore the probability of a height in the range

Table 3.1. (*b*) Ordinate (which is the value of *Y* in the equation, but which can also be called *f*(*X*) meaning *a function of X*) of the standard normal curve which has a mean of 0 and a standard deviation of 1 and thus *X* in this table is the unit normal deviate. For a general normal curve the unit normal deviate is calculated by $(X - M)/S$, referring to the earlier notation. (From Fisher and Yates, *Statistical Tables for Biological, Agricultural and Medical Research* (6th edn, 1974, table 11, p 44). Courtesy Longman Group UK Limited.)

X	Y	X	Y	X	Y	X	Y
0.00	0.39894	0.36	0.37391	0.72	0.30785	1.40	0.14972
0.02	0.39886	0.38	0.37115	0.74	0.30338	1.50	0.12952
0.04	0.39862	0.40	0.36827	0.76	0.29887	1.60	0.11092
0.06	0.39822	0.42	0.36526	0.78	0.29430	1.70	0.09405
0.08	0.39766	0.44	0.36213	0.80	0.28969	1.80	0.07895
0.10	0.39695	0.46	0.35889	0.82	0.28503	1.90	0.06562
0.12	0.39608	0.48	0.35553	0.84	0.28034	2.00	0.05399
0.14	0.39505	0.50	0.35206	0.86	0.27561	2.20	0.03547
0.16	0.39386	0.52	0.34849	0.88	0.27086	2.40	0.02239
0.18	0.39253	0.54	0.34481	0.90	0.26608	2.60	0.01358
0.20	0.39104	0.56	0.34104	0.92	0.26128	2.80	0.00792
0.22	0.38940	0.58	0.33717	0.94	0.25647	3.00	0.00443
0.24	0.38761	0.60	0.33322	0.96	0.25164	3.50	0.00087
0.26	0.38568	0.62	0.32918	0.98	0.24680	4.00	0.00013
0.28	0.38360	0.64	0.32506	1.00	0.24197	4.50	0.00002
0.30	0.38138	0.66	0.32086	1.10	0.21785		
0.32	0.37903	0.68	0.31659	1.20	0.19418		
0.34	0.37653	0.70	0.31225	1.30	0.17136		

Table 3.1. (*c*) Probability related to multiples of standard deviations (*S*) or standard errors for a normal distribution.

Number of SDs	Probability of an observation showing *at least as large a deviation* from the normal population mean
0.674*S*	$(1 - 0.5) = 0.5$
1*S*	$(1 - 0.683) = 0.317$
1.645*S*	$(1 - 0.90) = 0.10$
1.96*S*	$(1 - 0.95) = 0.05$
2*S*	$(1 - 0.954) = 0.046$
2.576*S*	$(1 - 0.99) = 0.01$
3*S*	$(1 - 0.9973) = 0.0027$

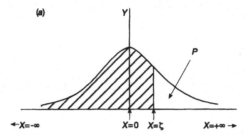

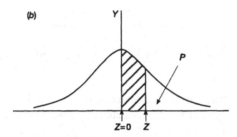

Figure 3.2. Schematic diagrams for the normal distribution curve. The unit normal deviate can be referred to as either ζ or as z in these diagrams. Note that in (a) the shaded area equals $1 - P$ and the white area P, which is given in Table 3.1(a). In (b), because the normal distribution is symmetrical, the area to the left of $z = 0$ to $-\infty$ is equal to 0.5 and is also equal to the area to the right of $z = 0$ to $+\infty$, the shaded area in (b) is $(0.5 - P)$ where P is given in Table 3.1(a).

$X = -1$ to $X = 0$ is $(0.5 - 0.15866) = 0.34134$.

For $X = +0.20$ the area P is 0.42074 and therefore the area for the range $X = 0$ to $X = +0.2$ is $(0.5 - 0.42074) = 0.07926$. We can now calculate, by addition, the probability that a height is in the range given by $X = -1$ to $X = +0.2$. This is $(0.34134 + 0.07926) = 0.4206$.

3.4 NORMAL PROBABILITY GRAPH PAPER

Special graph paper called arithmetic probability or normal probability graph paper exists for the purpose of testing data to see if it is normally distributed. However, a more rigorous statistical test, the chi-squared test of statistical significance should be used to demonstrate normality, but the graphical test is the quickest to apply for an approximate data check and a first estimate of the parameters M and S of the normal curve which can provide a fit to the data.

To illustrate the use of normal probability graph paper the observations in

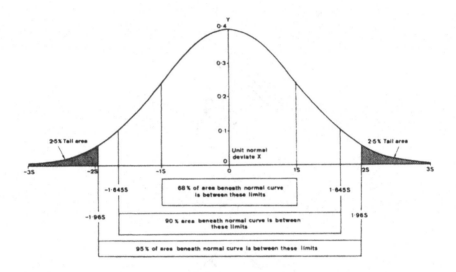

Figure 3.3. The standard normal distribution curve and values of the area beneath the curve in terms of multiples of standard deviations or standard errors.

histogram format in Figure 3.4 will be studied. These data refer to 96 monthly patient workload statistics for an eight-year period for a radiotherapy linear accelerator.

The total annual patient workload was 349, 527, 455, 464, 501, 618, 656 and 629, the range of monthly patient workloads was from 18 to 69 and, from Figure 3.4, the modal monthly workload is between 40 and 45 patients. Visually, Figure 3.4 looks approximately symmetrical and therefore might be normally distributed.

Figure 3.5 is the normal probability graph plot for the data, see Table 3.2, where the Y-raxis is a linear scale for patient workload from 20 to 70. The X-axis is the probability scale corresponding to the cumulative percentage of monthly workloads less than or equal to Y patients. If the observations are normally distributed, all the points will lie on a straight line, or at least most of them will, and the others will not be too far distant—except perhaps at the very far ends where X is less than 2% or greater than 98%. The mean of the normal, M, is the monthly workload which corresponds to the $X = 50\%$ cumulative value. This is because for the symmetrical normal curve, the mean and median coincide. $M = 43.5$ patients (the 0.5 patient is statistically acceptable although not practically possible!) in Figure 3.5.

Another property of the normal distribution curve is that 10% of the observations always lie at a distance from the mean which is greater than 1.645 standard deviations, see Figure 3.3, and half ot these observations are smaller than M and half are greater than M. The standard deviation, S, can thus be

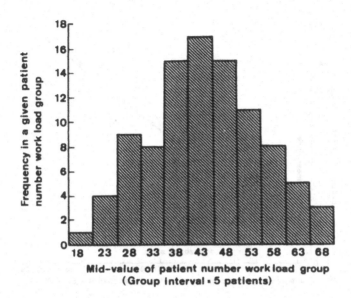

Figure 3.4. Patient workload statistics for a linear accelerator. This type of information is of interest to hospital planners who try and determine how many linear accelerators are required when building a new radiotherapy department. The lower workloads will in part be due to accelerator breakdowns and servicing when no patients can be treated, and the higher workloads will in part be due to the accelerator having to cope with an influx of patients when another treatment machine in the department cannot be used.

calculated from the formula

$$S \times 1.645 = [(Y \text{ value corresponding to } 95\% \text{ cumulation})$$
$$-(Y \text{ value corresponding to } 50\% \text{ cumulation})]$$

Hence $S = (Y_{95\%} - M)/1.645$

$S = 11.6$ for the data in Figure 3.5. To draw the shape of the normal distribution curve with $M = 43.5$ and $S = 11.6$ (Table 3.1(*b*)) can be used, computing Y-values for X-values of 43.5, 48, 53, 58, 63, 68, 73 and 78. This is illustrated in Table 3.3 and the curve is drawn in Figure 3.6.

It is *not* correct merely to superimpose this curve in Figure 3.6 on the histogram of Figure 3.4. The curve in Figure 3.6 has the ordinate Y in the formula for the general normal curve but the histogram blocks in Figure 3.4 refer to *areas beneath the normal curve* between defined limits. Thus in order to compare the observed data in Figure 3.4 directly with data *expected* from a

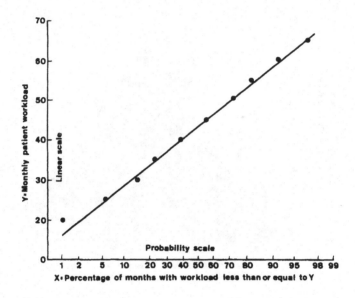

Figure 3.5. Graphical demonstration of normality using the data in Table 3.2.

normal curve with $M = 43.5$ and $S = 11.6$, data such as those in Table 3.1 must be used. This is shown in Table 3.4. Summation of the *observations* is 96 but summation of the *expectations* is 94.3. This is because there is still a small tail of the normal curve beyond the X-values of 15 and 70.

3.5   TESTING FOR NORMALITY USING THE CHI-SQUARED TEST

In the first paragraph of this section it was stated that a more rigorous test than visual assessment of a normal probability graph plot should be used. Table 3.4 illustrates the use of the chi-squared test to determine if observation and expectation are significantly different.

The information we have from Table 3.4 on page 43 to test the hypothesis that there is no difference between observation and expectation (this is called the *null* hypothesis) is

No of degrees of freedom DF $= (7 - 3) = 4$
Sum $\chi_i^2 = 1.79$

The critical value of χ^2 from statistical tables (see Table 9.1) for DF $= 4$ and $P = 0.05$, is 9.49. We conclude from this information that since 1.79 is less than 9.49, there is no significant difference, at the $P = 0.05$ level of significance, between observation of patient workload statistics and expectation from a normal distribution.

Table 3.2. Data requirements prior to a graphical demonstration of normality, see Figure 3.5.

Workload range (patient numbers)	Frequency	Cumulative frequency	Percentage cumulative frequency
16–20	1	1	1.0
21–25	4	5	5.2
26–30	9	14	14.6
31–35	8	22	22.9
36–40	15	37	38.5
41–45	17	54	56.3
46–50	15	69	71.9
51–55	11	80	83.3
56–60	8	88	91.7
61–65	5	93	96.9
66–70	3	96	100
Total = 96			

Estimation of M from Figure 3.5
$Y = 43.5$ when $X = 50\%$ therefore $M = 43.5$.
Estimation of S from Figure 3.5
$Y = 62.5$ when $X = 95\%$
therefore $S \times 1.645 = 62.5 - 43.5 = 11.6$

Table 3.3. Calculations required to be able to draw the normal curve with $M = 43.5$ and $S = 11.6$.

Workload X	$(X - 43.5)$	Unit normal deviate $(X - 43.5)/11.6$	Y-value for given unit normal deviate	$Y/11.6$
43.5	0	0	0.399	0.0344
48	4.5	0.39	0.369	0.0318
53	9.5	0.82	0.285	0.0246
58	14.5	1.25	0.183	0.0158
63	19.5	1.68	0.097	0.0084
68	24.5	2.11	0.043	0.0037
73	29.5	2.54	0.016	0.0014
78	34.5	2.97	0.005	0.0004

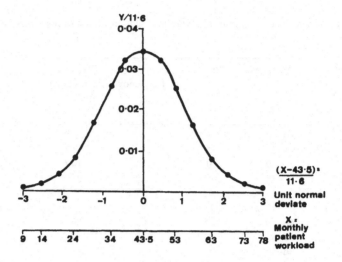

Figure 3.6. The normal curve with $M = 43.5$ and $S = 11.6$.

The area $\{1 - P(\zeta_i)\}$ in the last column of the top part of Table 3.4 denotes the area beneath the normal curve from $-\infty$ to $+\zeta_i$ where the P in Figure 3.2(a) and Table 3.1(a) is the same as $P(\zeta_i)$.

Thus for $\zeta = -2.46$ the value of $1 - P$ is given by $P(\zeta = +2.46)$ which from Table 3.1(a) is 0.007 and for $\zeta = +1.42$ the value of $1 - P$ is given by $1 - \{P(\zeta = +1.42)\}$ which from Table 3.1(a) is $1 - 0.077804$ which equals 0.922.

3.6 THE LOGNORMAL CURVE

The lognormal curve is the logarithmic transformation of the normal curve when X becomes $\log_e X$ and it is of interest since it has many applications. For instance, the following can be represented by lognormal distributions: relation between time and the death rate of bacteria; survival times subsequent to treatment of certain groups of cancer patients who die with their disease present (e.g. cancer of the cervix, cancers of the head and neck); cancer symptom duration times; distribution of sensitivities to drugs among individual animals of the same species, as measured by the dose required to cause some definite effect.

In addition, some investigators with more time than sense have also found that the following are lognormally distributed: size of foreheads of crabs; ages at second marriage; number of buttercup petals; number of words in a sentence by George Bernard Shaw.

Table 3.4. Calculations required to derive normal distribution expectations for a curve with $M = 43.5$ and $S = 11.6$.

i	X_i	$(X_i - M)$	Unit normal deviate $(X_i - M)/S = \zeta_i$	Area beneath normal curve from $-\infty$ to ζ_i $= \{1 - P(\zeta_i)\} = Q(\zeta_i)$
1	15	−28.5	−2.46	0.007
2	20	−23.5	−2.03	0.022
3	25	−18.5	−1.59	0.056
4	30	−13.5	−1.16	0.123
5	35	−8.5	−0.73	0.233
6	40	−3.5	−0.30	0.382
7	45	1.5	0.13	0.551
8	50	6.5	0.56	0.712
9	55	11.5	0.99	0.839
10	60	16.5	1.42	0.922
11	65	21.5	1.85	0.968
12	70	26.5	2.28	0.989

j	Interval $X_{J+1} - X_J$†	Observed frequency O_j	Expected frequency $N\{Q(\zeta_{J+1}) - Q(\zeta_J)\}$ $= E_J$	$(O_J - E_J)$	$(O_J - E_J)^2$	$(O_J - E_J)^2/E_J$ $=\chi_J^2$
1	15–30†	14	11.1	2.9	8.41	0.76
2	30–35	8	10.6	−2.6	6.76	0.64
3	35–40	15	14.3	0.7	0.49	0.03
4	40–45	17	16.2	0.8	0.64	0.04
5	45–50	15	15.5	−0.5	0.25	0.02
6	50–55	11	12.2	−1.2	1.44	0.12
7	55–70†	16	14.4	1.6	2.56	0.18
	Sum = N = 96		Sum = 94.3			Sum χ_J^2 = 1.79

† The intervals for the chi-squared (χ^2) test are chosen so that the individual O_J values are not too small. For this reason the observations for 15–30 and 55–70 have been grouped together.

Interval	Expected frequency
15–20	1.4
20–25	3.3
25–30	6.4
55–60	8.0
60–65	4.4
65–70	2.0

To illustrate the properties of the lognormal distribution curve, data for cancer patient symptom durations, that is the delay time between patients first noticing symptoms and receiving treatment, will be used. If the symptom time is denoted by T. then the general lognormal curve can be stated mathematically as

$$f(T) = \frac{1}{TS\sqrt{2\pi}} \exp\{-[\log_e(T/M)]^2/2S^2\}$$

where the symbol $f(T)$ is used to denote that the expression on the right-hand side of the formula is a function of T. For the lognormal:

$M = \exp(\bar{T})$ and is the median of the lognormal distribution ($\bar{T} = \log_e M$), S is the standard deviation;

$$S = \frac{1}{n-1} \sum_{1=1}^{n} [\log_e(T_i - \bar{T})]^2$$

and

$$\bar{T} = \frac{1}{n} \sum_{i=1}^{n} \log T_i$$

the generalised variable T_i being the individual symptom times in our example.

The properties of the lognormal distribution are such that:

the value of T at which the mean occurs is $M \times \exp(\frac{1}{2}S^2)$
the value of T at which the mode occurs is $M/\exp(S^2)$
the value of T at which the median occurs is M.

Using, as an example, data for 783 cancer of the stomach cases, a lognormal curve which will fit these data is one with $M = 4.1$ months and $S = 0.6$. Since $e^{0.36} = 1.433$ and $e^{0.18} = 1.197$,

the value of T at which the mean occurs is $4.1 \times 1.197 = 4.91$ months
the value of T at which the mode occurs is $4.1/1.433 = 2.86$ months
the value of T at which the median occurs is 4.1 months.

$\bar{T} = \log_e M$ and is the *mean log survival time* and the logarithm on the right-hand side of the formula for $f(T)$ can be written alternatively as $\log_e(T/M)$ or $[\log_e T - \log_e M]$ or $[(\log_e T) - \bar{T}]$. For the stomach cancer symptom time example, $\bar{T} = \log_e(4.1) = 1.411$ and the values of T at which the mean, mode and median occur can alternatively be stated in terms of $\bar{T}$ rather than in terms of M. Thus the relative positions of mean, median and mode are at

$$T_{\text{mean}} = \exp(\bar{T} + \tfrac{1}{2}S^2), \quad T_{\text{median}} = \exp(\bar{T}) \quad \text{and} \quad T_{\text{mode}} = \exp(\bar{T} - S^2)$$

which for this example gives the values $\exp[1.411 + \frac{1}{2}(0.36)] = \exp(1.591) = 4.91$ months, $\exp(1.411) = 4.10$ months, and $\exp(1.411-0.36) = \exp(1.051) = 2.86$ months. The answers are the same as before but the two presentations are included to emphasise that there is a possibility of making mistakes if absolute values of times and logarithms of times are mixed up.

Table 3.5. Calculations required to be able to draw the lognormal curve with $M = 4.1$ and $S = 0.6$. The constants are: $2S^2 = 0.72$ and $1/S\sqrt{2\pi} = 0.665$. The mode, median and mean are indicated in the footnote.

T	$T/4.1$	$\log_e(T/M)$	$\log_e(T/M)^2/0.72$ $= A$	e^{-A}	$0.665/T$	$0.665e^{-A}/T$ $= f(T)$
1.0	0.244	−1.411	2.765	0.063	0.665	0.042
2.0	0.488	−0.718	0.716	0.489	0.332	0.163
2.86†	0.698	−0.360	0.180	0.835	0.233	0.194
4.10‡	1.000	0	0	1.000	0.162	0.162
4.91§	1.198	0.180	0.045	0.956	0.135	0.129
5.5	1.342	0.294	0.120	0.887	0.121	0.107
6.0	1.463	0.381	0.201	0.818	0.111	0.091
8.0	1.951	0.668	0.621	0.538	0.083	0.045
12.0	2.927	1.074	1.602	0.202	0.055	0.011
16.0	3.902	1.362	2.575	0.076	0.042	0.003

† Mode.
‡ Median.
§ Mean.

Table 3.5 illustrates the arithmetical calculation steps required to compute $f(T)$ for the stomach cancer symptom time data curve with $M = 4.1$ and $S = 0.6$. and the curve is drawn in Figure 3.7 with the positions of the mode, median and mean shown for this positively skewed (see also Figure 2.1) curve.

As well as *normal probability graph paper* for demonstrating normality, Figure 3.5, *logarithmic probability graph paper* is available to demonstrate lognormality. Figure 3.8 illustrates the method, see also Table 3.6. In a similar manner to that described on page 40, we have for the lognormal distribution data in a log-probability plot where M corresponds to the 50% cumulation

$$S \times 1.645 = (\log_{10} T \text{ value corresponding to 95\% cumulation})$$
$$- (\log_{10} T \text{ value corresponding to 50\% cumulation})$$

hence
$$S = \frac{\log_{10}(T_{95\%}) - M}{1.645}$$

The lognormal curve with $M = 4.1$ and $S = 0.6$ is shown in Figure 3.7, represented as a negatively skewed curve drawn on a horizontal *linear time* axis. If the horizontal time axis is transformed to a logarithmic scale, say to base 2, with areas beneath the lognormal curve between defined limits on a logarithmic

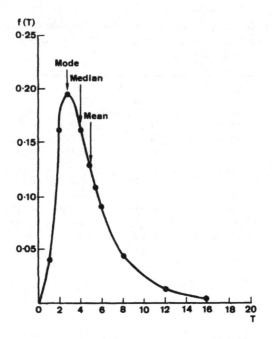

Figure 3.7. Lognormal curve with $M = 4.1$ and $S = 0.6$.

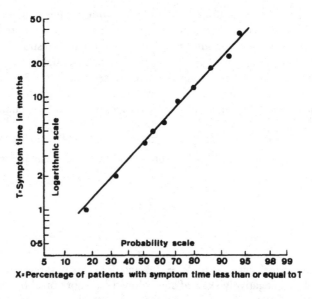

Figure 3.8. Graphical demonstration of lognormality using the data in Table 3.6.

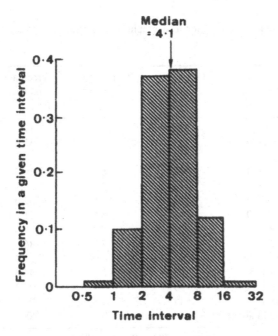

Figure 3.9. Lognormal distribution with $M = 4.1$ and $S = 0.6$ illustrated with a logarithmic scale for the horizontal axis (logarithms to base 2).

scale (e.g. 0.5, 1, 2, 4, 8, 16, 32, 64,....,) illustrated as histogram blocks, then the symmetrical bell-shaped pattern of the normal distribution is demonstrated, Figure 3.9 (see also Table 3.7). This transformation property gives the lognormal distribution curve its name.

The area $\{1 - P(\zeta_i)\}$ in the fourth column of Table 3.7 denotes the area beneath the normal curve from $-\infty$ to ζ_i where the P in Figure 3.2(a) and Table 3.1(a) is the same as $P(\zeta_i)$. A similar notation is used in Table 3.4.

Table 3.6. Data requirements prior to a graphical demonstration of lognormality, see Figure 3.8.

Symptom time range (months)	Frequency	Cumulative frequency	Percentage cumulative frequency
0–1	138	17.6	17.6
1.1–2	111	14.2	31.8
2.1–3	85	10.9	42.7
3.1–4	63	8.0	50.7
4.1–5	42	5.4	56.1
5.1–6	51	6.5	62.6
6.1–9	67	8.6	71.2
9.1–12	66	8.4	79.6
12.1–18	56	7 2	86.8
18.1–24	36	4.6	91.4
24.1–36	23	2.9	94.3
36.1–48	7	0.9	95.2
48.1–60	11	1.4	96.6
60.1 and above	27	3.4	100

Total = 783   Total = 100%

Estimation of M from Figure 3.8
$Y = 4.1$ when $X = 50\%$
Therefore $M = 4.1$.
Estimation of S from Figure 3.8
$Y = 39$ when $X = 95\%$
$\log_{10}(4.1) = 0.6128$ and $\log_{10}(39) = 1.5911$.
Therefore $S = (1.5911 - 0.6128)/1.645 = 0.595$.

Table 3.7. Data requirements for Figure 3.9.

i	T_i	Unit normal deviate $[\log_e(T_i/4.1)]/S = \zeta_i$	Area beneath normal curve from $-\infty$ to $\zeta_i = \{1 - P(\zeta_i)\} = Q(\zeta_i)$	$\{Q(\zeta_{j+1}) - Q(\zeta_j)\}$
1	0.5	−3.507	0.0002	0.009
2	1	−2.352	0.009	0.106
3	2	−1.196	0.115	0.369
4	4	−0.041	0.484	0.382
5	8	1.114	0.886	0.122
6	16	2.269	0.988	0.012
7	32	3.425	0.9997	0.0003
				Sum= 1.000†

† The total area beneath the normal curve is 1 and this summation provides a check on the arithmetic so long as the limits of the unit normal deviate are approximately ±3.

The above has been computed manually from tables of areas beneath the normal curve. However, in practice, if many of these calculations were required then a computer program would be most appropriate. In such a program, instead of storing a look-up table of values, the area $P(\zeta = x)$ can be computed directly from one of several available polynomial expressions which give good approximations, e.g.

$$P(x) = 1 - \tfrac{1}{2}(1 + d_1x + d_2x^2 + d_3x^3 + d_4x^4 + d_5x^5 + d_6x^6)^{-16}$$

$$d_1 = 0.04986\ 73470 \qquad d_4 = 0.00003\ 80036$$
$$d_2 = 0.02114\ 10061 \qquad d_5 = 0.00004\ 88906$$
$$d_3 = 0.00327\ 76263 \qquad d_6 = 0.00000\ 53830$$

THE PSYCHOANALYSIS OF MICKEY MOUSE

Chapter 4

Introduction to Sampling, Errors, Accuracy and Precision

4.1 INTRODUCTION

In statistical parlance the term population refers to the group of objects, events, results of procedures or observations (rather than the geographical connotation of population relating only to persons in a country or state etc) which is so large a group that usually it cannot be given exact numerical values for statstics such as the population mean μ or the population standard deviation σ. These statistics therefore can only be estimated.

To obtain for example, an estimate of the population mean μ of a certain characteristic x of the population, *sampling* must first take place because all the values of x for the entire population cannot be measured. Only a small part of the population can be surveyed and that part is called a *sample*.

There are various methods of sampling, including *random sampling*, which for clinical trials is discussed in a later chapter as simple randomisation, stratified randomisation and balanced randomisation. A type of sampling which is not appropriate is the so-called *judgement sampling*, in which the investigator makes his choice by personal whim, which can thus lead to biased results.

From a knowledge of this sample the sample mean x_m can be found and a *statistical inference* (i.e. drawing a conclusion about a population from a sample) can be drawn about 'how good' is this value x_m as an estimate of the true population mean μ. The phrase 'how good' can be stated in terms of *confidence limits*.

The standard deviation s_m of the sample mean x_m tells you about the spread of the measured sample values $x_1, x_2, \ldots x_i \ldots,$. The method of calculation of a standard deviation such as s_m is described in section 2.4. If the *sampling experiment* to measure x_m is then repeated N times, with the sample size n always remaining the same, a total of N values of x_m will be obtained. If these are then averaged, then M, which is the *mean of means* or *grand mean* is obtained.

The standard deviation of the mean of means M is given a special name: *standard error of the mean*, where

$$SE = \text{Sample standard deviation}/\sqrt{n}.$$

This value SE tells how accurately we know the mean of means M.

Accuracy and *precision* (which are also sometimes termed *validity* and *reliability*) should not be confused. Accuracy can be described as how closely the result of an experiment or trial agrees with the true or most probable value, whereas precision can be described as how closely different measurements of the same quantity agree with each other. Standard deviation and coefficient of variation are measures of precision. Figure 4.1 illustrates this difference between accuracy and precision.

The word *error* has been used above in standard error of the mean, but there are also other types of error which will be discussed in this chapter: gross accidental errors, systematic errors, errors of interpretation and random errors.

In Chapter 8 when introducing statistical significance we will also encounter type I and type II errors which are associated respectively with α risks (which can be chosen in a clinical trial to be $P = 0.05$) and β risks which in clinical trial design (as well as for other topics) are related to statistical power $(1 - \beta)$.

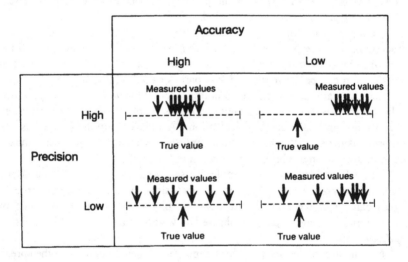

Figure 4.1. A high *accuracy* (i.e. high validity) means that in repeated measurements the results fall very close to each other; conversely, a low *precision* (i.e. low reliability) means that they are scattered. Accuracy determines how close the mean of repeated results is to the true value. A low accuracy will produce more problems when interpreting results than a low reliability. (Courtesy: World Health Organisation.)

4.2 SAMPLE DISTRIBUTION AND SAMPLING DISTRIBUTION OF THE SAMPLE MEAN

The title of this section is a real mouthful of words with sample/sampling mentioned three times! To help understand what is meant by these two distributions we have Figures 4.2 and 4.3 but first to make matters as easy as possible let us define the general notation which is used.

Sample mean from a random sample $= x_m$
Sample standard deviation $= s_m$
Mean of means $= M$
Standard deviation of the mean of means (i.e. the standard error of the mean) $= SE$
Population mean $= \mu$
Population standard deviation $= \sigma$
Size of a random sample $= n$

Specifically for the example in Figures 4.2 and 4.3 let us assume that $\mu = 2$ and the size of the total population is 20 000 and the size of a random sample is $n = 100$ (i.e. 0.5% of the population) and also assume that 4000 of the population have a value of x between 1.5 and 2.0. In this case we would expect some 20% of 100, i.e. 20 of the x values in the random sample lying between these limits of 1.5 and 2.0. We would indeed, expect the sample frequency curve (the lower curve in Figure 4.2) to have a similar frequency curve to that of the population (the upper curve in Figure 4.2).

The *sample distribution* is the lower curve in Figure 4.2 but a *sampling distribution* is different, and is a distribution of values from a mass of samples, one value per sample. This is seen as the lower curve in Figure 4.3 which is symmetrical and bell-shaped.

In general this *sampling distribution* can be imagined as being obtained after taking thousands of samples (**not** samples of thousands) and selecting one item of information from each sample, which in this example is x_m.

4.3 CENTRAL LIMIT THEOREM

Following on from the end of the previous section, the frequency curve of the sampling distribution of the sample mean x_m is very different from the population frequency curve but it clusters about the population mean μ and is, as mentioned above, both symmetrical and bell-shaped. These properties are recognised in what is known as the *central limit theorem* which can be stated as follows.

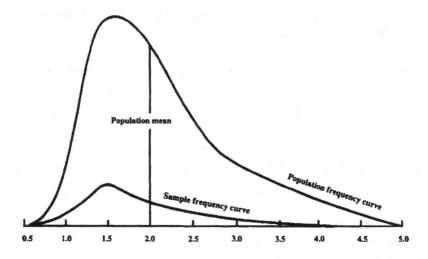

Figure 4.2. Frequency distributions of an item x (described on the horizontal axis) for a population and a sample drawn from that population. The population mean is μ and the sample mean is x_m. Note that the sample mean and dispersion will be approximately the same as the mean and dispersion of the population.

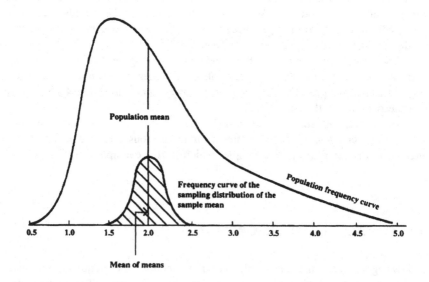

Figure 4.3. Frequency distributions of the population with mean μ and the sampling distribution of the sample mean (x_m) of all the samples taken from the population. The mean of means is M.

If x is a random variable with mean μ and standard deviation σ then if M is the mean of a random sample of size n chosen from the distribution of x_m values, then the distribution of

$$(M - \mu)/\{\sigma/\sqrt{n}\}$$

tends towards the *standard normal distribution* as n tends towards infinity.

As an example consider values of x_m for N random samples (where N is very large) of size $n = 64$ (thus obtaining N mean values of x_m equal to $x_1, x_2, \ldots x_i, \ldots x_N$) from a much larger population which is known to have a mean $\mu = 100$ and a standard deviation $\sigma = 16$. The question we wish to answer is: 'What is the probability of an observation x_m as high as 104 occurring when we draw N samples of size $n = 64$?'.

Assuming a mean of means $M = 100$ and a sample standard deviation (i.e. a standard error of the mean) of $SE = \{16/\sqrt{64}\} = 2$, then since $(104-100) = 4$ an observation of $x_m = 104$ differs from M by two standard errors.

From Table 3.1(c) it is seen that the area under the normal curve above the limit of two standard deviations is 0.023 which is the probability of obtaining an observation of 104 or greater. This is an example of a *one-tail test* situation. A *two-tail test* situation would be if we were interested in having an observation of *either* 104 or greater, *or* 96 or less. From Table 3.1(c) we this time obtain the probability of 0.046.

Referring again to Figure 4.3, the sampling distribution of the sample mean is only drawn schematically whereas Figure 4.4 shows actual data for a distribution of means x_m from 2000 samples of 5 random digits generated on a computer with the approximate normal distributions superimposed on the frequency distribution of x_m values. The mean $\mu = (0 + 1 + 2 + 3 + \ldots + 9)/10 = 4.5$ and the standard deviation can be calculated as in Chapter 2 as $\sigma = \sqrt{([\{1^2 + 2^2 + 3^2 + \ldots + 9^2\} - \{4.5\}^2]/9)} = \sqrt{(82.5/9)} = 3.03$. The standard error of the mean is thus $SE = 3.03/\sqrt{5} = 1.30$. The approximating normal distribution therefore has a mean of 4.5 and a standard deviation of 1.30.

[NOTE Some authors make no distinction between a divisor of n and a divisor of $(n - 1)$ when computing standard deviations for probability distributions. If in this example we used n rather than the $(n - 1)$ of the examples in Chapter 2 the value for S would have been $S = 2.87/\sqrt{5} = 1.28$.]

4.4 FORMULAE FOR STANDARD ERRORS

[1] Standard error of the mean

$$SE = \text{Standard deviation of the sample}/\sqrt{\text{Sample size}}$$

This formula $SE = s_m/\sqrt{n}$ has already been quoted in section 4.1.

[2] Standard error of the difference between two means M_1 and M_2 of populations 1 and 2 where the sample sizes are n_1 and n_2 and the standard

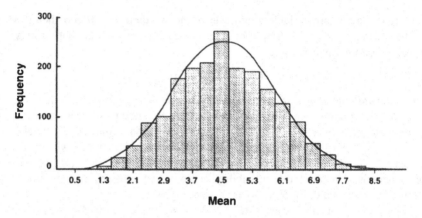

Figure 4.4. Distribution of means from 2000 samples of 5 random number digits with an approximating normal distribution.

deviations are S_1 and S_2

$$SE = \sqrt{\{[S_1^2/n_1] + [S_2^2/n_2]\}}$$

[3] Standard error of a percentage $P\%$

$$SE = \sqrt{\{[P(100 - P)]/n\}}$$

[4] Standard error of the difference between two percentages $P_1\%$ and $P_2\%$

$$SE = \sqrt{\{[P_1(100 - P_1)/n_1] + [P_2(100 - P_2)/n_2]\}}$$

4.5 GROSS ACCIDENTAL ERRORS

These are irregular errors, which could have been avoided and were caused by incorrect technique. Examples include transcription errors, errors in calculation and mismatching units of measurement.

4.6 SYSTEMATIC ERRORS

These consistently give a result which is wrong by a fixed amount and may be due to instrument errors such as zero adjustment or to calibration errors, faulty primary standards and personal errors in reading instrument measurements.

4.7 ERRORS OF INTERPRETATION

Many examples could be given of these types of errors, but that described by Altman[1] is a particularly good example. The story is as follows. In 1949 a divorce case was heard in which the sole evidence of adultery was that a baby was born almost 50 weeks after the husband had gone abroad on military service. A divorce was not granted. On appeal the judges agreed that the limit of credibility had to be drawn somewhere, but on medical evidence 349 days, whilst improbable, was scientifically possible. The appeal failed. The judges apparently did not look at the distribution of length of gestation, Figure 4.5, which shows that, although scientifically possible, a pregnancy lasting 349 days is an extremely unlikely occurrence.

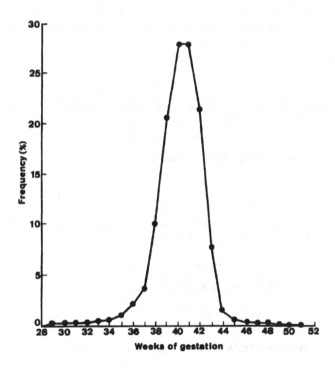

Figure 4.5. Distribution (shown in the form of a frequency polygon) of length of gestation measured using the standard convention of counting in completed weeks from the first day of the last menstrual period with conception assumed to have occurred 14 days later.

4.8 RANDOM ERRORS

If the same observer repeatedly makes the same measurements under apparently identical conditions, slight variations, called random errors, will occur. If a sufficiently large number of measurements are taken, the mean value will be close to the true value since positive and negative values are equally likely.

Also, the distribution of random errors will be approximately normal. Thus 68.3% of the errors lie within ±1 standard deviation, 95.4% within ±2 standard deviations, and 99.7% within ±3 standard deviations.

4.9 COMBINATION OF ERRORS

If there are two terms A and B which are used to calculate a final result and the error of the A-readings is $2e_A$ and of the the B-readings is $2e_B$ then the values of A and B can be quoted as $A \pm e_A$ and $B \pm e_B$. The final result in terms of whether A and B are added, subtracted, multiplied or divided, determines the way in which the errors e_A and e_B are combined.

When A and B are added together,

$$(A + B) \pm (e_A + e_B)$$

is the result and the combined error when B is subtracted from A is also as above:

$$(A - B) \pm (e_A + e_B)$$

However, when A is multiplied by B we have

$$A \cdot B \pm (A \cdot e_B + B \cdot e_A)$$

e_A and e_B are absolute errors, but the relative errors $E_A = |e_A/A|$ and $E_B = |e_B/A|$ can also be used in an alternative formula to calculate the absolute errors. $A \cdot B \pm (A \cdot e_B + B \cdot e_A)$ can be expressed as

$$A \cdot B \pm A \cdot B \cdot (E_A + E_B).$$

When A is divided by B, the formula is

$$A/B \pm A \cdot (E - A + E_B)/B.$$

4.10 ROOT MEAN SQUARE ERROR

A percentage root mean square (RMS) error can be used to compare, for example, two sets of experimental data for which there are N values for each set. One set can be used as a reference set, $R_1, R_2, R_3, \ldots R_i \ldots R_N$. The second set is $X_1, X_2, X_3, \ldots X_i \ldots X_N$. The formula for the percentage RMS is given below

$$\sqrt{\frac{\sum_{i=1}^{i=N}[(X_i - R_i)/R_i]^2}{N}} \times 100\%$$

One example of the use of a percentage RMS error is when four different measurement systems are used to determine the source strength of an ^{192}Ir high dose rate brachytherapy source. The term $[(X_i - R_i)/R_i]$ is sometimes termed the *fractional error*. A full description of the ^{192}Ir source measurement errors would be the percentage RMS together with the range of fractional errors.

4.11 CONFIDENCE LIMITS

As a first illustration of the use of confidence limits (others will follow in later chapters) consider the example given by Swinscow[2]. If the mean diastolic blood pressure of 72 people in a defined population is 88 mm Hg with a standard deviation of 4.5, then the standard error of the mean is $4.5/\sqrt{72} = 0.53$ mm Hg.

Confidence limits are related to standard errors. Since the distribution of means is normal, using the mean of 88 mm Hg and standard error of 0.53 mm Hg it can be stated that the sample mean ±1.96 times its standard error gives the following:

$$88 + (1.96 \times 0.53) = 89.04 \text{ mm Hg}$$
$$88 - (1.96 \times 0.53) = 86.96 \text{ mm Hg.}$$

These are called the 95% confidence limits and they are interpreted by saying that there is only a 5% probability that the range from 86.96 to 89.04 mm Hg excludes the population mean. With small samples (say 30) larger multiples of the standard error are required to compute the confidence limits and these are obtained from tables of the t-distribution rather than from tables of the normal distribution. Confidence limits are useful in deriving a range within which the true result is 95% likely to fall.

AUTHOR ERRORS
MAKING TOO MANY CHANGES

A pictogram, with apologies to Clint Eastwood, which is suitable for editors to send to authors when the latter make too many changes at final page proof stage!

Go ahead...

*Make one
more
change!*

Chapter 5

Introduction to Probability

5.1 DESCRIPTIONS OF PROBABILITY

The French mathematician Pierre Simon Laplace (1749–1827) described the theory of probability as only common sense reduced to calculation, in that it exhibits with accuracy what reasonable minds feel by a kind of instinct, without being able to describe it for themselves.

Another description could be the numerical expression of uncertainty which can take any value between 0 and 1 where 0 equates to definitely improbable (never happens) and 1 equates to definitely probable (always happens). The numerical values between 0 and 1 express degrees of belief for an eventuality. Probability is often introduced in terms of the likelihood of events when throwing one or more dice (dice is the plural for die) since probability is a study of random or non-deterministic experiments. Thus if a single die is thrown into the air, it is certain that it will come down, but not certain that a 6 will appear. However, if this die-throwing experiment is repeated n times and the number of times a 6 appears (call this a success) is s, then it has been empirically observed that the ratio $f = s/n$ will approach a limiting value as $n \to \infty$, That is, it will become stable. f is called a *relative frequency*. In probability theory, the probabilities associated with experimental events are the limiting values of the relative frequencies. It is also necessary in probability theory to distinguish between two types of event, Figure 5.1.

An example of two *independent* events is the die throwing of a 6 followed by a 5, since the outcome of the first throw (the 6) does not affect the outcome of the second throw (the 5). An example of two dependent events would be with a pack of cards, drawing first a 6 of any suit (hearts, spades, clubs, diamonds) and then a 5 of any suit, *without replacement of drawn cards*. After the 6 has been drawn from the full pack of 52 cards, there will only be 51 cards left from which the 5 is then drawn. The probability in this situation of drawing the 6 is 4/52 and of the 5 is 4/51. The probability of drawing at the third opportunity a second 6, would be 3/50. If, though, the card experiment was modified such that after each draw, the drawn card was replaced in the pack, then each draw,

Those which can occur SIMULTANEOUSLY. Two events are INDEPENDENT if the occurrence or non-occurrence of one event does not affect the probability of the occurrence or non-occurrence of the other.

Those which can only occur CONSECUTIVELY, since one of them PRECLUDES the other. These events are termed "Mutually Exclusive" events.

Figure 5.1. Types of event.

first, second, third, etc, would be independent and the probabilities previously quoted as 4/52, 4/51 and 3/50 would all be 4/52 (= 0.077). The probabilities are *a priori probabilities* in that they are calculated from a knowledge of the experimental conditions before the experiment is actually performed. An *empirical probability*, using the deck of cards, with card replacement after each draw, would be 0.070 if, in 1000 draws, 70 outcomes were cards with a 6 of any suit. Returning to the example of card drawing without replacement of drawn cards, the probabilities would be

$$\text{Pr(6 on the first draw)} = 4/52$$
$$\text{Pr(6 on the second draw)} = 3/51.$$

In this instance, 3/51 is a *conditional probability* since the probability of drawing the second 6 is conditional on the first draw having also been a 6 and there being no replacement of drawn cards.

A simple demonstration of *a priori* and *empirical* probabilities can be made with the throw of a die. The experimental outcomes of 50 throws were

5	5	3	1	4	6	6	1	4	5
2	6	5	1	2	1	1	1	6	5
5	5	4	3	6	5	5	4	1	3
2	4	5	3	2	2	5	2	4	3
1	3	5	4	2	3	3	1	4	6

which, for 1, 2, 3, 4, 5 and 6 give empirical probabilities of 9/50, 7/50, 8/50,

8/50, 12/50 and 6/50, which in decimals are 0.18, 0.14, 0.16, 0.16, 0.24 and 0.12. The *a priori* probability is $1/6 = 0.167$.

The empirical probabilities are in fact relative frequencies and these will stabilise as the number of throws, n, increases. 50 is a relatively small value of n for an experiment with 6 possible outcomes but, even so, it can be seen that they are more stable for $n = 50$ than for $n = 30$ (0.23, 0.07, 0.10, 0.13, 0.30 and 0.17).

5.2   THE TWO LAWS OF PROBABILITY

The two fundamental laws of probability are the multiplication law and the addition law, Figure 5.2. An example of an application of the *multiplication law* is that when throwing two dice there are 36 equally likely possibilities, but only one of these possibilities is a double 6. The probability of a double 6 is $1/6 \times 1/6 = 1/36 = 0.028$.

As a second example, for a pack of cards with replacement of drawn cards, the probability of a 6 card in the first draw and a 6 card (each of any suit) in the second draw is $4/52 \times 4/52 = 16/2704 = 0.006$.

An example of an application of the *addition law* is when using a single die: a 5 and a 6 cannot turn up together in a single throw but each number has an equally likely probability of $1/6$. The probability of either a 5 or a 6 turning up is $1/6 + 1/6 = 1/3 = 0.33$. As a second example of the application of this law: the probability of drawing from a pack of cards either a 6 or a 5, which are mutually exclusive outcomes since both cannot occur together in a single 1-card draw, is $4/52 + 4/52 = 8/52 = 0.154$.

5.3   BAYESIAN PROBABILITY

The probability we have been speaking of in the previous two sections is *classical probability* and its notion is based historically on the well known ratio

Number of favourable events/Number of possible events

which is a definition of probability credited to Jakob Bernoulli (1654–1705), and to Laplace who was mentioned in the first sentence of this chapter. All possible events can occur and every probability P is assigned a number between zero and one, $0 \leqslant P \leqslant 1$, and an impossible outcome has a probability of zero.

Bayesian statistics are named after an 18th century mathematician, the Reverend Thomas Bayes (1702–1761) who was interested in *conditional probability*, i.e. the probability that an event would occur under a given condition. An example of their use is given in section 21.12.2 for treatment optimisation, where it is seen that a Bayesian probability is a statement of a *personal probability* (i.e. expressing a degree of belief) and as such focuses as

Figure 5.2. The two fundamental laws of classical probability.

much attention on the decision maker as on the process of phenomenon under study.

Bayes' theorem is given below, where the vertical line between D and T^+ is read as 'given' or 'conditional on'. In an example relating to *diagnostic tests*, D is the disease and T^+ signifies a positive test result. The conditional probability $\text{Prob}(D|T^+)$ is the probability of having the disease D given that the result of the diagnostic test is positive.

Bayes' theorem applied to diagnostic tests is given below as a formula. $\text{Prob}(D)$ is the probability of having the disease D, $\text{Prob}(T^+)$ is the probability of having a positive test result, and $\text{Prob}(T^+|D)$ is the conditional probability of a positive test result given that the person has the disease.

These probabilities for a diagnostic test will dealt with in more detail in Chapter 15 where it is seen that they have special names. For example, $\text{Prob}(D|T^+)$ is called the *positive predictive value* and $\text{Prob}(T^+|D)$ is a true positive rate where the *sensitivity* of the test is the proportion of truly diseased persons in the screened population who are identified as diseased by the screening test. Specificity and negative predictive values are also given in Chapter 19.

$$\text{Prob}(D|T^+) = \{\text{Prob}(T^+|D) \cdot \text{Prob}(D)\}/\{\text{Prob}(T^+)\}$$

Stating this formula in words, Bayes' theorem says that the probability of disease, given a positive test, is the product of the probability of a positive test result given disease, multiplied by the probability of disease, with the product of these two probabilities divided by the probability of a positive test.

The probability Prob(D) is a *prior probability* because it is the probability that a test subject has the disease before (*a priori*) the test result is known. Prob(T^+) can also be expressed as the sum of two conditional probabilities since any subject must either be diseased (D) or disease-free (D^F):

$$\text{Prob}(T^+) = \{\text{Prob}(T^+|D) \cdot \text{Prob}(D) + \text{Prob}(T^+|D^F) \cdot \text{Prob}(D^F)\}$$

AN UNLIKELY PROBABILITY

Computers are useful, especially to take the labour out of lengthy calculations. However, care must be taken in writing computer programmes and in keying in data. In 1984 in Boraas county, Sweden, the county's computer was being used to update population statistics by taking into account recent deaths. Personal registration numbers in Sweden are based on a 6-digit number linked by a hyphen to a 4-digit number. The computer operator keyed in various data but unfortunately the programme was such that the hyphen was interpreted as a minus sign. The computer subtracted the second number from the first and interpreted the resultant number as obsolete. It then erased the records of several thousands of Boraas citizens from its memory: death by computer! Several million pounds and hundreds of hours were required to resurrect the dead Swedes.

Chapter 6

Binomial Probabilities

6.1 PERMUTATIONS AND COMBINATIONS

A knowledge of *permutations* and *combinations* is necessary before the binomial distribution is discussed, since they are needed for the calculations involved with binomial problems. The difference between the two terms is that for *combinations* the order of objects *is not* important, whereas for *permutations* the order *is* important.

To illustrate this, consider three objects A, B and C. The group of three will be called a *set*, and a *permutation* is an arrangement of a set or part of a set. Thus, permutations of *two from three* are

$$AB \; BA \; AC \; CA \; BC \; CB$$

which are written as $^3P_2$, that is, $^3P_2 = 6$.

If we now consider a set of four objects A, B, C and D, then permutations of *three from four* are

$$
\begin{array}{cccccc}
ABC & ACB & BAC & BCA & CAB & CBA \\
ABD & ADB & BAD & BDA & DAB & DBA \\
ACD & ADC & CAD & CDA & DAC & DCA \\
BCD & BDC & CBD & CDB & DBC & DCB
\end{array}
$$

which are written as $^4P_3$.

A general formula for obtaining permutations of r *from* n is

$$^nP_r = \frac{n!}{(n-r)!}$$

where $n!$ is called n *factorial*. $(1!) = 1$; $(2!) = 1 \times 2 = 2$; $(3!) = 1 \times 2 \times 3 = 6$; $(4!) = 1 \times 2 \times 3 \times 424$; $(n!) = (1 \times 2 \times 3 \times 4 \times 5 \times \cdots \times n)$.

This formula can be used for $^3P_2$ where $n = 3$ and $r = 2$:

$$^3P_2 = \frac{3!}{(3-2)!} = 6$$

For $^4P_3$ where $n = 4$ and $r = 3$,

$$^4P_3 = \frac{4!}{(4-3)!} = 24$$

The answers 6 and 24 can be checked by counting the permutations in the boxes.

For *combinations*, the order is not important, and thus, in our first example, although AB and BA count as *two* permutations, they only count as *one* combination; similarly, in the second example, ABC, BCA and CAB count as *three* permutations, but only as *one* combination. Thus, the combinations of *three from four* are

ABC
ABD
ACD
BCD

which are written as $^4C_3 = 4$.

The general formula for obtaining combinations of r *from* n is

$$^nC_r = \frac{n!}{r!(n-r)!}$$

There is an extra term $(r!)$ in the denominator of the above formula compared with the formula for ^nP_r, since there are always fewer combinations than permutations for the same r *from* n. An alternative symbolism for ^nC_r, is

$$\binom{n}{r}$$

In practice, the formulae for ^nP_r and ^nC_r are useful, since it is often too laborious to tabulate all the possible combinations or permutations.

For example, the number of ways in which two cards can be dealt from a pack of 52, when order does not matter, is $^{52}C_2$ or using the formula $52!/(2! \times 50!)$. Many terms cancel out in the numerator and denominator, leaving

$$^{52}C_2 = \frac{(52 \times 51)}{(2 \times 1)} = 1326.$$

If we are interested in drawing the ace of hearts and king of hearts, either ace first *or* king first, then we have one chance of success in 1326 attempts. A more complicated example is the chance that the first two cards dealt are a pontoon (an ace and a face card) and this is calculated as follows. There are twelve face cards and four aces in the pack, thus there are 48 different combinations which give a pontoon. The chance of drawing a pontoon is, therefore, 48 in 1326, i.e. 48/1326, or *one in 27.6*, or *a probability of 0.036.*

6.2 THE BINOMIAL DISTRIBUTION

The *binomial distribution* refers to simple yes or no, black or white, 0 or 1, dead or alive situations where there are only two alternatives, and it can be used to determine whether the results observed in a *trial-experiment* situation could have occurred *randomly.* This is important because if the results could quite easily have occurred by chance, *no significant conclusions can be drawn.*

Suppose that we have a trial in which the outcome can only be one of two events, A or B, and let the *probability* of event A be (p). Since the *total probability* is 1, the probability of event B is $(1 - p)$, and could be denoted (q). We could, for example, say that event A was success and event B failure, and so the binomial probability for success is p and the binomial probability of failure is q, where $p + q = 1$. Now suppose that the identical trial is conducted n times. This can be referred to as a *sample* of n trials. The problem which can be solved by *binomial* theory is:

What is the probability distribution
of the numbers of successes (As)
in the sample of n trials?

If the number of As (successes) is equal to r, then the number of Bs (failures) will be equal to $(n - r)$. The *binomial distribution* gives the probability that the sample of n contains (r) As and $(n - r)$ Bs.

The binomial distribution is written below where the first factor ^nC_r, can be recognised as a *combination* with formula ^nC_r.

Binomial probability of r successes in n trials
$$= {}^nC_r \cdot p^r \cdot (1 - p)^{n-r}$$

This combination ^nC_r is called the *binomial coefficient*, and can be obtained using a diagram called *Pascal's triangle*, Figure 6.1, without having to use any formula. In the triangle, the numbers can be extended *downwards.* Each entry is obtained as the sum of the two adjacent numbers of the line above. Thus, in the row for $n = 2$, the middle number 2 in the row 1 2 1 is obtained by $(1 + 1)$ from row $n = 1$. Similarly, the 5 in the row for $n = 5$ is obtained by $(1 + 4)$ from the row with $n = 4$. Each row contains the binomial coefficients ^nC_r for that particular value of n. Thus, for the row $n = 3$, there are four entries 1 3 3 1 which correspond to $^3C_0$, $^3C_1$, $^3C_2$, $^3C_3$, respectively.

Before any examples of the use of the binomial distribution are given, Table 6.1 summarises the three conditions under which a trial can be considered to be a binomial situation.

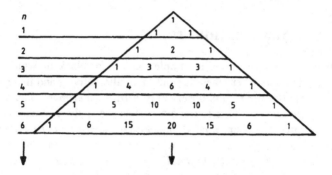

Figure 6.1. Pascal's triangle.

Table 6.1. Conditions for a binomial trial.

1	The experiment consists of a fixed number of trials, n
2	Each trial has only two possible outcomes, usually called success and failure.
3	The outcome of any trial is independent of the outcome of any other trial.

Condition 3 in Table 6.1 for a trial with a pack of cards where success was signified by drawing an ace, would be satisfied if, after each draw, the card chosen was replaced in the deck of cards before the next draw. Otherwise the trials would *not* be independent, since the second trial would be related to the first because of the removal of the first card. If on the first draw the card was an ace, the initial probability of success, p, would be 4/52 and if there was no replacement, the probability of an ace for a second draw would be $p = 3/51$. This would not be a trial when the binomial distribution could be used since p must be the same throughout all the n binomial trials.

6.3 EXAMPLES OF BINOMIAL PROBLEMS

To illustrate the use of the binomial probability formula and of Pascal's triangle, Figure 6.1, consider the simple experiment of Table 6.2. From Pascal's triangle for $n = 3$, the binomial coefficients ${}^nC_r$ are 1, 3, 3 and 1 for $r = 0, 1, 2$ and 3, respectively. The method of solution and answers to the problem are given in Table 6.3 and Figure 6.2.

A binomial situation of historical importance is the work of Sir Edward Jenner on smallpox vaccination (an enquiry into the causes and effects of the variolae vaccinae, 1798). A sample of 23 people was infected with cowpox ($n = 23$). The probability of contracting smallpox when inoculated with the virus was some 90% ($p = 0.9$), but none of the previously vaccinated 23 people did in fact contract smallpox ($r = 0$). The binomial probability of such an event occurring is exceedingly small, and the observations are therefore definitely not random.

Table 6.2. Description of a binomial problem.

Experiment	5 balls in an opaque box, 3 are red and 2 are black. Three balls are drawn successively, and after each has been drawn and recorded, it is replaced prior to the next draw.
Problem?	What is the probability that 0,1, 2 or 3 red balls will be drawn?
Number of trials, n	$n = 3$
Probability of success, p	Pr(of a red outcome) $= p = \frac{3}{5}$ Thus, $1 - p = \frac{2}{5}$

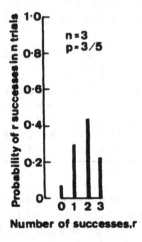

Figure 6.2. Binomial probabilities for the problem stated in Table 6.2.

Table 6.3. Calculation of binomial probabilities.

Value of r, the number of successes	^nC_r	p^r where $p = \frac{3}{5}$	$(1-p)^{n-r}$ where $n = 3$	Binomial probability
0	1	1	8/125	8/125 = 0.06
1	3	3/5	4/25	36/125 = 0.29
2	3	9/25	2/5	54/125 = 0.43
3	1	27/125	1	27/125 = 0.22

NOTE $p^0 = (1-p)^0 = 1$.

A binomial problem, relevant to clinical trials, is stated in Table 6.4. The method of solution and answer to the problem are given in Table 6.5. The binomial probability for 5 or more successes is 0.11 or 11%, that is, 11 chances per 100 experiments that the results are random—each experiment consisting of six matched pairs.

Table 6.4. Description of a binomial problem.

Experiment	Testing whether one of two drugs, A or B is more *effective*. Pairs of patients matched for sex and age are placed in the trial and one of each pair is randomly assigned to drug A or B.
Problem?	There are 6 matched pairs and in 5 cases drug A was found to be more *effective* than drug B. Is this result likely to have occurred quite randomly?—or can it be assumed that there is a *real* difference between A and B?
No of trials, n	$n = 6$. In this problem n will equal the number of matched pairs, each pair being regarded as a single trial.
Probability of success, p	What do we choose for p, the probability that A is more effective than B? Since we have no concrete evidence beforehand, choose in the first instance $p = 0.5$, $1 - p = 0.5$. That is, it is equally likely either that A is more effective than B or that B is more effective than A.

We now pass to the subject of *statistical significance*, which will be covered in Chapter 8. It is the *investigator who must fix the value of the critical probability level*. This is the level below which he will accept that the results were *not* obtained by chance, that is were *not* random. For this example, Tables 6.4 and 6.5 assume that the level fixed was 0.10. Since 0.11 is greater than 0.10, he would not accept that A is a more effective treatment than B and would have to accept that no difference had been demonstrated between *A* and *B* at the 0.10 level. The investigator's follow-up step would then be to repeat the experiment with a greater number of matched pairs, that is, *n* greater than 6, in the hope of a clear answer.

Table 6.5. Calculation of binomial probabilities. In this example, we only need to calculate the probabilities for $r = 5$ and $r = 6$.

Value of r, the number of successes (i.e. *A* more effective than *B*)	^nC_r	p^r where $p = 0.5$	$(1 - p)^{n-r}$ where $n = 6$	Binomial probability
0	1			
1	6			
2	15			
3	20			
4	15			
5	6	$(\frac{1}{2})^5 = \frac{1}{32}$	$(\frac{1}{2})^1 = \frac{1}{2}$	$\frac{3}{32} = 0.094$
6	1	$(\frac{1}{2})^6 = \frac{1}{64}$	$(\frac{1}{2})^0 = 1$	$\frac{1}{64} = 0.016$

When we assess the significance of a result we require under the null hypothesis the probability of obtaining the observed data *or more extreme results*. In the above example, for drug *A* better than drug *B*, we need a 1-tailed test finding the probability of obtaining 5 or 6 successes, if our action level is to be $r = 5$. The probability of 5 or more successes will be $0.094 + 0.016 = 0.11$. For a 2-tailed test, which would be relevant if we were interested in *A* being better than *B*, or in *B* being better than *A*, this probability will need to be doubled.

As a further example consider the disease Huntington's chorea, a fatal degenerative brain disorder in which the inheritance pattern is autosomal dominant and therefore the probability of any one child in the family developing the disease is 0.5. The answer to the binomial question in Table 6.6 is therefore that of all such 10 children families, the probability that there will be three affected children is 0.117. Figure 6.3 is the complete binomial distribution for this problem giving the probabilities of numbers of affected children in families of 10 with one parent having Huntington's chorea.

6.4 THE NORMAL APPROXIMATION TO THE BINOMIAL

Figure 6.2 is a *binomial probability distribution* with $n = 3$ and $p = 0.6$ and thus $np = 1.8$. As the value of *n* increases, so does the range of values of *r*, and the distribution takes on a bell-shaped pattern, as can be seen in

Table 6.6. Description of a binomial problem.

Basic information	The pattern of inheritance of Huntington's chorea
Problem?	What is the probability of having 3 affected children in a family of 10?
No. of trials, n	The value of n will be the number of children in the family, $n = 10$
Probability of any child eventually developing the disease	$p = 0.5$

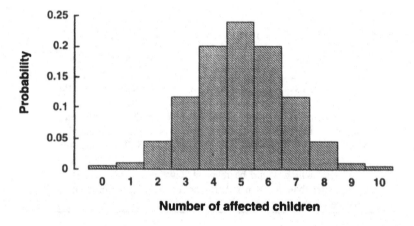

Figure 6.3. Probabilities of affected children in families of 10 children with one parent having Huntington's chorea.

Figure 6.3. This can be approximated to a *normal* probability distribution under certain circumstances which are usually taken to be: large n and p not close to zero, where the product np is greater than 5. The advantage of this normal approximation to the binomial is that the standard tables available for the normal distribution curve may be used for binomial problems when np is greater than 5.

Comparisons of binomial and normal distributions are given in Figures 6.4 and 6.5. In the former the mean is 4 and the standard deviation is $\sqrt{2} = 1.41$, whereas in the latter the mean is 6 and the standard deviation is $\sqrt{3} = 1.73$. The mean of the binomial is given by the product np and the standard deviation of $\sqrt{np(1-p)}$.

As a final example of binomial probabilities, suppose that a die is thrown

Table 6.7. Calculation of binomial probabilities.

Value of r the number of affected children	^nC_r	p^r where $p = 0.5$	$(1 - p)^{n-r}$ where $n = 10$	Binomial probability
3	120	0.125	0.0078125	0.117

Pascal's triangle in Figure 6.1 does not extend to $n = 10$ but it can easily be calculated by obtaining each new entry as the sum of the two adjacent numbers in the row above. For the row $n = 10$ the numbers $^{10}C_0$, $^{10}C_1$, ... $^{10}C_{10}$ are:

1 10 45 120 210 252 210 120 45 10 1

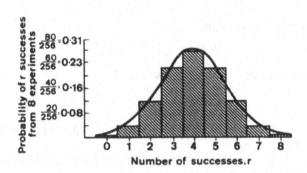

Figure 6.4. Comparison of the binomial and normal distributions for $n = 8$ and $p = (1 - p) = 0.5$. The histogram is the binomial and the curve is the normal. For the binomial the mean is 4 and the standard deviation is $\sqrt{2} = 1.41$.

180 times ($n = 180$) and that a gambler is betting that the 6 (probability of success, $p = \frac{1}{6}$) will appear between 29 and 32 times. What is the probability that he will win his bet?

$$\text{Mean } M = np = 180/6 = 30$$

$$\text{Standard deviation } S = \sqrt{np(1 - p)} = \sqrt{180 \times \tfrac{1}{6} \times \tfrac{5}{6}} = 5.$$

If r is the number of successes, the gambler will want to know the probability of r in the range 29–32, including both limits. The normal approximation to the binomial can be used to solve this problem. However, the normal is a

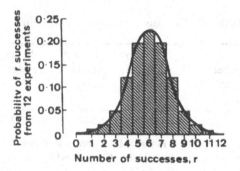

Figure 6.5. Comparison of the binomial and normal distributions for $n = 12$ and $p = (1 - p) = 0.5$. A probability of $p = 0.5$ could be for an experiment with the tossing of a coin when a head is a success and a tail is a failure. The histogram is the binomial and the curve is the normal. For the binomial the mean is 6 and the standard deviation is $\sqrt{3} = 1.73$.

continuous distribution and thus

$$\Pr(28.5 \leqslant r \leqslant 32.5)$$

has to be found. The unit normal deviate formula is

$$\frac{r - M}{S}$$

and for 28.5 and 32.5 is therefore $(28.5 - 30.0)/5 = -0.3$ and $(32.5 - 30.5)/5 = +0.5$ respectively.

Using the notation in Figure 3.2(a) and Table 3.1(a) which gives the area P under the normal curve from $+\zeta$ to $+\infty$ we must use these data to determine the area beneath the standard normal curve between the limits $\zeta = -0.3$ to $\zeta = +0.5$.

From $\zeta = +0.5$ to $+\infty$ the area is 0.30854, therefore from $\zeta = 0$ to $+0.5$ the area is $(0.5 - 0.30854) = 0.19146$. Because of the symmetrical properties of the normal curve the area from $\zeta = +0.3$ to $+\infty$ is the same as the area from $-\infty$ to $\zeta = -0.3$ which is 0.38209 from Table 3.1(a). Therefore from $\zeta = -0.3$ to 0 the area is $(0.5 - 0.38209) = 0.11791$. Therefore the area between $\zeta = -0.3$ and $+0.5$ is $(0.19146 + 0.11791) = 0.30937$.

We can therefore say the the probability $\Pr(28.5 \leqslant r \leqslant 32.5) = 0.309$ which implies that the gambler will only have a 30.9% chance of winning the bet.

6.5 GAMBLING WITH DICE

As a historical anecdote, it it noted that it was in fact gambling problems which first led to classical probability theory in the 17th century. The Chevalier de

Méré acquired a substantial amount of money by betting that he would get at least one six in a sequence of 4 tosses of a die: and then lost it all by betting that he would get at least one double-six in a sequence of 24 tosses with two dice.

When the Chevalier won, we can talk about getting a six *at least* once or getting no sixes. The probability of getting no sixes with a single toss is 5/6 and with four tosses it is $(5/6)^4$ and thus the probability of getting at least one six with four tosses is $1 - (5/6)^4 = 0.518$ which is calculated using the formula

Probability of *at least* one successful result $= 1 - (1 - p)^n$

where $(1 - p)$ is the probability of failure. This Chevalier de Méré example can also be computed using binomial probabilities, arriving at the same answer of 0.518 quoted above, Table 6.8.

Table 6.8. Calculation of the probability of the Chevalier de Méré obtainining a six at least once with four tosses of a die. The probability is the sum of the individual probabilities of obtaining either 1, 2, 3 or 4 sixes (but not no sixes) in the four tosses.

No. of sixes r	^4C_r	p^r where $p = \{1/6\}$	$(1 - p)$ since $n = 4$	Binomial probability
1	4	0.1667	0.5786	0.3858
2	6	0.0278	0.6944	0.1158
3	4	0.0046	0.8333	0.0154
4	1	0.0008	1.0000	0.0008
				$\sum = 0.5178$

If the bet had been obtaining *only one six* and not *at least one six*, then the probability would have been 0.386.

The result of a probability of 0.518 can be said to predict a profitable outcome for anyone who has patience, money and an honest die and bets on the appearance of at least one six in four tosses, since the probability 0.518 is greater than 0.5. However, if the bet had been for getting four sixes simultaneously when four unbiased dice were tossed, this probability would be only $(1/6)^4 = 0.00132$ and consequently would be a very poor betting proposition. When the Chevalier lost, the probability of getting at least one double-six in 24 sequences with two dice is $1 - (35/36)^{24} = 0.491$, which is less than 0.5.

The exchange of letters between Pierre de Fermat (1601–1665) and Blaise Pascal (1623–1662), which had been requested by the Chevalier de Méré in order to solve the problem described above, established in 1654 a foundation for probability theory which was later developed by Jakob Bernoulli into a mathematical theory of probability. For this reason, binomial trials are sometimes referred to as Bernoulli trials.

$$\pi$$

The most famous irrational number is π and is a number which can never be expressed as a fraction and whose series of numbers after the decimal point is infinite. However, a knowledge of π to 39 decimal places is sufficient to calculate the circumference of the universe to give an answer which is accurate to the radius of a hydrogen atom. Nevertheless, the current (1996) record is held by Yasumasa Kanada of the University of Tokyo who has calculated π to six billion decimal places.

THE VALUE OF π TO OVER 1500 DECIMAL PLACES

3.14159265358979323846264338327950288419716939937510582
09749445923078164062862089986280348253421170679821480 86
51328230664709384460955058223172535940812848111745028 41
02701938521105559644622948954930381964428810975665933 44
61284756482337867831652712019091456485669234603486104 54
32664821339360726024914127372458700660631558817488152 09
20962829254091715364367892590360011330530548820466521 38
41469519415116094330572703657595919530921861173819326 11
79310511854807446237996274956735188575272489122793818 30
11949129833673362440656643086021394946395224737190702 17
98609437027705392171762931767523846748184676694051320 00
56812714526356082778577134275778960917363717872146844 09
01224953430146549585371050792279689258923542019956112 12
90219608640344181598136297747713099605187072113499999 98
37297804995105973173281609631859502445945534690830264 25
22308253344685035261931188171010003137838752886587533 20
83814206171776691473035982534904287554687311595628638 82
35378759375195778185778053217122680661300192787661119 59
09216420198938095257201065485863278865936153338182796 823
03019520353018529689957736225994138912497217752834791 31
51557485724245415069595082953311686172785588907509838 17
54637464939319255060400927701671139009848824012858361 60
35637076601047101819429555961989467678374494482553797 74
72684710404753464620804668425906949129331367702898915 21
04752162056966024058038150193511253382430035587640247 49
64732639141992726042699227967823547816360093417216412 19
92458631503028618297455570674983850549458858692699569 09
27210797509302955321165344987202755960236480665491198 81
83479775356636980742654252786255181841757467289097777 27
93800081647020016145249192173217214772350141441973 5....

Chapter 7

Poisson Probabilities

7.1 THE POISSON DISTRIBUTION

In the binomial distribution situation there is a calculable probability of the occurrence or non-occurrence of the event in question. This is not possible in a Poisson distribution situation and the Poisson is therefore very useful for calculating the probabilities of rare events for which no *a priori* estimate can be made of the probability that they will occur.

Simeon Denis Poisson (1781–1840) was a French mathematical genius who originally intended to be a doctor but who gave up his studies at the age of 17 because of an upset about a patient who died. His law of the probability of occurrence of rare events states that the probability of occurrence of n such events, $Pr(n)$ in calculated from the formula

$$Pr(n) = \frac{e^{-m}m^n}{n!}$$

where m is the mean number of events and $n!$ is factorial n. That is, $n! = 1 \times 2 \times 3 \times 4 \times \cdots \times n$. For the Poisson distribution both the mean and the variance are equal to m. The standard deviation thus equals $\sqrt{n}$. This probability distribution is, however, only applicable if the following four conditions are met.

1 Discontinuous (i.e. discrete) data.
2 When the chance of a result is small.
3 When the chance of a result is independent of previous results.
4 When a large number of tests can be performed.

The distribution for selected values of m is shown in Figure 7.1 and it is seen that for small values of m the distribution is positively skewed. A table of Poisson probabilities for the three Poisson means $m = 0.5, 1$ and 5, and for a number of occurrences $n = 0, 1, 2, 3, 4$ and 5 is given in Table 7.1.

It should also be noted that although the Poisson is a distribution in its own right, the binomial equates to it under certain conditions. A binomial distribution may be imagined in which the probability of success, p, is very nearly equal to 1, and the probability of failure $1 - p$ is hence very nearly equal to 0, so that the product $n(1 - p)$ is small. If we also have a large number, n, of experiments, then $n - r$, where r refers to the probability of r successes in n experiments, can be assumed to be equal to n since r will be negligible compared with n. Under these conditions the formula for binomial probabilities becomes

$$\mathrm{e}^{-m}\frac{m^r}{r!}$$

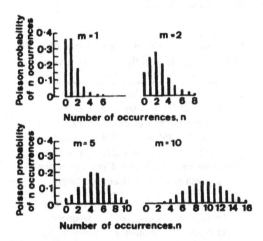

Figure 7.1. Poisson distributions for mean values $m = 1, 2, 5$ and 10.

Table 7.1. Poisson probabilities.

Number of occurrences n	Formula for Poisson probability	Poisson probabilities for different values of m		
		$m = 0.5$	$m = 1$	$m = 5$
0	e^{-m}	0.607	0.368	0.007
1	$m\mathrm{e}^{-m}$	0.303	0.368	0.034
2	$0.5m^2\mathrm{e}^{-m}$	0.076	0.184	0.084
3	$0.17m^3\mathrm{e}^{-m}$	0.013	0.061	0.140
4	$0.042m^4\mathrm{e}^{-m}$	0.002	0.015	0.174
5	$0.0083m^5\mathrm{e}^{-m}$	0.000	0.003	0.174

7.2 EXAMPLES OF POISSON PROBLEMS

The classic example of the Poisson distribution is that of the number of cavalrymen kicked to death in 10 Prussian army corps over a period of 20 years. There are 200 *observation units* (10 corps ×20 years) and the total number of deaths observed was 122. The mean of the Poisson distribution is thus $m = 122/200 = 0.61$ and Table 7.2 gives the number of expected deaths in the corps assuming a Poisson distribution with $m = 0.61$ and the number of deaths actually observed. There is very good agreement. The closeness of agreement between observed and expected values can be tested for statistical significance using the chi-squared test.

Table 7.2. Poisson distribution data showing the chance of a 19th century Prussian cavalryman being killed by a horse kick in the course of a year.

Number of years for which data are
available for a cavalry corps = 200
Total number of deaths = 122
$\sum\{nf\} = \{65 + 44 + 9 + 4\} = 122$
Poisson mean $m = 122/200 = 0.61$
Frequencies are number of years in which
a given number of deaths occurred

No. of deaths by horse kick per corps per year (n)	(f) Observed frequency	Expected frequency, Poisson mean $m = 0.61$
0	109	108.7
1	65	66.3
2	22	20.2
3	3	4.1
4	1	0.6
5	0	0.1
Totals	200	200

Two other examples of Poisson distributions are those of drownings and of suicides in Malta, as reported in the *St. Luke's Hospital Gazette*. The data for drownings were from a consecutive 355-month period when 167 drowning deaths were reported, giving a mean of 0.47 deaths per month. The data for suicides were for a consecutive 216-month period when there were 141 suicides, giving a mean of 0.65. Table 7.3 gives the observed and Poisson-expected deaths. There is an impressive agreement between observation and expectation, although some argue that it is invalid to apply Poisson to the number of suicides per year in a given community because the temptation to commit suicide varies with the stress of the times, such as the Wall Street crash of 1929 which preceded the American slump of the 1930s. However, it can be applied to an adequate period of observation where no such undue extraneous factors may have occurred. Indeed, no departure from Poisson may be a valuable indication that no such factors did exist during the period of observation.

Table 7.3. Poisson distribution data for Malta.

Number of consecutive months observed $= 216$
Total number of deaths by drowning $= 167$
$\sum\{nf\} = \{102 + 46 + 15 + 4\} = 167$
Poisson mean $m = 167/355 = 0.65$
Frequencies are the number of months in which
a given number of deaths occurred

No. of deaths by drowning per month (n)	(f) Observed frequency	Expected frequency, Poisson $m = 0.47$
0	224	221.9
1	102	104.4
2	23	24.5
3	5	3.9
4	1	0.4
Totals	355	355.1

Number of consecutive months observed $= 216$
Total number of suicides $= 141$
$\sum\{nf\} = \{76 + 50 + 15\} = 141$
Poisson mean $m = 141/216 = 0.65$
Frequencies are the number of months in which
a given number of suicides occurred

No. of deaths by suicide per month (n)	(f) Observed frequency	Expected frequency, Poisson $m = 0.65$
0	110	112.3
1	76	73.3
2	25	24.0
3	5	5.2
Totals	216	214.8

Examples of other rare events which have been shown to be represented by a Poisson distribution include the number of patients admitted with acute poisoning to Patna Medical College Hospital on days of full moon; the number of goals in a football match (see section 9.4); the number of printer's errors per page in a book; the number of faulty electrical components in a batch; and radioactive decay.

7.3 THE POISSON APPROXIMATION TO THE BINOMIAL

The printer's error problem has been described in the following manner[1] and illustrates a situation when the Poisson approximation to the binomial can be used.

Suppose 300 misprints are distributed randomly throughout a book of 500 pages. Find the probability P that a given page contains [1] exactly 2 misprints, [2] 2 or more misprints.

We view the number of misprints on one page as the number of successes in a sequence of Bernoulli trials. Here $n = 300$ since there are 300 misprints, and $p = 1/500$, the probability that a misprint appears on the given page. Since p is small, we use the Poisson approximation to the binomial distribution with $np = 0.6$.

[1] $P = p(2; 0.6) = \frac{(0.6)^2 e^{-0.6}}{0!} = (0.36)(0.549)/2 = 0.0988 \approx 0.1$.

[2] $P(0 \text{ misprints}) = \frac{(0.6)^0 e^{-0.6}}{0!} = e^{-0.6} = 0.549$

$P(1 \text{ misprint}) = \frac{(0.6)e^{-0.6}}{1!} = (0.6)(0.549) = 0.329$

Then $P = 1 - P(0 \text{ or } 1 \text{ misprint}) = 1 - (0.549 + 0.329) = 0.122$.

7.4 THE NORMAL APPROXIMATION TO THE POISSON

The Poisson distribution is not simply more symmetrical as its mean m increases (see Figure 7.1). It also approaches a normal distribution with mean m and standard deviation $\sqrt{m}$. This is shown in Figure 7.2 for a Poisson distribution with $m = 10$. There still remains a slight asymmetry and a rather better fit around the peak is obtained if the normal distribution is displaced with a slightly lower mean value, 9.5 rather than 10. For mean values in excess of 30, the asymmetry is negligible. The advantage of this normal approximation to the Poisson is that standard tables for the normal curve may be used for Poisson problems with large values of m.

7.5 RADIOACTIVE DECAY

A practical use of the normal distribution as an approximation to the Poisson is in the measurement of radioactive disintegrations. For example, when this is performed during nuclear medicine diagnostic tests by counting with radiation detection equipment such as scintillation counters or Geiger counters. If it is assumed that the radioactive sample is counted a number of times for 1 minute

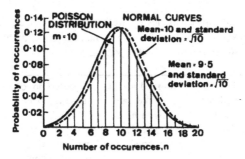

Figure 7.2. Comparison of the Poisson distribution with $m = 10$ and two normal curves.

and a mean value of 1000 counts is found, the probability of observing a count, N, will be given by a normal distribution of mean $M = 1000$ and standard deviation $S = \sqrt{1000} = 31.62 \approx 32$. This is the normal approximation to the Poisson distribution

$$\frac{e^{-1000}(1000)^N}{N!}$$

Recalling Figure 3.3 and the interpretation of an area beneath the normal curve between limits expressed in terms of standard deviation, S, we can state the following for the radioactive sample counting problem with a mean count rate of 1000 counts per minute.

There is a 68% probability of finding a measurement in the range 968–1032, i.e. $M \pm S$.

There is a 90% probability of finding a measurement in the range 947–1053, i.e. $M \pm 1.645S$.

There is a 95% probability of finding a measurement in the range 937–1063, i.e. $M \pm 1.96S$.

The term *probable error* is also sometimes used and this is equal to $0.67 \times$ standard deviation. For the counting problem $0.67S = 21$ and hence there is a 50% probability of finding a measurement in the range 979–1021, i.e. $M \pm 0.67S$.

Some historical data on radioactivity counting are given in Table 7.4 which reproduces[2] the counts of alpha particles by Sir Ernest Rutherford and Hans Geiger which were published in their 1910 paper in the *Philosophical Magazine*.

Figure 7.3 plots $\{E\}$ in the last column of Table 7.4, using a similar vertical bar format to Figure 7.1. Of the two scales on the vertical axis the Poisson probability scale can be calculated using the formulae in Table 7.1. Thus for example, for $m = 3.87$ and $n = 2$, the Poisson probability for two α particles to be observed in an interval equals $0.5 \cdot (3.87)^2 e^{-3.87} = 0.15619$. Since $\sum \{E\} = 2608$, this particular expected Poisson frequency for the Rutherford and Geiger experiment is $2608 \times 0.15619 = 407.3$.

Table 7.4. Counting of alpha particles by Rutherford and Geiger. The time interval for the measurements was 7.5 seconds. Total number of α particles $= 2608$ and $\sum\{nf\} = 10,094$. Poisson mean $= 10,094/2608 = 3.87$.

No. of α particles (n)	Observed no. of time intervals with n α particles per interval $\{f\}$	Expected no. of time intervals with n α particles per interval $\{E\}$ (rounded down)
0	57	54
1	203	210
2	383	407
3	525	526
4	532	509
5	408	394
6	273	254
7	139	140
8	45	68
9	27	29
10	10	11
11	4	4
12	2	1
$\geqslant 13$	0	1
	$\sum\{f\} = 2608$	$\sum\{E\} = 2608$

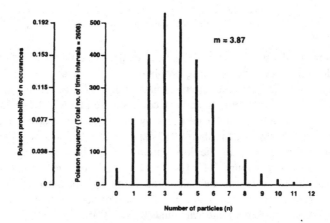

Figure 7.3. Poisson probabilities for the Rutherford and Geiger α particle counting experiment.

Chapter 8

Introduction to Statistical Significance

8.1 THE NULL HYPOTHESIS

One of the problems encountered by those involved with statistics is how, and with what accuracy, inferences can be drawn about the nature of a population when the only evidence which exists is that from samples of the population. In order to solve this problem an understanding of *statistical significance* is essential and it should be immediately recognised that this is not necessarily the same as *clinical significance* when the statistics refer to medicine. It should also be noted that editors of journals containing results using statistical tests always insist on a statement of a P-value, and that this is regarded by some readers and authors as having mystical properties more commonly associated with those of a fortune teller's crystal ball. Statistical inference is not such a simple matter that if P is less than 0.05 then results are worth publishing as they demonstrate statistical significance. It is an absolute priority for those using tests for statistical significance that they understand the conditions which must apply for a particular test to be valid and that they have a clear understanding of the hypotheses which are being tested.

The starting point of any practical use of a test for statistical significance is the *null hypothesis*, often denoted by H_0, to distinguish it from alternative hypotheses such as H_1 and H_2. This will in effect state that 'there is no difference between ... and ...' and one example is that 'there is no difference between the effect of treatment A and the effect of treatment B'. In a clinical trial where the investigator is hoping that a new treatment is better than an old one, the starting point is still the null hypothesis: an alternative hypothesis is that 'the effect of treatment A is better than the effect of treatment B'.

The null hypothesis is therefore usually set up to be rejected when testing for statistical significance in medical research, since it is usually not the one the researcher hopes to be true, one exception being curve fitting in section 9.5. The logic of this is that if the hypothesis of no relationship, H_0, can be rejected, the rational conclusion is that there is a relationship between the phenomena or the attributes stated in the null hypothesis. This logic would be compelling if there were a null hypothesis, H_0, and a single alternative hypothesis, H_1, which were mutually exclusive. However, in most medical situations there will be a series of possible alternative hypotheses, H_1, H_2,..., and the outcome of a single test of statistical significance seldom totally solves a research project. Examples of null hypotheses are:

H_0: The observations come from a Poisson distribution.

H_0: The mean age of a population of patients with cancer of the ... is 60 years.

Examples of alternative hypotheses for the above are:

H_1: The observations do not come from a Poisson distribution.

H_1: The mean age of a population of patients with cancer of the ... is 50 years.

H_2: The mean age of a population of patients with cancer of the ... is 70 years.

It should also be noted that statistical significance tests can be classed as 1-tailed or 2-tailed depending on the form of the alternative hypothesis. Taking an opinion poll H_0 as an example, if we are interested in testing to see if the proportion is different from 0.7 either on the high side or on the low side, then this is a *2-tailed test*, since we are interested in values both higher than 0.7 and lower than 0.7. If, though, the alternative hypothesis we are interested in is only for values higher than 0.7 then this is a *1-tailed* test since we are not, in this instance, interested in values lower than 0.7. This is a trivial example, but it demonstrates the meaning of 1- and 2-tailed tests.

8.2   NULL, POSITIVE AND NEGATIVE RESULTS OF STUDIES

In practice, one should sometimes guard against introducing bias by choosing only a 1-tailed test in, for example, a drug trial. If drug A is a placebo and drug B the new drug of interest, then it might be thought that only

H_0: There is no difference in the effects of drugs A and B

and

H_1: Drug B is *better* than drug A

should be considered, as a result of a pilot study involving only a small number of patients and the *hunch* of the medical researcher. (*Better* must of course be clearly defined.) However, a full-scale trial might show that there is a second alternative hypothesis, albeit one the researcher would prefer to ignore:

H_2: Drug A is *better* than drug B.

This is particularly true if drug A is a *placebo*. In this case if H_0 is accepted (*statistical acceptance* is now defined) this is a null result. If H_0 has not been rejected at a given level of probability then we must assume that it has been *accepted*. However, this really implies *not proven*—because we are working in terms of probability. What *acceptance* means is that *if the hypothesis is in fact false, then the experiment was not able to detect it at the established level of significance.*

However, if H_1 is accepted then this is a positive result but if H_2 is accepted this is a negative result. In the literature on clinical trials there are seldom any negative results reported, although two chemotherapy examples are a sequential analysis clinical trial for head and neck cancer[1] and a Medical Research Council clinical trial on advanced carcinoma of the cervix[2].

A review[3] entitled *When is a negative study not negative?* emphasises that the jargon used should be consistent throughout the literature (null, positive, negative). This is very important when in many instances a null result is interpreted as negative.

8.3   PROBABILITY LEVELS AND SIGNIFICANCE

8.3.1 $P < 0.05$

Returning to the expression '*P is less than 0.05*' mentioned in the first paragraph of section 8.1, it must be emphasised that it is most important to distinguish between a *probability level derived* from using a statistical test and a *significance level chosen* by the investigator. The P-value derived from a test is the chance of observing the result in the present trial or a more extreme result (i.e. more contradictory to the null hypothesis), given that the null hypothesis is true.

Using the clinical trial as an example, the explanation of '*P is less than 0.05*' is well described by Peto *et al*[4] as follows: 'The patients in one treatment group have fared better than the patients in the other. If there is no difference between the medical effects of the two treatments and the only cause of difference between the treatment groups is the chance allocation of more good-prognosis patients to one group than to the other, **then** the chance of one treatment group faring **at least** this much better than the other group would be less than 0.05, i.e. less than a 1 in 20 chance'.

Hence when we are talking of '*P is less than 0.05*', the chance that we would have obtained the same set of results (or a more extreme set) purely randomly is *less than 1 in 20*, but if we are talking of '*P is less than 0.01*', this chance would be *less than 1 in 100*. For '*P is less than 0.10*' this chance would be *less than 1 in 10*.

A chosen significance level such as 0.05 is one chosen by the investigator after due thought (it is not always automatically 0.05), and is the level with which the P-value derived from the statistical test is compared.

8.3.2   Clinical and Statistical Significance

Statistical significance does not necessarily equate to *clinical significance* and when looking for trends, and when using tests of statistical significance to help towards decision making, it is not necessarily true that if the P-value is 0.06 'you can forget all about it'.

When $P < 0.05$ the chance we would have obtained our results (or a more extreme set) purely randomly is, as stated above, less than 1 in 20 or, turning this around, the chance that we would *not* have obtained our results purely randomly is greater than 19 in 20. If $P < 0.06$ then the chance that we would not have obtained our results purely randomly is still greater than 17 in 18. Commonsense should therefore tell us that this is not all that different from $P < 0.05$ and therefore the results do not warrant 'forgetting'; rather, they warrant further investigation.

We have to set boundaries at some point, such as $P < 0.05$, but although theoretically this leads to an acceptance/rejection boundary, clinically if the results fall *only just over* the boundary into the region of rejection, we cannot, as it were, give a conclusion set in stone for ever after, since medicine is not such a precise science. Certainly we reject at $P < 0.05$, as this is what we have decided *a priori*, but commonsense must then be used to answer the question 'What now?'.

8.4   TYPE-I AND TYPE-II ERRORS AND ALPHA AND BETA RISKS

There are two types of error which can be made in arriving at a decision about the null hypothesis, H_0. A type-I error is to *reject H_0 when in fact it is true* and a type-II error is to *accept H_0 when in fact it is false*. By convention the probability of a type-I error is usually denoted by α and the probability of a type-II error by β

Table 8.1. Error types and associated risks.

Error	Definition of error type	Associated risk
Type I	Wrongly rejecting H_0 when H_0 is true	Probability of a Type-I error $= \alpha$
Type II	Wrongly accepting H_0 when H_0 is false	Probability of a Type-II error $= \beta$

The β-risk is a function of an alternative hypothesis, and since H_0 is false, H_1 or H_2 must be true. The probability $(1 - \beta)$ is defined as the *power* of the test of the hypothesis H_0 against an alternative hypothesis.

The α-*risk* is the chance of wrongly rejecting H_0 and acting upon the premise *that at the α-level there is a difference*. The consequence of this in treatment evaluation, for example, is the discontinuation of one particular treatment in preference to another.

The β-*risk* is the chance of wrongly accepting the null hypothesis when it is false. We then accept *that there is no difference*. The consequence which follows could be either that we attempt further investigations or discontinue the trial without reaching any conclusion as to whether one treatment is better than the other.

α- and β-risks are related; they are *not independent* of each other. If the α-risk is increased, then the β-risk is decreased, and *vice versa*.

α and β are also related to the size of the sample, N, which for a clinical trial would be the number of patients. Ideally, the values of α and β would be specified by the researcher before the trial or experiment, and these values would then determine the size, N, of the sample to be drawn for computation of the test chosen for statistical significance. To reduce the possibility of both types of error, I and II, the value of N must be increased.

Different types of catastrophe will follow type-I and -II errors. As an example, consider a clinical trial of an existing drug A and a new drug B where the null hypothesis, H_0, is that there is no difference and the alternative hypothesis, H_0, is that the drug B is better than drug A (ignore H_0: drug A better than drug B for this example). If a type-I error occurs then H_0 is wrongly rejected and following the inference that H_0 should thus be accepted, drug A will be abandoned and drug B now given to patients. What are the consequences? Since H_0 is really true and there is no difference between A and B then the patients will not have suffered since they have only been placed on another drug of similar efficacy.

Now consider a type-II error when H_0 is wrongly accepted. In this instance, the truth is that H_0 is correct and the consequences to the patient are that a drug with a better efficacy is denied to them, since the acceptance of H_0 would mean that the current drug A should remain in use and research be abandoned on the new drug B (or perhaps continued more stringently, if the value of N is relatively small). Of the two types of error in this situation, a type-II error would seem to be worse than a type-I error since it could mean rejection of a new and useful drug—if the trial is not carefully planned.

In conclusion, the information in table 8.1 is presented in a slightly different way to that in table 8.2.

Table 8.2. Error types and α and β risks in terms of clinical trial result and actual truth. $\{1 - \beta\}$ is termed power but there is no terminology for $\{1 - \alpha\}$. *P*-values relate to α risks.

Actual truth	Clinical trial result	
	Treatment benefit	No treatment benefit
Treatment benefit	Correct result $\{1 - \beta\}$	Type-II error $\{\beta\}$ False −ve result
No treatment benefit	Type-I error $\{\alpha\}$ False +ve result	Correct result $\{1 - \alpha\}$

8.5   A GENERALISED SCHEDULE FOR SIGNIFICANCE TESTING

The following flow pattern of actions represents a generalised schedule which is applicable for many test situations.

Clear conception of the problem to be studied, of the possible outcomes of planned experiments or trials, and of future possible actions which can be taken should the null hypothesis be rejected or accepted at the chosen level of significance.

↓

Statement of the null hypothesis, H_0.

↓

Ensure that the test of statistical significance to be used is valid for testing H_0.

↓

Choose a significance level, e.g.. 0.05.

↓

Calculate the test statistic (e.g., chi-squared, χ^2) using the appropriate formulae into which observations expressed as numbers can be inserted. These observations will already have been obtained from the appropriate sampling distribution.

↓

Consult a table of values of the test statistic stated in terms of probability (P levels and degrees of freedom (see section 8.6). As an example for chl-squared:

↓

Chi-squared values for $P = 0.05, 0.01$ and DF $= 1, 2, 3, 4$.

Degrees of	Probability levels	
freedom (DF)	$P = 0.05$	$P = 0.01$
1	3.84	6.64
2	5.99	9.21
3	7.82	11.34
4	9.49	13.28

↓

Assume for this example that the test statistic was calculated to be 10.5 and that the degrees of freedom are DF $= 2$. The derived probability level from the test is less than 0.05 because 10.5 is greater than 5.99.

↓

From a comparison of the probability level derived from the test and the significance level chosen, accept or reject H_0.

For the example in the schedule on page 90: degrees of freedom = 2, critical value of the test statistic (χ^2) from standard tables for $P = 0.05$ is 5.99, derived value of the statistic $\chi^2 = 10.5$. 10.5 is greater than 5.99, therefore $P < 0.05$ and therefore we conclude that there is a significant difference at the $P = 0.05$ level, and therefore we reject the null hypothesis H_0.

If, however, the derived value of the statistic had been 3.1, which is less than 5.99, the null hypothesis would not have been rejected at the $P = 0.05$ level. Alternatively, though, if the derived value of the statistic had been 8.1, which is greater than 5.99 but less than 9.21, the null hypothesis would be rejected at the $P = 0.05$ level but not rejected at the $P = 0.01$ level.

PARROT FASHION STATISTICAL TESTING

FORMULAE FOR TESTS OF STATISTICAL SIGNIFICANCE SHOULD NOT BE USED PARROT FASHION WITHOUT CAREFUL THOUGHT AS TO WHETHER A PARTICULAR TEST IS APPROPRIATE

I was once asked the following question. 'On my statistical computer software I have 10 tests of statistical significance. I fed my data through all 10 and obtained a significant result $P < 0.05$, in one test and no significant results in nine tests. Can I use the one test which gave me $P < 0.05$ for my publication?'

8.6 DEGREES OF FREEDOM

Teaching experience has shown that the concept of degrees of freedom is rather difficult to explain in understandable simple English and therefore three explanations have been chosen. However, when examples of tests of statistical significance (e.g. χ^2 and t-tests) are illustrated in the later chapters, the concept of degrees of freedom (often termed DF or sometimes v) may become clearer.

[1] The size of the degrees of freedom reflects the number of observations that are free to vary after certain restrictions have been placed on the data. These restrictions are not arbitrary, but rather are inherent in the organisation of the data. Examples in the following chapter on the chi-squared test, $\chi^2 = \{(Observed-Expected)^2/Expected\}$, include table 9.2 where there are 10 pairs of O and E values (often referred to as *cells*) for which there are no assumptions made with regard to E.

$$DF = (10 - 1) = 9.$$

There will always be in a situation like table 9.2, the -1 term subtracted from the total number of cells. However, in table 9.5 the expected values E are calculated from the Poisson formula for which an assumption has to be made about the Poisson mean, m, and this has to be taken into account such that the total number of cells, nine in this example, are reduced by -2 and not only by -1. Thus:

$$DF = (9 - 2) = 7$$

Then in table 9.6 where the expected values E are calculated from the lognormal formula for which an assumption has to be made about the lognormal mean M and the lognormal standard deviation S, the total number of cells, 11 in this example, are reduced by -3 and not by -1. Thus:

$$DF = (11 - 3) = 8$$

For a generalised contingency table with n rows and m columns, see section 9.6 for the 2×2 contingency table, there is a general formula which should be remembered, although there is no need to know how it is derived, which gives:

$$DF = (\text{Number of rows} - 1) \times (\text{Number of columns} - 1)$$

which for a 2×2 table, as in table 9.7, is:

$$DF = (n - 1) \times (m - 1) = 1$$

[2] DF is linked with *constraints* (i.e. restrictions) on the expectations E, such as the Poisson mean and the lognormal mean and standard deviation, as mentioned above in **[1]**. In the language of physics, a point that can move freely in three-dimensional space has DF= 3 and three variable coordinates X, Y and Z which can take on different values independently. If we constrain the point to move only in a plane (i.e. two-dimensional space) the point then has DF= 2.

[3] Imagine trying to decide which chocolate to choose from a box of N chocolates. Every time we choose a chocolate we have a choice, until we come to the last chocolate and then we have no choice. We thus have $N - 1$ choices: in other words DF= $N - 1$. This contains the same -1 term as commented upon above in **[1]**.

Chapter 9

The Chi-Squared Test

9.1 INTRODUCTION

The chi-squared test (the Greek symbol for chi is χ) is probably the most commonly used test of statistical significance. It is a non-parametric test, unlike the t-test in Chapter 11 which is a parametric test, since for the chi-squared test there are no underlying assumptions that must be made about a normally distributed population before the test can be considered to be appropriate.

The χ^2-statistic is

$$\frac{(\text{Observation} - \text{Expectation})^2}{\text{Expectation}}$$

where Observation is the observed frequency and Expectation is the expected frequency of the quantity being studied. Two important applications of the chi-squared test are goodness of fit (sections 9.2–9.5 and Table 3.4) and contingency tables (sections 9.6–9.8). The basic form of the test determines whether the observed frequencies of a particular parameter value or range of values differ significantly from the frequencies which would be expected under some theory or hypothesis. When the test is applied there will be a number n of observed and expected values and each of these cells will give a corresponding value for

$$\chi_i^2 = (O_i - E_i)^2/E_i$$

where $i = 1, 2, 3, 4, \ldots, n$. The χ^2-statistic is the summation of all the $(O_i - E_i)^2/E_i$ values.

Table 9.1 is a table of χ^2-values in terms of degrees of freedom (DF) and probability levels P. Some of these data are presented graphically in Figure 9.1. The derived probability level from the derived summation $\Sigma\chi^2$ of all the χ_i^2 values for each cell i can be obtained from Table 9.1. This derived probability level can then be compared with the chosen level of significance, as described in section 8.5, and a decision on acceptance or rejection of the null hypothesis, H_0, can then be made.

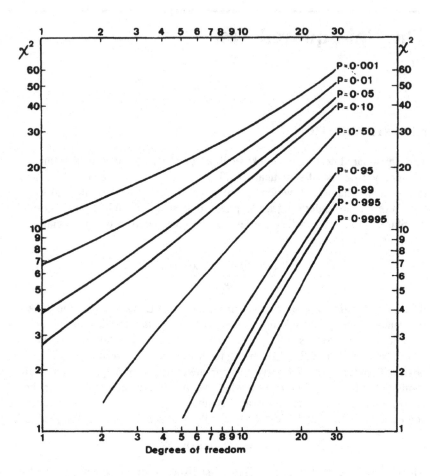

Figure 9.1. χ^2 as a function of degrees of freedom and probability levels, P. Some statistical tables give P as a percentage (i.e. 5% as distinct from 0.05) and the term P as percentage points of the χ^2-distribution; others tabulate a probability in percent which is $100 - P\%$ in the notation and term this fractiles of the χ^2-distribution (i.e. a fractile of 95% corresponds to $P = 0.5$ or 5%). Thus when consulting tables or graphs of χ^2 ensure first that the notation is understood.

Table 9.1. Values of χ^2 for selected probabilities and degrees of freedom. The rows in the table indicate the degrees of freedom (DF) and the columns indicate probability levels (P). The values in the body of the table are the values of χ^2 and the probabilities given are for values of χ^2 with the larger values being less probable. For example, the value corresponding to $P = 0.05$ for DF $= 6$ is 12.592. Values larger than 12.592 thus have a probability of less than 5%. (From Fisher and Yates, *Statistical Tables for Biological, Agricultural and Medical Research* (6th edn, 1974, table IV, p 47). Courtesy Longman Group UK Limited.)

DF	$P = 0.99$	0.95	0.10	0.05	0.01	0.001
1	0.000157	0.00393	2.706	3.841	6.635	10.827
2	0.0201	0.103	4.605	5.991	9.210	13.815
3	0.115	0.352	6.251	7.815	11.345	16.266
4	0.297	0.711	7.779	9.488	13.277	18.467
5	0.554	1.145	9.236	11.070	15.086	20.515
6	0.872	1.635	10.645	12.592	16.812	22.457
7	1.239	2.167	12.017	14.067	18.475	24.322
8	1.646	2.733	13.362	15.507	20.090	26.125
9	2.088	3.325	14.684	16.919	21.666	27.877
10	2.558	3.940	15.987	18.307	23.209	29.588
11	3.053	4.575	17.275	19.675	24.725	31.264
12	3.571	5.226	18.549	21.026	26.217	32.909
13	4.107	5.892	19.812	22.362	27.688	34.528
14	4.660	6.571	21.064	23.685	29.141	36.123
15	5.229	7.261	22.307	24.996	30.578	37.697
16	5.812	7.962	23.542	26.296	32.000	39.252
17	6.408	8.672	24.769	27.587	33.409	40.790
18	7.015	9.390	25.989	28.869	34.805	42.312
19	7.633	10.117	27.204	30.144	36.191	43.820
20	8.260	10.851	28.412	31.410	37.566	45.315
21	8.897	11.591	29.615	32.671	38.932	46.797
22	9.542	12.338	30.813	33.924	40.289	48.268
23	10.196	13.091	32.007	35.172	41.638	49.728
24	10.856	13.848	33.196	36.415	42.980	51.179
25	11.524	14.611	34.382	37.652	44.314	56.620
26	12.198	15.379	35.563	38.885	45.642	54.052
27	12.879	16.151	36.741	40.113	46.963	55.476
28	13.565	16.928	37.916	41.337	48.278	56.893
29	14.256	17.708	39.087	42.557	49.588	58.302
30	14.953	18.493	40.256	43.773	50.892	59.703
40	22.164	26.509	51.805	55.759	63.691	73.402
50	29.707	34.764	63.167	67.505	76.154	86.661
60	37.485	43.188	74.397	79.082	88.379	99.607

9.2 GOODNESS OF FIT: PRECISION OF NUCLEAR MEDICINE COUNTING INSTRUMENTS

Measuring instruments must be precise and with nuclear medicine instrumentation, which has to record radioactive isotope count rates for diagnostic or therapeutic purposes, precision is most important. The International Atomic Energy Agency[1] recommends that the chi-squared test be used for the following test of precision.

(a) A standard caesium-137 source is positioned in the counter.

(b) A radiation counting time is fixed to give counts of at least 10 000.

(c) Ten replicate counts are recorded.

(d) A table is constructed as in Table 9.2 and the data are analysed using the chi-squared test.

For a chi-squared goodness of fit test, the *number of degrees of freedom* is the number of cells, n, minus 1 for the total (since the total number n is a fixed constraint), minus 1 for the mean if it is fixed *a priori* (and is therefore a second constraint), and minus 1 for the variance if this has also been fixed *a priori* when the expected frequencies are calculated. For the radioactivity counter no *a priori* assumptions are made concerning mean or variance and therefore

$$DF = n - 1.$$

In Table 9.2, O_i is an individual count, $i = 1, 2, 3, \ldots, 10$, $n = 10$ and the expected number of counts, E, is taken to be the mean of the ten O_i values, 7632 in this example. DF$= n - 1 = 9$ and therefore from Table 9.1, for $P = 0.05$, $\chi^2 = 16.92$; and for $P = 0.95$, $\chi^2 = 3.32$.

A value of the derived $\Sigma\chi^2$ statistic greater than 16.92 would indicate that the variation in counts, at the chosen level of significance of 0.05, is greater than can be plausibly attributed to chance alone. A value of less than 3.32 for the derived $\Sigma\chi^2$ statistic will similarly indicate that the results cannot be expected to occur by chance alone—in this case they are too good to be true! Thus, if the derived $\Sigma\chi^2$ statistic is greater than 16.92 or less than 3.32, the test should be repeated according to the IAEA protocol and if the second test also gives $\Sigma\chi^2$ statistic outside the range of acceptance, this may be taken to indicate faulty performance.

In practice, it has been found in some laboratories that the technician has switched the counting instrument controls to the electronic test position for 10 000 counts rather than for a sample count. In this case, the $\Sigma\chi^2$ statistic is always less than 3.32. If the $\Sigma\chi^2$ statistic is greater than 16.92, this may be due to spurious counts from random electrical noise, from an unstable power supply, from temperature changes or from electronic faults. In Table 9.2 the derived $\Sigma\chi^2$ statistic is 6.34 which is taken from real data for a counting system in a Bogota hospital during an IAEA Training Workshop on Nuclear Medicine Instrumentation.

For this example, where the counting equipment was an activity meter, the null hypothesis is

H_0: There is no difference in the expected number of counts and those measured, and any observed differences are merely chance variations to be expected from the random nature of radioactive decay.

Table 9.2. Test of precision of a counting system for radioactive diagnostic test measurements.

Measurement number	$O =$ counts in a given time	$E =$ expected counts	$O - E$	$(O - E)^2$
1	7589	7632	−43	1849
2	7687	7632	55	3025
3	7660	7632	28	784
4	7658	7632	26	676
5	7592	7632	−40	1600
6	7728	7632	96	9216
7	7551	7632	−81	6561
8	7534	7632	−98	9604
9	7744	7632	112	12544
10	7581	7632	−51	2601

Sum of all O-values $= 76324$.
Expected number of counts, $E = 76324/10 = 7632$.
Since E is a constant, the simplest method of calculating the sum of all $(O - E)^2/E$ values is to calculate the sum of all $(O - E)^2$ values and divide this by 7632.
Number of cells $= n = 10$.
Number of degrees of freedom $= n - 1 = 9$.
The derived $\Sigma \chi^2$ statistic is the summation of all the $(O - E)^2$ values divided by 7632.
$\Sigma \chi^2 = 48460/7632 = 6.34$.

9.3 GOODNESS OF FIT: A RACING PROBLEM

Another example of the chi-squared test is quoted by Siegel[2] for the benefit of horse racing fans, from data published in the *New York Times* of 30 August 1955. This gave results by track number position where Track 1 was the inside track and Track 8 the outside track of a circular course. The total number of races was 144. The null hypothesis is

H_0: There is no difference in the expected number of winners starting from each of the tracks, and any observed differences are merely chance variations to be expected in a random sample from a population of horse races.

Table 9.3. The chi-squared test for the horse racing problem.

Track position	$O =$ observed number of wins	$E =$ expected number of wins	$O - E$	$(O - E)^2$	$(O - E)^2/E$
1	29	18	+11	121	6.72
2	19	18	+1	1	0.06
3	18	18	0	0	0.00
4	25	18	+7	49	2.72
5	17	18	−1	1	0.06
6	10	18	−8	64	3.56
7	15	18	−3	9	0.50
8	11	18	−7	49	2.72

Total number of races = 144.
Number of track positions = 8.
Expected number of wins per track if H_0 is true = $144/8 = 18$.
$\Sigma\chi^2$ = sum of all $(O - E)^2/E$ values = 16.3.
Number of degrees of freedom = $8 - 1 = 7$.
(In a similar manner to Table 9.1, some computation time can be saved in practice since E is a constant: $\Sigma\chi^2$ = sum of all $(O - E)^2$ values$/18 = 294/18 = 16.3$.)

Table 9.3 gives the calculation procedure and the derived $\Sigma\chi^2$ statistic is 16.3. From Table 9.1, for 7 degrees of freedom (DF = number of tracks -1) and $P = 0.05$, the χ^2 statistic is 14.07. The derived $\Sigma\chi^2$ of 16.3 is greater than 14.07 and therefore H_0 is rejected at the $P = 0.05$ level of significance. If, however, the gambler is cautious and does not want to bet all his money on a particular track, such as the inside track, he will further consider

H_1: *Alternative hypothesis*: Significantly more winners occur from the inside track position.

So as to be as certain as possible not to be on a losing streak, he will take the $P = 0.01$ level as his chosen level of significance. In this case, for DF = 7 and $P = 0.01$, the χ^2 statistic is 18.48 and the derived $\Sigma\chi^2$ of 16.3 is less 4 than 18.48 and therefore the gambler cannot now reject the null hypothesis, H_0, at this chosen 0.01 level of significance.

Since the myth of the *inside track* is bound to persist in the minds of some, I have repeated the problem illustrated for horses in New York, by considering data for greyhound racing in London. The raw information is that for the annual Greyhound Derby finals, 1927–1985. In these finals there are 6 and not 8 traps, no Greyhound Derby was held during the war years 1941–1944, and thus 55 results are available for analysis. The expected number of wins per trap is $55/6 = 9.17$ and the observed numbers of wins per trap are

Trap	Number of wins
1	13
2	7
3	9
4	11
5	7
6	8

These data give a derived $\Sigma\chi^2$ statistic of 3.1. From Table 9.1 for DF = 5 and $P = 0.05$ the χ^2 statistic is 11.07. Since 3.1 is less than 11.07, the null hypothesis, H_0, cannot be rejected at the 0.05 level of significance.

For the 1986 Greyhound Derby races (total of 58) it was also interesting to note that the favourites were not drawn significantly more often (0.05 level of significance) in any one trap position. For the 58 races, the expected number of favourites in any trap position is $58/6 = 9.67$ and the observed number of times a favourite was drawn in any trap position was

Trap	Number of favourites
1	10
2	$11\frac{1}{2}$
3	6
4	$7\frac{1}{2}$
5	$11\frac{1}{2}$
6	$11\frac{1}{2}$

Note: $\frac{1}{2}$ denotes a joint favourite.

The derived value of $\Sigma\chi^2$ is 2.9 which is less than 11.07 and therefore it cannot be said, at the 0.05 level of significance, that the favourite greyhound is drawn more often in any particular trap position.

9.4   GOODNESS OF FIT: A POISSON PROBLEM

In Chapter 7 it was stated that the Poisson distribution can be shown to represent the distribution of rare events, such as goals in football matches. Whether this is true or not can be tested using the chi-squared test. The basic data chosen were the 636 football matches played in the English Football Association Cup of 1985–1986 with Liverpool eventually beating Everton in the final at Wembley. These 636 matches included replays after initially drawn games and all the matches of the preliminary rounds of the FA Cup. A total of 1959 goals were scored and therefore the null hypothesis is

H_0: The probability of occurrence of goals in the 1985–1986 FA Cup is the same as that given by a Poisson distribution with a mean, $m = 3.1$.

The mean m is calculated as 1959/636. Table 9.4 shows the computations for deriving the distribution of the expected number of goals using the Poisson formula, and Table 9.5 shows the chi-squared test calculations. The derived $\Sigma\chi^2$ statistic is 5.6. From Table 9.1 for DF $= 9 - 2 = 7$ and $P = 0.01$, the χ^2 statistic is 18.48. Since 5.6 is less than 18.48 the null hypothesis, H_0, cannot be rejected at the 0.01 level of significance.

Table 9.4. Expected number of goals derived from the Poisson distribution with $m = 3.1$. Poisson probabilities are given by $\Pr(n) = e^{-m}m^n/n!$ (see section 7.1) where m is the mean of the distribution and $\Pr(n)$ is the probability of n goals in a football match—in this problem. $m = 3.1$ and $e^{-3.1} = 0.0450$.

n	$n!$	$(3.1)^n$	$(3.1)^n/n$	$\Pr(n)$	Expected goals $= 636 \times \Pr(n)$
0	1	1	1	0.0450	28.62
1	1	3.1	3.1	0.1395	88.72
2	2	9.61	4.805	0.2162	137.52
3	6	29.79	4.965	0.2234	142.10
4	24	92.35	3.848	0.1732	110.13
5	120	286.29	2.386	0.1074	68.28
6	720	887.50	1.233	0.0555	35.28
7	5040	2751.26	0.546	0.0246	15.62
8	40320	8529.91	0.212	0.0095	6.05
9	362880	26439.62	0.073	0.0033	2.09
10	3628800	81962.83	0.023	0.0010	0.65

Table 9.5. Chi-squared test for the football match problem.

Number of goals in an FA Cup match	$O =$ observed number of goals	$E =$ expected number of goals from Poisson, $m = 3.1$	$O - E$	$(O - E)^2$	$(O - E)^2/E$
0	32	28.6	3.4	11.56	0.40
1	92	88.7	3.3	10.89	0.12
2	141	137.5	3.5	12.25	0.09
3	132	142.1	−10.1	101.00	0.71
4	111	110.1	0.9	0.81	0.01
5	70	68.3	1.7	2.89	0.04
6	32	35.3	−3.3	10.89	0.31
7	12	15.6	−3.6	12.96	0.83
8–10	14	8.8	5.2	27.04	3.07

Total number of football matches = 636.
Total number of goals = 1959.
Mean of the Poisson distribution, $m = 3.1$.
$\Sigma\chi^2 = $ sum of all $(O - E)^2/E$ values = 5.6.
Number of degrees of freedom = $9 - 2 = 7$.

9.5 GOODNESS OF FIT: A LOGNORMAL CURVE FITTING PROBLEM

In Table 3.4 the chi-squared test is used as a goodness of fit test of observational data to the normal distribution curve. There were 7 data cells and the number of degrees of freedom was therefore $7 - 3 = 4$. In this section a similar problem will be described, the difference being that it is the *lognormal distribution curve* which is being tested for goodness of fit. Table 3.7 shows the method by which the area beneath the lognormal curve can be computed between defined limits. The initial stage of this computation involves specifying the unit normal deviate for the lognormal curve, $[\log_e(T/M)]/S$, so that standard data tables for the normal curve can be used. This method is used to calculate the expected number of deaths, E, in Table 9.6. The null hypothesis for the current lognormal problem is

H_0: There is no difference between the distribution of observed values and the distribution of the expected values calculated from a lognormal distribution with a specified mean M and standard deviation S.

The observations used for this example are the survival times of 583 stage 2 cancer of the cervix patients who died with their disease present[3] and Figure 9.2 presents the data in histogram format. The *areas of the histogram blocks* represent the number of deaths in a given interval and this is taken into account when some blocks, such as for 0–6 and 6–12 months, have block widths half the value of the remaining histogram blocks. Also incorporated into Figure 9.2 is the logarithmic probability graph plot, which demonstrates visually that the data are reasonably lognormal. Lognormality is also demonstrated graphically in Figure 3.8 with the associated data requirements given in Table 3.6. From the straight line in Figure 9.2,

Estimation of M
$T = 23$ when $X = 50\%$
therefore $M = 23$

Estimation of S
$T = 160$ when $X = 95\%$
$\log_{10}(23) = 1.3617$ and $\log_{10}(160) = 2.2041$
therefore $S = (2.2041 - 1.3617)/1.645 = 0.51$.

The problem we have is to decide whether at the chosen 0.05 level of significance we can accept H_0 and therefore show that the lognormal curve is a good approximation to the observed distribution of cancer of the cervix survival times. Table 9.6 gives the computation method, with the expected number of deaths, E, being calculated for a lognormal distribution with $M = 23$ and $S = 0.51$. The derived $\Sigma\chi^2$ statistic is 12.3. From Table 9.1, for DF $= (11 - 3) = 8$ and $P = 0.05$, the χ^2 statistic is 15.51. Since 12.3 is less than 15.51 the null hypothesis, H_0, cannot be rejected at the 0.05 level of significance.

9.6 THE 2 × 2 CONTINGENCY TABLE

The application of the chi-squared test in a 2 × 2 contingency table format (also sometimes termed a 2-way table or a fourfold table) is a test of association between mutually exclusive categories of one variable (given in the rows of the table) and mutually exclusive categories of another variable (given in the columns of the table). It is a table of frequencies showing how the total frequency is distributed among the four cells in the table. The null hypothesis which is tested is

H_0: No relationship (i.e. association) exists between the two variable classifications.

It is a test for a *comparison of two proportions*, but it is actual numbers, that is, the actual frequencies, which are in the cells of the 2 × 2 table, and not percentage values. The degrees of freedom for the chi-squared test are

$$DF = (\text{Number of rows } - 1) \times (\text{Number of columns } - 1)$$

Table 9.6. Calculations required for the chi-squared test for the lognormal curve fitting problem.

j	Survival time interval (years), $T_{j+1} - T_j$	Observed number of deaths, O_j	Expected number of deaths, E_j	$O_j - E_j$	$(O_j - E_j)^2$	$(O_j - E_j)^2/E_j$ $= \chi_j^2$
1	0–0.5	56	57.1	−1.1	1.21	0.02
2	0.5–1	107	98.1	8.9	79.21	0.81
3	1–2	155	146.7	8.3	68.89	0.47
4	2–3	78	87.9	−9.9	98.01	1.11
5	3–4	62	54.6	7.4	54.76	1.00
6	4–6	50	60.1	−9.9	98.01	1.63
7	6–8	30	29.8	0.2	0.04	0.00
8	8–10	27	16.5	10.5	110.25	6.68
9	10–12	10	9.9	0.1	0.01	0.00
10	12–14	5	6.3	−1.3	1.69	0.27
11	14–16	3	4.2	−1.2	1.44	0.34
		Sum = 583	Sum = 571.2†			Sum = 12.3

†11.8 deaths are expected beyond 16 years.   Derived $\Sigma \chi^2$ statistic = 12.3.

which for a 2 × 2 table gives DF = 1. If the observed frequencies in the table are a, b, c and d

	Column 1	Column 2	Marginal totals
Row 1	a	b	$a + b$
Row 2	c	d	$c + d$
Marginal totals	$a + c$	$b + d$	Grand total $N = a + b + c + d$

then the expected frequencies are calculated from the row and column *marginal totals* and the *grand total*, N, as follows:

	Column 1	Column 2
Row 1	$(a + b) \times (a + c)/N$	$(a + b) \times (b + d)/N$
Row 2	$(c + d) \times (a + c)/N$	$(c + d) \times (b + d)/N$

The $\Sigma \chi^2$ statistic is then computed by the summation of the four $(O - E)^2/E$ values. This yields a formula for $\Sigma \chi^2$ which sometimes enables

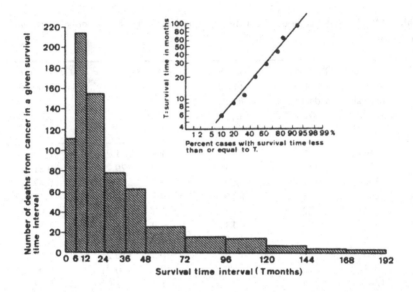

Figure 9.2. Distribution of survival times of 583 stage 2 cancer of the cervix patients who died with their disease present. Graphical demonstration of lognormality for these data.

a quicker computation to be made:

$$\Sigma \chi^2 = \frac{N(ad - bc)^2}{(a + b)(c + d)(a + c)(b + d)}$$

9.7 A 2 × 2 CONTINGENCY TABLE: A CHOLERA EPIDEMIC PROBLEM

Data from de Lorenzo *et al* in the *Lancet* of 1974 are used to illustrate a practical application of the 2 × 2 contingency table. From August to October 1983 there was an outbreak of cholera in Naples and the scattered geographical distribution of cases initially indicated that water was not the cause of the outbreak. This was later confirmed by laboratory tests.

Histories were taken from 911 patients admitted to one particular hospital during the epidemic and the following points were found. 31.5% of 130 patients who had eaten raw mussels in the 5 days before hospital admission were affected by cholera, but only 10.9% of the 781 patients who had not eaten raw mussels were found to suffer from cholera. A 2 × 2 contingency table can be used to test the null hypothesis

H_0: There is no association between the cholera infection and a diet including raw mussels in the 5-day period preceding the cholera outbreak.

The two variables in the contingency table are cholera infection and diet (see Table 9.7).

Table 9.7. Basic data for the cholera problem.

Variable = Diet	Variable = cholera infection		
	Column 1 Infected	Column 2 Not infected	Marginal totals
Row 1 Mussel eaters	41	89	130
Row 2 Not mussel eaters	85	696	781
Marginal totals	126	785	Grand total = 911

Table 9.8 shows the procedure to calculate the $\Sigma \chi^2$ statistic from the data in the 2×2 contingency table. The derived $\Sigma \chi^2$ statistic is 39.8. From Table 9.1 for DF = 1 and $P = 0.05$, the χ^2 statistic is 3.84. 39.8 is very much larger than 3.84 and therefore the null hypothesis, H_0, is rejected at the 0.05 level of significance. Indeed, it would even be rejected at the 0.001 level of significance since from Table 9.1 the χ^2 statistic is 10.83. This result points to the very probable role of mussels in this epidemic. This was confirmed by the striking reduction in cholera cases once the sale of mussels was forbidden.

9.8 THE GENERALISED $r \times c$ CONTINGENCY TABLE

The general form of an $r \times c$ contingency table is given in Table 9.9.

The expected frequency for the cell in the ith row and jth column is $(R_i C_j)/N$. The χ^2 statistic is the sum of all $(O - E)^2/E$ values for all the O_{ij}, E_{ij} cells.

An example of an application of a contingency table larger than a 2×2 would be for three different treatments A, B and C, where the patients are assessed for treatment success as either complete success, partial success, or no change. In this case, the table would be a 3×3 contingency table and the number of degrees of freedom is 4. The null hypothesis is

H_0: There is no difference in the pattern of outcome (i.e. in terms of response) between the treatments A, B and C.

Table 9.8. Calculation of the $\Sigma\chi^2$ statistic from a 2×2 contingency table.

Cell position	$O =$ observed frequency	$E =$ expected frequency	$O - E$	$(O - E)^2$	$(O - E)^2/E$
R1C1	41	18	23	529	29.39
R1C2	89	112	−23	529	4.72
R2C1	85	108	−23	529	4.90
R2C2	696	673	23	529	0.79

R = row, C = column.
Total number of cases = 911.
$\Sigma\chi^2$ = sum of all $(O - E)^2/E$ values = 39.8.
Number of degrees of freedom = 1.

Table 9.9. Form of a generalised $r \times c$ contingency table with the number of degrees of freedom DF $= (r - 1) \times (c - 1)$.

		Variable y					Row marginal totals
		y_1	y_2	$\cdots$	y_j	$\cdots$ y_c	
	x_1				.		R_1
	x_2				.		R_2
	$\vdots$				$\vdots$		$\vdots$
Variable x	x_i	.	.	$\cdots$	O_{ij}	$\cdots$.	R_i
	$\vdots$				$\vdots$		$\vdots$
	x_r				.		R_r
Column marginal totals		C_1	C_2	$\cdots$	C_j	$\cdots$ C_c	N

O_{ij} is the frequency for the ith row and jth column.
$R_i = \Sigma_{j=1}^{c} O_{ij}$ is the row marginal frequency for the ith row.
$C_j = \Sigma_{i=1}^{r} O_{ij}$ is the column marginal frequency for the jth column.
$N = \Sigma_{i=1}^{r}\Sigma_{j=1}^{c} O_{ij}$ is the total frequency.

9.9   YATES CORRECTION FOR SMALL SAMPLES

It should be noted that contingency tables and the chi-squared test are not always appropriate if the numbers involved are too small. However, opinion differs on what is too small but the recommendations below are a good guideline.

Table 9.10. Recommendations for small samples.

Use Yates correction: 2 × 2 contingency table	When sample size is less than 100 or with any cell less than 10
2 × 2 contingency table not to be used	If the smallest expected value is less than 5 and the sample size 20–40
	If the sample size is less than 20
r × c contingency table with DF> 1 not to be used	If more than 1/5th of expected values are less than 5
	If any expected value is less than 1
When 2 × 2 contingency table cannot be used	Use the Fisher exact probability test. (This was specifically designed for use with small samples when a 2 × 2 table cannot be used, see Chapter 10)

When a Yates correction (sometimes termed a *continuity* correction) is used all $|O - E|$ values for the four cells of the 2x2 table are reduced by 0.5 before calculating $\Sigma \chi^2$, thus

$$\Sigma \chi^2 = \Sigma \{|O - E - 0.5|\}^2 / E$$

The mathematical symbol | | is termed *modulus* and means that the numerical value of $O - E$ is reduced by 0.5 regardless of whether $O - E$ is negative or positive. Thus $+11.6$ would reduce to 11.1 and -11.6 would reduce to -11.1.

The formula in section 9.6 when the Yates correction is applied should be modified to become:

$$\Sigma \chi^2 = \frac{N(|ad - bc| - \frac{1}{2}N)^2}{(a + b)(c + d)(a + c)(b + d)}$$

One example of when the 2 × 2 contingency table was not appropriate is in the trial by Chain and colleagues of penicillin for *staphyloccus aureus* infection in the mouse, for which the trial results are shown in Table 9.11. Indeed, statistics are hardly required at all with such results

Table 9.11. Basic data for the pencillin trial.

	No. of mice dying	No.of mice surviving
Treated	3	21
Not treated	24	0

9.10   DID MENDEL CHEAT?

The experimental data for Mendel's experiment with peas, intended to verify his theory of inheritance, is a good example for a workshop seminar on the chi-squared test since teaching experience has shown that many approach the setting up of the chi-squared test from the wrong point of view: by drawing up a 2×2 contingency table.

According to Mendel's theory the numbers for the different (shape and colour) classes of pea should be in the proportions 9:3:3:1 for the experimental results (given in brackets) with 556 peas.

{315} Round and Yellow
{108} Round and Green
{101} Wrinkled and Yellow
{32} Wrinkled and Green

The following 2×2 table (DF = 1) is incorrect because it implies testing for an association between shape (round/wrinkled) and colour (yellow/green) when in fact the testing should be for experimental results *versus* theoretical expectation.

	Round	Wrinkled
Yellow	315	101
Green	108	32

A goodness of fit chi-squared should be used in a manner similar to the problems in sections 9.4 and 9.5.

The critical value of χ^2 for DF= 3 and $P = 0.05$ from Table 9.1 is 7.815. This therefore looks like a very good fit. Indeed from Table 9.1 and Figure 9.1 the value of P corresponding to $\chi^2 = 0.47$ and DF= 3 is between $P = 0.90$ (for which $\chi^2 = 0.584$) and $P = 0.95$ (for which $\chi^2 = 0.352$). Mendel's results are too good to be true and the famous statistician Sir Ronald Fisher said that the experimental results were so close to the expected that it would have taken 'an absolute miracle of chance' to produce them.

Table 9.12.   Chi-squared test of Mendel's experimental results and his theory. DF = (4 − 1) = 3. Derived $\Sigma \chi^2 = 0.470$.

i	Observed O_i	Expected E_i	$(O_i - E_i)$	$(O_i - E_i)^2$	$(O_i - E_i)^2/E$
1	315	$(9/16) \times 556 = 312.75$	+2.25	5.0625	0.0162
2	108	$(3/16) \times 556 = 104.25$	+3.75	14.0625	0.1349
3	101	$(3/16) \times 556 = 104.25$	−3.25	10.5625	0.1013
4	32	$(1/16) \times 556 = 34.75$	−2.75	7.5625	0.2176

He uses statistics as a
drunken man uses
lamp-posts, for support
rather than for illumination.
Andrew Lang (1844–1912)

Chapter 10

The Fisher Exact Probability Test

10.1 INTRODUCTION

The Fisher exact probability test for 2×2 tables is very useful when samples are small. The test consists of finding the exact probability of the observed occurrences by taking the ratio of the product of the factorials of the four marginal totals to the product of the cell frequencies multiplied by $N!$ (using the same notation for a 2×2 table as in section 9.6, that is a, b, c, d, N). P is the probability of the observed distribution of frequencies under the null hypothesis, H_0, where

$$ P = \frac{(a+b)!(c+d)!(a+c)!(b+d)!}{N!a!b!c!d!} $$

Table 10.1 is a table of $N!$ values which, if known, make the computations quicker. They can also be simpler when one of the cell frequencies, a, b, c or d, is 0; but generally the factorials can be cancelled down to make the arithmetic easier.

For any 2×2 table of observed frequencies, the probabilities of all tables with the same marginal totals can be calculated and a derived probability level for the test can be calculated by summation.

This is because the basic assumption underlying the Fisher's exact test is that the row and column totals (i.e. the *marginal totals*) can be treated as fixed quantities. The individual cells within the table, however, are free to vary subject to the constraint that these marginal totals remain constant.

10.2 ONE-TAILED EXAMPLE

Figure 10.1 illustrates the post-Chernobyl iodine-131 thyroid uptake measurements of 31 people living in embassies in Warsaw who were sent to the Westminster Hospital for thyroid monitoring[1]. Potassium iodide tablets were only given to the children, not to the adults, and whereas in the United Kingdom,

Table 10.1. Table of $N!$ for $1 \leqslant N \leqslant 20$.

N	$N!$
0	1
1	1
2	2
3	6
4	24
5	120
6	720
7	5040
8	40320
9	362880
10	3628800
11	39916800
12	479001600
13	6227020800
14	87178291200
15	1307674368000
16	20922789888000
17	355687428096000
18	6402373705728000
19	121645100408832000
20	2432902008176640000

when a thyroid needs to be blocked, 100 mg of potassium iodate (more stable than the iodide) would be given, the Warsaw group of children only received 10 mg of the iodide. Nevertheless, for the Malaysian group who remained in Warsaw for 28 days after the iodine dispensation, it would appear from Figure 10.1 that, even though the iodine was given 3 days after the radioactive cloud passed over Warsaw, it was still effective to a certain extent. Is there a significant difference between the Malaysian adults and children for this small population of 15 persons? The sample size is too small for a chi-squared test to be appropriate, but the Fisher exact probability test can be used. The null hypothesis is

H_0: There is no difference in iodine uptake in the thyroid between Malaysian adults and children.

a	b	$a+b$
c	d	$c+d$
$a+c$	$b+d$	N

The basic data for this problem is given in Table 10.2. There are 5 possible tables with the same marginal totals as those observed (i.e. 9, 6, 4, 11) since neither a nor c can fall below 0 or exceed 4, which is the smallest marginal total.

Table 10.2. Basic data for the one-tailed Fisher exact probability test.

Variable = Age group	Variable = I-131 thyroid uptake level		
	Column 1 Activity > 2 KBq	Column 2 Activity < 2 KBq	Marginal totals
Row 1 Children	0	9	9
Row 2 Adults	4	2	6
Marginal totals	4	11	15

These 5 tables are each labelled as a set, numbered according to the frequency $a = 0, 1, 2, 3, 4$

0	9	9		1	8	9		2	7	9		3	6	9		4	5	9
4	2	6		3	3	6		2	4	6		1	5	6		0	6	6
4	11	15		4	11	15		4	11	15		4	11	15		4	11	15
Set 0				Set 1				Set 2				Set 3				Set 4		

Sets 0 and 4 are the extreme frequency distributions.
Using the formula in section 10.1, the probability of set 0 is

$$P_0 = \frac{9! \times 6! \times 4! \times 11!}{15! \times 0! \times 9! \times 4! \times 2!} = 0.0110$$

and of set 1 is

$$P_1 = \frac{9! \times 6! \times 4! \times 11!}{15! \times 1! \times 8! \times 3! \times 3!} = 0.1319$$

and, similarly, P_2, P_3 and P_4 can be calculated. The total probability is of course 1, since the 5 sets represent all the possible alternatives ($P_0 + P_1 + P_2 + P_3 + P_4 = 1$).

In the Fisher exact probability test, the investigator must calculate the probability of the observed table of frequencies *or of one (table) which is more extreme*. In this example, set 0 which is the observed set is in fact one of the two most extreme tables. The other is set 4. Therefore, for a 1-tailed test§

§ In this example there is only interest in a 1-tailed test and not in a 2-tailed test. We are only interested in the adults having significantly higher activities than the children, as the Malaysian adults received no iodine. From Figure 10.1 it would obviously be a nonsense to consider a 2-tailed test since the children did not have higher thyroid activities than the adults.

the derived probability level is 0.011, which is less than the chosen level of significance of 0.05. The null hypothesis, H_0, is therefore rejected.

The implication of this rejection of H_0 is that even though the iodine was given after the radioactive cloud had passed over Warsaw and consequently after the maximum hazard had occurred, the exercise was still worthwhile for these children.

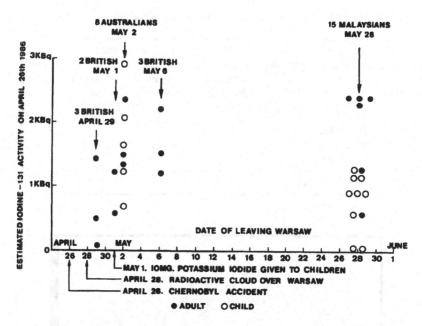

Figure 10.1. Post-Chernobyl accident iodine-131 uptake measurements on 31 people living in embassies in Warsaw. The apparent difference for Malaysian and Australian children in the effectiveness of the potassium iodine may be due to the biological loss of some of the iodine-131 in the Malaysian children after the Australians had left Warsaw Further uptake of iodine-131 in the period 2–28 May might then have taken place in the Malaysian adults, but would have been blocked in the children[1].

10.3   TWO-TAILED EXAMPLE

There is some argument as to whether, and also as to how, the derived probability level should be calculated for both tails of the distribution curve. Armitage[2] suggests doubling the derived level of probability for a l-tailed test on the grounds that a significant result (i.e. rejection of the null hypothesis, H_0), is interpreted as strong evidence for a difference in the observed direction. He illustrates the use of the test with data on malocclusion of the teeth of infants, Table 10.3. Using the previous notation there will be 6 possible sets: 0, 1, 2,

3, 4 (the observed set) and 5. The probabilities are $P_0 = 0.0310$, $P_1 = 0.1720$, $P_2 = 0.3440$, $P_3 = 0.3096$, $P_4 = 0.1253$, $P_5 = 0.0182$; total $= 1.0001$. For a 1-tailed test the probability of set 4 or a more extreme set (i.e. set 5) is $0.1253 + 0.0182 = 0.1435$, which is not significant at the 0.05 level of significance. The 2-tailed test would give 0.2870.

Table 10.3. Basic data on breast feeding and malocclusion of teeth.

Variable = Feeding technique	Variable = Teeth assessment		
	Column 1 Normal	Column 2 Malocclusion	Marginal totals
Row 1 Breast fed	4	16	20
Row 2 Bottle fed	1	21	22
Marginal totals	5	37	42

Chapter 11

The t-test

11.1 INTRODUCTION

There are two types of significance test and they are described as parametric and non-parametric. For the parametric tests, certain conditions should exist for the population being tested. For non-parametric tests, no such conditions are laid down. The chi-squared test is non-parametric, but the t-test and the F-test are parametric. One *condition* which must hold before a t-test or F-test can be used is that the population from which the sample under observation is drawn must be normally distributed. That is, the population distribution can be fitted by a normal curve. In practice, however, for some tests the investigator is allowed a certain latitude and, as long as the population is approximately normal, the test may be applied. In statistical terminology, it is said that the test is *robust* when we can accept approximate normality. The t-test is a robust test and the t-statistic is

$$t = \frac{\text{Difference in means}}{\text{Standard error of the difference in means}}$$

Formulae for standard errors of a mean or of a percentage and of the difference in means or the difference in percentages have already been given in section 4.4.

The conditions, including normality, which must be satisfied in order for the t-test to be used, are given in Table 11.1. The term *interval* scale is referred to in this table and Figure 11.1 illustrates the three possible types of measurement: *interval*, *nominal* and *ordinal*.

The t-distribution was first published by W S Gosset in 1908, who was a brewer working for the Guinness Company and who used the pseudonym Student. Hence, the distribution is often called Student's t-distribution. It describes the variability of the mean and standard deviation of small samples taken from a normal population and the distribution curve is similar in shape to the normal curve. However, it has fewer observations in the mode and more in the tails of the distribution, see Figure 11.2. The t-distribution solves

Table 11.1. Conditions which must be satisfied in order for a t-test to be appropriate.

1	The observations must be *independent* in order to avoid bias.
2	The observations must be drawn from *normal populations*.
3	These normal populations must have the same *variance* (or in special circumstances, a known ratio of variances).
4	The variables involved must have been measured in an *interval* scale, so that it is possible to use arithmetical operations (e.g. add, divide, obtain means) on the values of the variables.

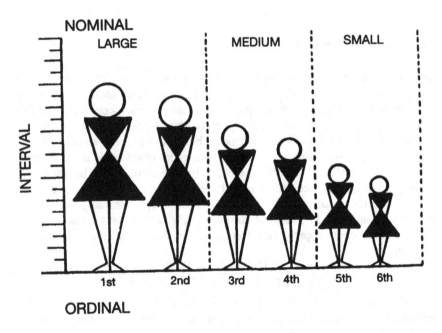

Figure 11.1. Illustration of three types of measurement.

the problem of working with small sample numbers of 30 or less, which are impossible to deal with using the normal distribution itself. Gosset tested his t-distribution in an interesting manner by obtaining data on the heights of 3000 criminals, writing each on a separate card, and sampling four cards 750 times so that each sample represented the heights of four criminals. He calculated the mean and standard deviation of each sample (x_m and s_m using the same notation as in section 4.2) and derived a value of t for each sample. t was calculated by taking the difference of the sample mean x_m and the population mean μ and dividing this by the standard error of the mean, SE, where

$$SE = s_m/\sqrt{n}$$

and n is the number in the sample:

$$t = \{(x_m - \mu)/(s_m/\sqrt{n})\}$$

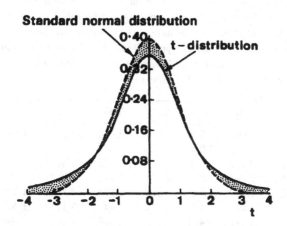

Figure 11.2. Comparison of the standard normal distribution curve (mean = 0, standard deviation = 1, infinite number of degrees of freedom) which is shown as a broken curve, and the t-distribution (n = 4 observations, degrees of freedom = 3) which is shown as a full curve. With a decreasing number of degrees of freedom the maximum of the t-distribution decreases and the shaded area between the two distribution curves increases.

Table 11.2 is a table of t-statistic values in terms of degrees of freedom (DF) and probability levels (P). Some of these data are presented graphically in Figure 11.3.

11.2   ESTIMATE OF THE POPULATION MEAN μ FROM THE SAMPLE MEAN x_m AND CALCULATION OF CONFIDENCE LIMITS

The solution is to first construct t using the formula given at the top of this page and then solve it to obtain the value of μ. Section 4.10 also describes the calculation of confidence limits but in that instance there is a defined population of 72 people and the population mean μ and standard deviation σ are known. In the example in this section the value of μ and σ are not known and moreover the sample size of $n = 10$ is too small for the normal distribution to be used instead of the t-distribution: that can only occur with large sample sizes as will be explained in section 11.6.

It must be remembered with t-tests that t can be either positive or negative, depending on whether $x_m > \mu$ or $\mu > x_m$ and that tables of the critical values of t, see Table 11.2, list only the positive values of t.

Table 11.2. Values of the t-statistic for selected probabilities and degrees of freedom. The rows in the table indicate the degrees of freedom (DF) and the columns indicate probability levels (*P*). The values in the body of the table are the values of t and the probabilities given are for given values of $+t$ (and of $-t$) with numerically larger values being less probable. The t-statistic corresponding to a probability of 0.05 for DF= 5 is 2.571. Values larger than $+2.57$ thus have a less than 5% probability and values less than -2.57 thus have a less than 5% probability. (From Fisher and Yates, *Statistical Tables for Biological, Agricultural and Medical Research* (6th edn, 1974, Table 111, p 46). Courtesy Longman Group UK Limited.)

DF	$P = 0.90$	0.50	0.20	0.10	0.05	0.02	0.01	0.001
1	0.158	1.000	3.078	6.314	12.706	31.821	63.657	636.619
2	0.142	0.816	1.886	2.920	4.303	6.965	9.925	31.598
3	0.137	0.765	1.638	2.353	3.182	4.541	5.841	12.924
4	0.134	0.741	1.533	2.132	2.776	3.747	4.604	8.610
5	0.132	0.727	1.476	2.015	2.571	3.365	4.032	6.869
6	0.131	0.718	1.440	1.943	2.447	3.143	3.707	5.959
7	0.130	0.711	1.415	1.895	2.365	2.998	3.499	5.408
8	0.130	0.706	1.397	1.860	2.306	2.896	3.355	5.041
9	0.129	0.703	1.383	1.833	2.262	2.821	3.250	4.781
10	0.129	0.700	1.372	1.812	2.228	2.764	3.169	4.587
12	0.128	0.695	1.356	1.782	2.179	2.681	3.055	4.318
15	0.128	0.691	1.341	1.753	2.131	2.602	2.947	4.073
20	0.127	0.687	1.325	1.725	2.086	2.528	2.845	3.850
25	0.127	0.684	1.316	1.708	2.060	2.485	2.787	3.725
30	0.127	0.683	1.310	1.697	2.042	2.457	2.750	3.646
40	0.126	0.681	1.303	1.684	2.021	2.423	2.704	3.551
60	0.126	0.679	1.296	1.671	2.000	2.390	2.660	3.460
120	0.126	0.677	1.289	1.658	1.980	2.358	2.617	3.373
∞	0.126	0.674	1.282	1.645	1.960	2.326	2.576	3.291

To illustrate this problem of estimating μ from a knowledge of x_m suppose we have a random sample of 10 males with the following diastolic blood pressure measurements

94	94
94	92
98	98
74	95
84	86

for which $x_m = 91$. The sample standard deviation s_m can be calculated as shown in section 2.4 on page 26 (using the notation $\bar{x}$ for x_m).

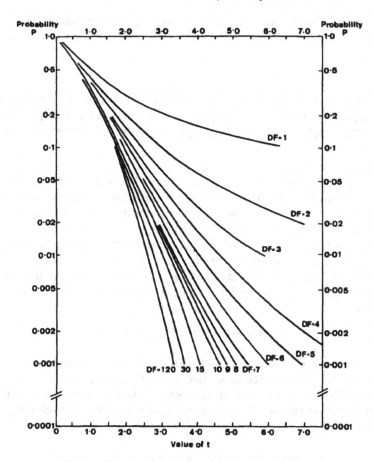

Figure 11.3. The t-statistic as a function of degrees of freedom and probability level.

Since $\Sigma(x_i - x_m)^2 = 514$ we have $s_m = \sqrt{(514/9)} = 7.56$. By rearranging the formula for t we have

$$\mu = x_m - t(s_m/\sqrt{n})$$

but since t can be either positive or negative this expression giving the 95% confidence limits (in a similar manner to section 4.10) becomes

$$\mu = x_m \pm t(s_m/\sqrt{n})$$

Using Table 11.2 for $P = 0.05$ and DF$= (10 - 1) = 9$ we have $t = 2.262$ and therefore

$$\mu = \{91 \pm 2.262(7.56/\sqrt{10})\} = \{91 \pm 5.41\} = 85.59 \text{ to } 96.41$$

and we can say that we are 95% confident that the population mean μ lies in the range 85.59 to 96.41 and that there is only a 5% probability that μ lies outside this range.

11.3 THE ONE-SAMPLE *t*-TEST: INFERENCE BASED ON A SINGLE SAMPLE MEAN x_m WHEN THE REFERENCE POPULATION FROM WHICH THE SAMPLE IS DRAWN IS KNOWN TO HAVE A MEAN OF μ

In this application of the t-test suppose that we know the mean μ_0 of some reference population of interest from which a random sample is drawn with a mean x_m and standard deviation s_m. The problem is to decide whether the difference between x_m and μ_0 is statistically significant. That is, whether there is only a small probability that a difference at least as large as the one observed, $(x_m - \mu_0)$, could have arisen by chance.

The null hypothesis H_0 states that $H_0 : \mu = \mu_0$ where the mean of the population from which the random sample was drawn is μ. The formula for the t-statistic is

$$t = \{(x_m - \mu_0)/(s_m/\sqrt{n})\}$$

To illustrate this consider again data for diastolic blood pressures and assume that for the reference population $\mu_0 = 93$ and suppose that a random sample of 20 males has a mean of $x_m = 98$ and standard deviation $s_m = 6.20$. Our problem is to estimate if the mean $x_m = 98$ is significantly higher than the mean $\mu_0 = 93$:

$$t = \{(98 - 93)/(6.20/\sqrt{20})\} = \{5/1.386\} = 3.61$$

Using Table 11.2 for DF= $(20-1) = 19$ we see that the one-sided P-value lies between 0.001 and 0.01. We therefore reject H_0 that our study sample of 20 males is a random sample of the reference population (with $\mu_0 = 93$) and conclude that it derives from a different population having a higher diastolic blood pressure, i.e $\mu > \mu_0$.

11.4 PAIRED *t*-TEST: DIFFERENCE BETWEEN MEANS

In this and succeeding sections in this chapter, the arithmetic becomes more complicated and to make understanding as easy as possible the t-test applications are illustrated for hypothetical data, for a quantity x measured for four patients before and after treatment where the means before and after treatment are given the notation x_{1m} and x_{2m} to be consistent with previous notation.

Patient	Before treatment	After treatment
	x_1	x_2
A	2	3
B	4	3
C	3	1
D	5	2

The two sets of measurements, before and after treatment, are paired for each patient and the null hypothesis H_0 is

H_0: There is no difference in quantity x before (x_1) and after (x_2) treatment, and that consequently both x_1 and x_2 are drawn from the same normal population and that ($x_{1m} - x_{2m}$) is normally distributed about a mean equal to 0.

The t-statistic in this situation is given by

$$t = \{(x_{1m} - x_{2m})/(s_{1m-2m}/\sqrt{n})\}$$

From the data we have $(x_{1m} - x_{2m}) = (3.5 - 2.25) = 1.25$ and the sample standard deviation of the difference (s_{1m-2m}) of $(x_1 - x_2)$ is given by the formula in section 2.4 on page 25 where the term (x_i) is replaced by $(x_1 - x_2)$ and the term $\bar{x}$ by $(x_{1m} - x_{2m})$. The standard error of the difference in means is then given by SE= $(s_{1m-2m}/\sqrt{n})$ where $n = 2$ since there is only a pair of data groups:

$$s_{1m-2m} = \sqrt{\{\Sigma(x_1 - x_2)^2 - [(\Sigma[x_1 - x_2])^2/n]\}}/\sqrt{(n - 1)}$$

From the data for these four patients we find that $\Sigma(x_1 - x_2)^2 = 15$ and $\sqrt{[x_1 - x_2]} = 5$. Therefore $s_{1m-2m} = \sqrt{\{15 - [25/4]\}}/\sqrt{3} = 1.71$. The derived t-statistic is therefore given by $t = 1.25/[1.71/\sqrt{2}] = 1.03$.

The degrees of freedom equal the number of pairs minus 1, i.e. DF= $(4 - 1) = 3$ and from Table 11.2 for DF= 3 and $P = 0.05$ the t-statistic is 3.18. Since 1.03 is less than 3.18 we cannot reject the null hypothesis H_0 at the 0.05 level of significance. We must therefore assume that there is no detectable change in the value of x before and after treatment.

11.5 UNPAIRED TWO-SAMPLE *t*-TEST

When the data are unpaired the formulae for the standard error of the difference in means is more complicated. To illustrate the calculation procedure we use the same two small numerical sets of data for x_1 and x_2 as in section 11.4, but we now assume that they are *not matched pairs* but two separate groups of patients with n_1 in one group and n_2 in the second group. It should be remembered that to perform the t-test it is not necessary that n_1 and n_2 are always equal.

Group 1	Group 2
x_1	x_2
2	3
4	3
3	1
5	2

We cannot now work directly with the differences $(x_1 - x_2)$ because the data are not matched pairs but we can of course still calculate the means x_{1m} and x_{2m} and determine the difference in means $(x_{1m} - x_{2m})$ which is the denominator for the t-statistic.

The formula for the standard error of the difference in means SE_p is calculated using a *pooled estimate* s_p of the standard deviations, of the two groups of observations, such that

$$SE_p = s_p \left(\sqrt{((1/n_1) + (1/n_2))} \right)$$

and the t-statistic is then calculated as

$$t = (x_{1m} - x_{2m})/SE_p$$

In order to calculate SE_p which is given by the formula

$$SE_p = \left[(((\Sigma x_1^2 - (\Sigma x_1)^2/n_1) + (\Sigma x_2^2 - (\Sigma x_2)^2/n_2)) \right.$$
$$\left. * ((1/n_1) + (1/n_2))/((n_1 - 1) + (n_2 - 1))) \right]^{1/2}$$

This looks complicated but from Chapter 2 it is seen that the formula for the standard deviation of x_1 (and similarly for x_2) is given by

$$SD_1 = \sqrt{((\Sigma(x_1 - x_{1m})^2)/(n_1 - 1))}$$

where x_{1m} is the mean value of all the n_1 observations x_1. Alternatively SD_1 may be written as

$$SD_1 = \sqrt{((\Sigma(x_1^2) - ((\Sigma x_1)^2)/n_1)/(n_1 - 1))}$$

and therefore the above formula for SE_p may also be written as

$$SE_p = \left[(((n_1 - 1) * SD_1^2) + ((n_2 - 1) * SD_2^2)) \right.$$
$$\left. * ((1/n_1) + (1/n_2))/((n_1 - 1) + (n_2 - 1))) \right]^{1/2}$$

using the notation $[\ ]^{1/2}$ for $\sqrt{[\ ]}$ and $*$ to symbolise multiplication.

In order to perform the calculations it is recommended that a list of the various individual terms required be written down since this will make it easier to insert them into the formula:

$$n_1 = 4 \qquad n_2 = 4$$
$$\Sigma x_1 = 14 \qquad \Sigma x_2 = 9$$
$$x_{1m} = 3.5 \qquad x_{2m} = 2.25$$

$$(x_{1m} - x_{2m}) = 1.25$$

$$\Sigma(x_1^2) = 54 \qquad \Sigma(x_2^2) = 23$$
$$(\Sigma x_1)^2 = 196 \qquad (\Sigma x_2)^2 = 81$$
$$((\Sigma x_1)^2/n_1) = 49 \qquad ((\Sigma x_2)^2/n_2) = 20.25$$
$$SD_1 = \sqrt{(54-49)/3} = 1.291 \qquad SD_2 = \sqrt{(23-20.25)/3} = 0.957$$

$$SE_p = \sqrt{((3*1.667)+(3*0.917))/(3+3))*((1/4)+(1/4))}$$
$$= \sqrt{7.751/6)*(0.5)} = 0.803$$

$$t = (m_1 - m_2)/SE_p = 1.25/0.803 = 1.56$$

The formula for the degrees of freedom for an unpaired two sample t-test is given by

$$\text{DF} = (n_1 - l) + (n_2 - l)$$

which for our example gives DF= 6.

From Table 11.2 for DF= 6 and $P = 0.05$ the t-statistic is 2.447 and since 1.56 is less than 2.447 we cannot reject the null hypothesis H_0 at the 0.05 level of significance. We must therefore assume that there is no detectable difference between the means x_{1m} and x_{2m} of our observations.

The sample sizes of 4 in the above example are too small in practice for a well planned study but it is emphasised that the t-test is for small samples although opinion varies as to *how small is small* with most recommendations taking 30 as the guideline for the upper limit. Above 30 the normal distribution can be used to determine if there is a significant difference between the means of two *large* samples, see section 11.6.

As a further example of the calculation of the t-statistic, assuming that the standard deviations of the two samples have been calculated already and also the means, we have the necessary terms for the calculation below, e.g. for a clinical trial that has taken place and for which there was no randomisation with respect to age. The t-test will test the null hypothesis H_0 that there is no difference in the true mean ages (i.e. the population means) of the two patient groups.

$$n_1 = 20 \qquad n_2 = 18$$
$$x_{1m} = 58 \text{ years} \qquad x_{2m} = 55 \text{ years}$$

$$(x_{1m} - x_{2m}) = 3.0 \text{ years}$$
$$SD_1 = 6.71 \text{ years} \qquad SD_2 = 8.06 \text{ years}$$

$$SE_p = \Big[((20-1)*(6.71)^2) + ((18-1)*(8.06)^2)$$
$$* ((1/20)+(1/18))/((20-1)+(18-1)) \Big]^{1/2}$$
$$SE_p = \sqrt{(19*45.02)+(17*64.96)*(0.1055)/(36)} = \sqrt{5.74} = 2.40$$

$$t = (x_{1m} - x_{2m})/SE_p = 3.0/2.40 = 1.25$$

The derived value of the t-statistic using the formulae above is 1.25 and from Table 11.2 it is seen that for DF= 36 and $P = 0.05$ the critical value of the t-statistic is 2.03. Thus since 1.25 is less than 2.03 ($P > 0.05$) we cannot reject the null hypothesis.

From Table 11.2 for DF= 36 and $P = 0.10$ the t-statistic is 1.7 and therefore we do not obtain a significant result even at this level ($P > 0.10$) since 1.25 less than 1.7.

In other words, the probability of obtaining a t-statistic of $+1.25$ or greater, or of -1.25 or less, is more than 0.20, that is, more than a 1 in 5 chance, and the data provide no evidence that the true mean ages differ between the two patient subgroups.

11.6 NORMAL TEST FOR THE DIFFERENCE BETWEEN MEANS OF LARGE SAMPLES

When a random sample is large and is from a normally distributed population with mean μ and a known standard deviation σ, (or if σ is not known, because it is a large sample ($n > 30$: see Figure 11.2) it can be estimated by using the sample standard deviation s_m) the t-test can be replaced by what is sometimes called the *normal test* or *z-test*, where

$$z = (x_m - \mu)/\{s_m/\sqrt{n}\}$$

The distribution of z is normal with a standard deviation equal to 1 because $\{s_m/\sqrt{n}\}$ is the estimate of the standard error of the mean.

As an example, the mean range of a rocket is 2000 metres and the range standard deviation is 120 metres. A total of 40 rounds are fired after a year's storage and give a mean range of 1863 metres. The problem is to determine at the $P = 0.05$ level of significance whether storage has changed the mean range.

Substituting $x_m = 1863$, $\mu = 2000$, $s_m = 120$ and $n = 40$ into the formula for z we find that $z = -7.22$. We now need to refer to Table 3.1(*c*) which is reproduced here in expanded form as Table 11.3, see also Figure 11.4 and Figure 3.3.

Table 11.3. Probability related to multiples of standard errors (SE) for a normal distribution; see also Figure 11.3.

Number of SEs	Probability of an observation showing *at least as large a deviation* from the normal population mean
0.25	0.80
0.50	0.62
0.67	0.50
1.00	0.32
1.50	0.133
1.645	0.10
1.96	0.05
2.00	0.046
2.58	0.01
3.00	0.0027
3.29	0.001

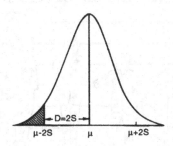

Figure 11.4. The probability that the observation falls in the shaded area, at a deviation (*D*) from μ which is greater than 2*S* is (0.046/2).

For a $P = 0.05$ level of significance, the two-tailed rejection region is $|z| > 1.96$. Since -7.22 is less than -1.96, it falls into this region of rejection and therefore we reject the null hypothesis that the mean after storage is 2000 metres. We conclude that the year's storage does change the mean range of this type of rocket.

A *two-tailed test* (two 0.025 tails) was used under the assumption that before the experiment was performed we were interested in a change in the range of this rocket in either direction. If we had been interested only in whether the range decreased, we would have used a *one-tailed test* (one 0.05 tail) for $z < -1.645$.

As another example, consider the mean ages of two large samples ($n_A = 4090$ and $n_B = 2214$) of cancer of the cervix patients, one group (A) with invasive cancer and one group (B) with *in situ* cancer[1] for which the means and standard deviations are $x_{Am} = 55.3$ years, $x_{Bm} = 38.4$ years, $s_{Am} = 13.7$ and $s_{Bm} = 10.6$.

The formula (see section 4.4) for the standard error of the difference in means is

$$SE = \sqrt{\{(s_{Am}^2/n_A) + (s_{Bm}^2/n_B)\}}$$

which for this example gives $SE = 0.03$. The difference in means is 16.9 years and the number of multiples of its SE that this difference in means represents is $(16.9/0.03) = 563$, which is far larger than 3.29 in Table 11.3. The result is therefore very highly significant and, indeed, we can assume that it is impossible to occur by chance. We can therefore assume that there are two different patient populations for newly registered cases of *in situ* and invasive cancers of the cervix.

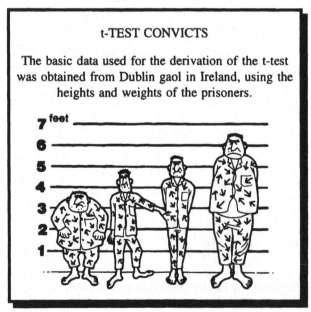

t-TEST CONVICTS

The basic data used for the derivation of the t-test was obtained from Dublin gaol in Ireland, using the heights and weights of the prisoners.

Chapter 12

Difference Between Proportions for Independent and for Non-Independent (McNemar's Test for Paired Proportions) Samples

12.1 INDEPENDENT SAMPLES FROM NORMAL POPULATIONS

We have already encountered the standard error of the difference between two proportions (when P_1 and P_2 are percentages then the factors $(1 - P_i)$ become $(100 - P_i)$ but the remainder of the formula remains unchanged) in section 4.4 on page 56. This is reproduced below and is for two *independent* samples of size n_1 and n_2 and the populations from which the samples are drawn are assumed to be normally distributed

$$SE = \sqrt{\{[P_1(1 - P_1)/n_1] + [P_2(1 - P_2)/n_2]\}}$$

The 95% confidence interval for the difference in proportions for the two independent samples from normal populations is $(P_1 - P_2) - 1.96 \times SE$ to $(P_1 - P_2) + 1.96 \times SE$ (see also section 4.10).

For the significance test for the difference $(P_1 - P_2)$ a slightly different formula (SE_p) is used for the SE compared to that given above. It is based on the null hypothesis H_0 that both samples n_1 and n_2 have the same proportion for which P is the estimate

$$SE_p = \sqrt{\{[P(1 - P)/n_1] + [P(1 - P)/n_2]\}}$$

To obtain P the two samples are combined and the proportion P is therefore that of $(n_1 + n_2)$. As an example suppose we have 76 women and 46 men in one sample, $n_1 = 122$, and 362 women and 280 men in the second sample, $n_2 = 642$. Let the proportions P be the proportions of women in the samples.

Thus $P_1 = 76/122 = 0.623$ and $P_2 = 362/642 = 0.564$; thus $(P_1 - P_2) = 0.059$, and for the combined sample $P = 438/764 = 0.573$:

$$
\begin{aligned}
SE_p &= \sqrt{\{([0.573 \times 0.427]/122) + ([0.573 \times 0.427]/642)\}} \\
&= \sqrt{\{(0.2447/122) + (0.2447/642)\}} \\
&= \sqrt{\{0.002006 + 0.000381\}} = \sqrt{\{0.002387\}} = 0.0489
\end{aligned}
$$

Since the difference $(P_1 - P_2) = 0.059$ this difference is now found to be $0.059/0.0489 = 1.21$ multiples of the standard error SE_p. Table 11.3 gives probability related to multiples of standard errors for a normal distribution and we find from this table that 1.21 standard errors gives a probability within the range 0.133–0.32 (since Table 11.3 only gives data for 1.00 and 1.50 multiples) of an observation showing *at least as large a deviation*. This P-value is larger than $P = 0.05$ (P now is the notation for the level of significance and not for a proportion) and so the difference between the percentages in the two samples could have been due to chance alone. The associated 95% *confidence interval* is obtained using the formula for SE described above:

$$
\begin{aligned}
SE &= \sqrt{\{([0.623 \times 0.377]/122) + ([0.564 \times 0.436]/642)\}} \\
&= \sqrt{\{(0.2349/122) + (0.2459/642)\}} \\
&= \sqrt{\{0.001925 + 0.000383)\}} = \sqrt{\{0.002308\}} = 0.0480
\end{aligned}
$$

The 95% *confidence intervals* are therefore $(0.059 - 1.96 \times 0.0480)$ to $(0.059 + 1.96 \times 0.0480) = -0.035$ to 0.153.

12.2 McNEMAR'S TEST FOR DIFFERENCE BETWEEN PAIRED PROPORTIONS

Studies are often performed where the patient acts as their own control and for example assesses the effect of a medication *before and after* treatment where the measurements are of the *strength* of either a nominal or interval (see Figure 11.1) scale. For example results might be assessed as:

Responded *or* did not respond
Improved *or* did not improve
Positive *or* negative

The populations from which the paired samples are drawn are therefore not independent of each other, as for example they are in section 9.7 for the 2×2 contingency table example of a cholera epidemic problem. It is also noted that the paired alternatives might not only be for the same patients acting as their control, but could also be for *matched pairs* of individuals.

The data for the McNemar test of paired proportions (sometimes called a test for the significance of changes) can be set out as in Table 12.1.

Table 12.1. Variables required for the McNemar test for two alternative scenarios: before and after treatment, and treatment A *versus* treatment B.

Pair result *before* treatment (or for *treatment A*)	Pair result *after* treatment (or for *treatment B*)	No. of paired results
+	+	n_a
+	−	$n_b(+ \rightarrow -)$
−	+	$n_c(- \rightarrow +)$
−	−	n_d

The rows for which there is no change, i.e. the results n_a and n_d, are ignored because the null hypothesis we are testing is:

H_0: The same proportion of patients changed result in in one direction $(+ \rightarrow -)$ as in the other direction $(- \rightarrow +)$

In practice the numbers involved in the type of study for which the McNemar test is appropriate are small, and therefore a *continuity correction* is included (in much the same way as Yates correction is used for the chi-squared test, section 9.9) and the test statistic (z) is calculated as follows using n_b and n_c, which we will now call n_1 and n_2 with n_1 defining the larger number of pairs and n_2 the smaller number of pairs:

$$z = \{[|n_1 - n_2| - 1]/\sqrt{[n_1 + n_2]}\}$$

This z statistic is normally distributed and its probability of occurrence can be obtained from Table 11.4 once the standard error of the difference SE_{MCN} has been calculated using the formula below, which is seen to be slightly different from the previous standard error formulae. This is because the groups of observations are *not* independent and SE_{MCN} cannot therefore be based simply on the variances of each proportion, but must take into account in some mannner the paired results. N is the total number of pairs where $N = (n_a + n_b + n_c + n_d)$

$$SE_{MCN} = \{1/N\}\sqrt{\{(n_1 + n_2) - ([n_1 - n_2]^2/N)\}}$$

The statistic z^2 may be regarded as a χ^2 statistic with DF= 1 and therefore critical values of z^2 can be found from the first row of figures in Table 9.1.

$$z^2 = \{[|n_1 - n_2| - 1]/\sqrt{[n_1 + n_2]}\}^2$$

To illustrate the arithmetical procedure for the McNemar test we will use real data[1] from a double-blind study of aspartame and the incidence of headaches. 40 patients were given aspartame and a placebo at different times and previously all of the patients had reported headaches after taking products containing aspartame. Table 12.2 is set out in the same manner as Table 12.1 and it is seen that $n_1 = n_c = 12$ and $n_2 = n_b = 8$. $N = 40$.

The null hypothesis H_0 is that there is the same proportion of headaches present with aspartame as with placebo. Thus H_0: there is no difference between $P_1 = 18/40 = 0.45$ (placebo) and $P_2 = 14/40 = 0.35$ (aspartame). The difference $(P_1 - P_2)$ is therefore 0.10. This difference in proportions can also be calculated by $\{(n_1/40) - (n_2/40)\} = \{(12/40) - (8/40)\} = 0.10$.

Table 12.2. Variables for the McNemar test. H = Headache.

Pair result after aspartame	Pair result after placebo	No. of paired results
H	H	6
H	No H	$8 = n_b$
No H	H	$12 = n_c$
No H	No H	14

The z statistic and its associated standard error SE_{MCN} are calculated using the formulae already described:

$$z = \{[12 - 8 - 1]/\sqrt{[12 + 8]}\} = \{3/\sqrt{20}\} = \{3/4.472\} = 0.671$$

From Table 11.4 it is seen that the probability value associated with $z = 0.671$ is about 0.50 and therefore we cannot reject the null hypothesis.

From the value of SE_{MCN} we can calculate the 95% confidence interval which is found to be $\{0.10 \pm 1.96 \times 0.111\}$ which is -0.118 to $+0.318$, which is seen to include zero and therefore not show a significant result at the $P = 0.05$ level of significance:

$$SE_{\text{MCN}} = \{1/40\}\sqrt{\{(12 + 8) - ([12 - 8]^2/140)\}} = \{0.025\}\sqrt{\{20 - (16/40)\}}$$
$$= \{0.025\}\sqrt{19.6} = \{0.025\} \times \{4.427\} = 0.111$$

Looking at this problem from a different point of view, $z^2 = 0.450$ and from Table 9.1 it is seen that for DF = 1 and $P = 0.05$ (two-tailed test) the critical value of χ^2 is 3.841 and since 0.45 is less than 3.84 we do not have a significant result. Although not shown in Table 9.1, the critical value of the χ^2

statistic for DF= 1 and $P = 0.50$ is $\chi^2 = 0.455$ which is approximately equal to the derived z^2. Thus again we end with a probability value of 0.50, showing that both approaches are consistent with each other.

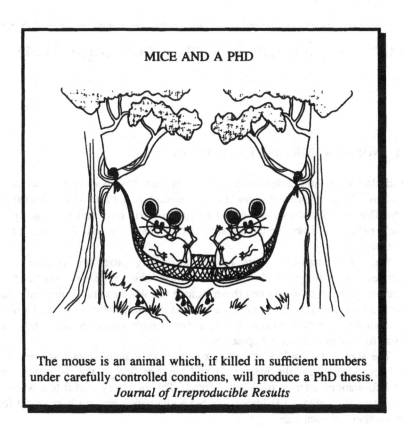

MICE AND A PHD

The mouse is an animal which, if killed in sufficient numbers under carefully controlled conditions, will produce a PhD thesis.
Journal of Irreproducible Results

Chapter 13

Wilcoxon, Mann–Whitney and Sign Tests

13.1 NON-PARAMETRIC RANK TESTS

Not all data are normally distributed and special tests have been devised to study situations where the data are non-normal, or indeed do not follow any other type of distribution. Such tests are called *non-parametric tests* and two examples already encountered are the chi-squared test in Chapter 9 and McNemar's test in Chapter 12.

Some non-parametric tests require a ranking procedure (Figure 11.1 illustrates the *ordinal* method of measurement in which *ranks* are used) as part of the computation schedule and these include the *Wilcoxon* rank sum tests (for paired and for unpaired data), the sign test and the *Mann–Whitney* U test which uses a similar approach to the Wilcoxon two-sample (unpaired data) test: as well as the *logrank* test of Chapter 15.

The sign test is unlike the Wilcoxon tests in that the sign test does not use information about the magnitude and the direction of the difference but only information in terms of + or − signs. This is a very useful test when a quantitative measurement is impossible but a qualitative ranking is practical for two members of each pair.

13.2 WILCOXON SIGNED RANKS TEST FOR MATCHED PAIRS

In this test for matched pairs a 6-column table is constructed for:

Pairs of patients, total number = N
Measurement of parameter X for group 1
Measurement of parameter X for group 2
Difference (some will be − and some will be +)
Rank in terms of numerical value where the smallest difference is rank 1 and the largest difference is rank N for a series of N patients
Signed rank (the + or − from the difference column is inserted before the rank).

132

The procedure is then to total the + ranks and the − ranks. Only the smaller of the two totals is used (regardless of whether it is + or −) and this is the derived Wilcoxon rank sum total statistic for this *matched-pairs signed-ranks* test. Table 13.1 gives the rank sum totals for selected numbers of patient-pairs and for 0.05 and 0.01 levels of significance. The relevant (in terms of patient-pairs) sum total from Table 13.1 for a chosen level of significance is compared with the derived statistic. If the value of the derived rank sum total is the larger, then no significant difference has been demonstrated at this level of chosen significance.

An example of a trial situation in which the Wilcoxon matched-pairs signed-ranks test was used[1] is for an assessment of conventional radiographic films (large films: 24×30 cm^2) and 100 mm^2 films (small films) for intravenous urogram imaging. Figure 13.1 summarises the results of 10 different radiologists reading 50 film pairs and assessing the images on a scoring scale of 0–5 for 5 particular radiographic features (0, all features invisible; 5, all features visible). The trial therefore consisted of 500 assessments of film pairs. Ten Wilcoxon tests were applied, one for each of the ten radiologists A–K. It was found that two radiologists (J and K) considered there was no significant difference between small and large films at the 0.05 level of significance; seven radiologists considered the differences to be significant in favour of the large film at the 0.01 level of significance; and one (F) considered that the difference in favour of the large film was at a level between 0.01 and 0.02.

As an example of the test procedure the data on 14 patients from a sequential analysis double blind clinical trial[2] for cancer of the head and neck will be used. The two treatment groups were *radiotherapy + drug* (B) and *radiotherapy + placebo* (A) and the tumour response within three months of completion of treatment was assessed for each patient in terms of complete regression (CR), partial regression (PR), no change (NC) and progression of the disease (P). For the purposes of this test example the data have been scored from 1 to 5 as follows:

Score 5 = CR with no recurrence subsequently up to 6 months
Score 4 = CR initially but with a subsequent recurrence within 6 months
Score 3 = PR
Score 2 = NC
Score 1 = P.

Table 13.2 summarises the test data in the format required for this Wilcoxon matched-pairs signed-rank test.

If two scores are tied, as for patients 11–14, then the pairs are omitted from the analysis. If the differences are the same for more than one pair, as for patients 2, 3, 6, 7, 8 and 10, then the same rank is given to each patient. These six patients all have differences of 1 and therefore the rank numbers 1, 2, 3, 4, 5 and 6 must be divided amongst them. That is, they all have a rank of $(1+2+3+4+5+6)/6 = 3\frac{1}{2}$. A similar method is used for patients 4 and 9

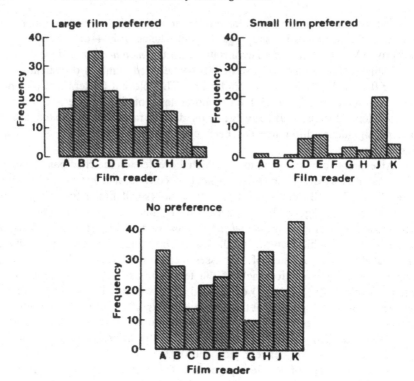

Figure 13.1. Radiologist reader performance for 50 film pairs and 10 readers, classified in terms of individual preferences for each reader. Frequency on the vertical axis is the *frequency of assessment preference for a film pair.*

where the difference is 3. Their ranks are $(8 + 9)/2 = 8\frac{1}{2}$.

The smaller of the two rank totals in Table 13.2 is used (this always occurs regardless of sign + or −) and this is 7, which is smaller than 8 in Table 13.1 for $N = 10$. The result is therefore significant at the 0.05 level of significance. It is not, though, significant at the 0.01 level of significance, since the appropriate sum in Table 13.1 is 3. The null hypothesis is

H_0: There is no difference between treatment A and treatment B at the 0.05 level of significance

and it is therefore rejected. It is emphasised that the data in Table 13.2 arose from a planned prospective sequential analysis clinical trial[2] and are only used to demonstrate the computations of this Wilcoxon text because they are good examples for tied scores. In a planned prospective trial using the Wilcoxon test it would be expected that the possible scores would have a wider range than 1–5 and that there would not be so many tied and similar scores.

Table 13.1. Wilcoxon matched-pairs signed-ranks test rank sum total statistic. Material from *Some Rapid Approximate Statistical Procedures*. Copyright ©1949, 1964, Lederle Laboratories Division of American Cyanamid Company, all rights reserved and reprinted with permission.

Number of pairs = N	Level of significance for 1-tailed test		
	0.025	0.01	0.005
	Level of significance for 2-tailed test		
	0.05	0.02	0.01
6	0	—	—
7	2	0	—
8	4	2	0
9	6	3	2
10	8	5	3
11	11	7	5
12	14	10	7
13	17	13	10
14	21	16	13
15	25	20	16
16	30	24	20
17	35	28	23
18	40	33	28
19	46	38	32
20	52	43	38
21	59	49	43
22	66	56	49
23	73	62	55
24	81	69	61
25	89	77	68

13.3 WILCOXON RANK SUM TEST FOR UNPAIRED DATA

For this test the two sample groups need not be of the same size. The observations from both samples are combined into a single series and ranked in order, using some symbol to distinguish one group from the other. The totals of the ranks are computed for both groups and, as a check for the arithmetic, the sum of ranks is $N(N+1)/2$ where the total number of ranks is N. The smaller total of ranks is then taken and compared with the Wilcoxon test statistic in Table 13.3 for the relevant numbers in each group, $N_1 + N_2 = N$ and for a chosen level of significance, usually either 0.05 or 0.01. This will determine if the two groups

Table 13.2. Test example data used in section 13.2.

Patient pair	Post-treatment follow-up score		Difference in score	Rank	Signed rank
	Treatment A	Treatment B			
1	3	5	2	7	−7
2	3	2	1	$3\frac{1}{2}$	$+3\frac{1}{2}$
3	2	1	1	$3\frac{1}{2}$	$+3\frac{1}{2}$
4	5	2	3	$8\frac{1}{2}$	$+8\frac{1}{2}$
5	5	1	4	10	+10
6	5	4	1	$3\frac{1}{2}$	$+3\frac{1}{2}$
7	2	1	1	$3\frac{1}{2}$	$+3\frac{1}{2}$
8	5	4	1	$3\frac{1}{2}$	$+3\frac{1}{2}$
9	5	2	3	$8\frac{1}{2}$	$+8\frac{1}{2}$
10	5	4	1	$3\frac{1}{2}$	$+3\frac{1}{2}$
11	1	1	0		
12	5	5	0		
13	5	5	0		
14	5	5	0		

No. of patient pairs for analysis $= 14 - 4 = 10$.
Total + ranks = 48.
Total − ranks = 7.

are significantly different. The test is illustrated below using data on funding for cancer diagnosis and treatment in different regions of the world[3]. The data were obtained from an international survey which, among other questions, posed one on budgets for separate fields of activity, including: cancer research; cancer diagnosis and treatment; public education in cancer; rehabilitation and facilities for the seriously ill and dying. Suppose that for those centres which have some diagnostic and treatment services, the null hypothesis is

H_0: There is no difference in the level of funding for cancer diagnosis and treatment, as a percentage of the total funding, for centres in Western and Eastern Europe.

The percentage fundings for Eastern Europe are 64, 75, 91, 93, 98 and for Western Europe are 30, 61, 68, 80, 82, 88, 93, 99.8. The numbers in these groups are $N_1 = 5$ and $N_2 = 8$. The data are ranked in Table 13.3.

For a chosen level of significance of 0.05 and for $N_1 = 5$ and $N_2 = 8$ in Table 13.4, the Wilcoxon statistic is 21. The derived Wilcoxon statistic is 39.5 (*the smallest total rank* [which is usually for the smallest group] *is always taken*) and since 39.5 is higher than 21 the result is not significant. Thus from

Table 13.3. Sample ranking for the Wilcoxon rank sum test for unpaired data. Group 1, Eastern Europe; $N_1 = 5$ (denoted by *). Group 2, Western Europe; $N_2 = 8$.

Funding (%)	Rank
40	1
61	2
64*	3*
68	4
75*	5*
80	6
82	7
88	8
91*	9*
93*	$10\frac{1}{2}$*†
93	$10\frac{1}{2}$
98*	12*
99.8	13

Total rank (group 1) = 39.5
Total rank (group 2) = 51.5

†If two values are the same, e.g., 93, then they are given equal ranks, e.g., $10\frac{1}{2}$

the questionnaire data it would appear that for those centres which took the trouble to reply (presumably the most active centres) there is no significant difference at the 0.05 level of probability between the percentage funding for cancer treatment and diagnosis in Eastern and Western Europe.

If group 1 is now changed to American centres, of whom 7 ($= N_1$) responding to the questionnaire with percentage data of 30, 42, 57, 82, 84, 89 and 96, the total ranks for the USA are 53 and for Western Europe are 67. Again choosing a 0.05 level of significance, we see that the Wilcoxon statistic in Table 13.4 is 38. Since 53 is higher than 38 there is again no significant difference: this time between the USA and Western Europe.

These data are used only to illustrate the computation schedule, and if, for example, a funding survey is required for Eastern Europe, Western Europe and the USA, a larger sample size than 5, 8 and 7 would be required by the survey organisers. The international survey[3] included 57 questions and covered several topics other than funding. Replies were received from 619 centres in 119 countries, including 16 from Eastern Europe, 151 from Western Europe and 90 from the USA. This shows, without any need for a test of statistical significance, how reluctant or unable are those completing questionnaires to include financial data.

Table 13.4. Wilcoxon rank sum test statistic for unpaired data. The numbers of samples in each to two groups are N_1 and N_2. (*a*) For a level of significance of 0.05. (*b*) For a level of significance of 0.01. Reproduced from C White, 'The use of ranks in a test of significance for comparing two treatments,' *Biometrics* **8** 33–41, 1952. With permission from The Biometric Society.

(*a*) $P = 0.05$

$N_1 =$	2	3	4	5	6	7	8	9	10	11	12	13	14	15
N_2														
4			10											
5		6	11	17										
6		7	12	18	26									
7		7	13	20	27	36								
8	3	8	14	21	29	38	49							
9	3	8	15	22	31	40	51	63						
10	3	9	15	23	32	42	53	65	78					
11	4	9	16	24	34	44	55	68	81	96				
12	4	10	17	26	35	46	58	71	85	99	115			
13	4	10	18	27	37	48	60	73	88	103	119	137		
14	4	11	19	28	38	50	63	76	91	106	123	141	160	
15	4	11	20	29	40	52	65	79	94	110	127	145	164	185
16	4	12	21	31	42	54	67	82	97	114	131	150	169	
17	5	12	21	32	43	56	70	84	100	117	135	154		
18	5	13	22	33	45	58	72	87	103	121	139			
19	5	13	23	34	46	60	74	90	107	124				
20	5	14	24	35	48	62	77	93	110					
21	6	14	25	37	50	64	79	95						
22	6	15	26	38	51	66	82							
23	6	15	27	39	53	68								
24	6	16	28	40	55									
25	6	16	28	42										
26	7	17	29											
27	7	17												
28	7													

For another illustration of this test, data are used on whether rats would generalise learned initiation under a new drive and in a new situation[4]. Five rats were trained to imitate leader rats in a T-maze, following the leader when hungry to reach a food incentive. Then the five rats were transferred to a shock avoidance situation where imitation of the leader rats would involve avoiding an electric shock. Their behaviour in the shock avoidance test was compared with that of four control rats who had no previous *follow your leader* training. The numerical comparison is in terms of how many trials each rat took to reach

(b) $P = 0.01$

$N_1 =$	2	3	4	5	6	7	8	9	10	11	12	13	14	15
N_2														
5				15										
6		10	16	23										
7		10	17	24	32									
8		11	17	25	34	43								
9	6	11	18	26	35	45	56							
10	6	12	19	27	37	47	58	71						
11	6	12	20	28	38	49	61	74	87					
12	7	13	21	30	40	51	63	76	90	106				
13	7	14	22	31	41	53	65	79	93	109	125			
14	7	14	22	32	43	54	67	81	96	112	129	147		
15	8	15	23	33	44	56	70	84	99	115	133	151	171	
16	8	15	24	34	46	58	72	86	102	119	137	155		
17	8	16	25	36	47	60	74	89	105	122	140			
18	8	16	26	37	49	62	76	92	108	125				
19	3	9	17	27	38	50	64	78	94	111				
20	3	9	18	28	39	52	66	81	97					
21	3	9	18	29	40	53	68	83						
22	3	10	19	29	42	55	70							
23	3	10	19	30	43	57								
24	3	10	20	31	44									
25	3	11	20	32										
26	3	11	21											
27	4	11												
28	4													

a criterion of 10 correct responses in 10 trials.

The two hypotheses of interest are:

H_0: The number of trials to the criterion in the shock-avoidance situation is the same for rats previously trained to follow a leader to a food incentive as for rats not previously trained

H_1: Rats previously trained to follow a leader to food incentive will reach the criterion in the shock-avoidance situation in fewer trials than will rats not previously trained

For a chosen level of significance of 0.05 and for $N_2 = 5$ and $N_1 = 4$, the Wilcoxon statistic in Table 13.4 is 11. The required total rank derived from the data in Table 13.5 is 19 (because this is the sum of ranks) and since that is higher than 11, the null hypothesis, H_0, cannot be rejected at this 0.05 level of significance. The *follow your leader* training has not led to this group reaching the criterion in the shock avoidance trial quicker than the control group.

Table 13.5. Ranking procedure for the Wilcoxon rank sum test for unpaired data. Group 2, trained rats; $N_2 = 5$ (denoted by *). Group 1, untrained rats (control group), $N_1 = 4$.

Test score	Rank
45*	1*
51	2
53	3
64*	4*
70	5
75*	6*
78*	7*
82*	8*
110	9

Total rank (Group 2) = 26
Total rank (Group 1) = 19

In some data there will be *ties* in which there are for example, two or more identical scores in a table such as Table 13.5, which will give identical ranks. In this situation averaging is required for the tied scores. Thus for example, if there were no ties and the data are . . . 24 (3rd rank), 25 (4th), 26 . . . there is no need for averaging. However, if the data are . . . 24, 24, 26 . . ., then the averaging of the ranks will give . . . 24 (score = 3.5), 24 (score = 3.5), 26 (score = 5) and the ranks (i.e. scores) are then summed in the manner previously described.

To recapitulate, suppose we have samples from two populations A and B. in which an observation from A is a, and an observation from B is b; then the null and alternative hypotheses can be stated as follows.

H_0: A and B have the same distribution
H_1 (Which is a *directional hypothesis*): Probability that a score from A is larger than a score from B is greater than 0.5
H_2 (Which is a *directional hypothesis*): Probability that a score from B is larger than a score from A is greater than 0.5

H_1 and H_2 are tested by *one-tailed tests* to determine respectively if:

$$\text{Prob}(a > b) > 0.5$$

or

$$\text{Prob}(a > b) < 0.5$$

If the evidence supports H_1, then this implies that the *bulk* of population A is higher than the *bulk* of population B. A *two-tailed test* is for a prediction of differences which does not state direction.

13.4 MANN–WHITNEY *U* TEST

There is a certain amount of confusion which can arise for the rank sum test for unpaired data which in section 13.3 we have called Wilcoxon. The problem is that sometimes it is called Wilcoxon, sometimes Mann–Whitney and sometimes Mann–Whitney–Wilcoxon, with tables such as Table 13.4 entitled accordingly! This is because variants of the test were developed by Wilcoxon and also by Mann and Whitney.

The best advice is to keep to one formulation, such as in section 13.3, and do not use a variety as this policy is a recipe for mistakes. This section is therefore included only for general information and to illustrate the problems and pitfalls which might occur if formulae are mixed up.

If we call the Wilcoxon statistic (section 13.3) T then the sum of ranks will be, using the data in Table 13.5, $T_1 = 19$ and $T_2 = 26$. Convention is usually to denote the group with the smallest sample size N, and whichever T-statistics is the smaller, T_1 or T_2, is used for the test with Table 13.4.

The Mann–Whitney U-statistic can take two values:

$$U = [\{N_1 N_2\} + \{N_1(N_1 + 1)/2\} - T_1]$$

and

$$U' = [\{N_1 N_2\} + \{N_2(N_2 + 1)/2\} - T_2]$$

which for the data in Table 13.5 are:

$$U = [\{20\} + \{10\} - 19] = 11$$

and

$$U' = [\{20\} + \{15\} - 26] = 9$$

where they are related by:

$$U = \{N_1 N_2\} - U'.$$

The rule for the Mann–Whitney test is that one takes the smallest value of U which in this case is 9. The original publication by Mann and Whitney[5] contained for a series of values of N_2 (the larger sample size), tables of the probabilities associated with values as small as the observed values of U for values of N_2 in the range 3–8. The relevant table for $N_2 = 5$ is reproduced as Table 13.6.

For $U' = 9$ for the data in Table 13.5 we see from Table 13.6 that $U \leq 9$ when $N_2 = 5$ and $N_1 = 4$ has a probability of occurrence of $P = 0.452$ (the ' is dropped from U' for Table 13.6) and since 0.452 is less than 0.5 the null hypothesis H_0 is not rejected. The same conclusion was also reached in section 13.3 using the Wilcoxon rank sum test for unpaired data.

Table 13.6. Table for $N_2 = 5$ of exact probabilities associated with values as small as the observed value of U. (For N_2 in the range 9–20 different format tables are available[6] which give *critical values* of U for various signficance levels $P = 0.001, 0.01, 0.025$ and 0.05 for a one-tailed test (and double these values of P for a two-tailed test) and do not give *exact probabilities*. These tables have later been extended in *Geigy Scientific Tables*).

$N_1 =$ U	1	2	3	4	5
0	0.167	0.047	0.018	0.008	0.004
1	0.333	0.095	0.036	0.016	0.008
2	0.500	0.190	0.071	0.032	0.016
3	0.667	0.286	0.125	0.056	0.028
4		0.429	0.196	0.095	0.048
5		0.571	0.286	0.143	0.075
6			0.393	0.206	0.111
7			0.500	0.278	0.155
8			0.607	0.365	0.210
9				0.452	0.274
10				0.548	0.345
11					0.421
12					0.500
13					0.579

13.5   KRUSKAL–WALLIS TEST

The Kruskal–Wallis test is a more general form of the Mann–Whitney test and the simplest version of the formula[8] for the Kruskal–Wallis statistic H is given by

$$H = \left[\{12/[N(N + 1)]\} \cdot \left\{ \sum (R_i^2)/n_i \right\} \right] - 3(N - 1)$$

When the null hypothesis is true, H follows a χ^2 distribution. Thus if there are k groups of observations (the Mann–Whitney test is a two-sample test) the statistic H is compared with a χ^2 distribution with $(k - 1)$ degrees of freedom. For a full description of the Kruskal–Wallis test see Altman[8].

13.6 SIGN TEST

The Wilcoxon matched pairs signed ranks test (section 13.2) uses information about the magnitude and the direction of the differences of the paired samples. The sign test uses only + or − signs, rather than magnitudes and is therefore very useful when a quantitative measurement is impossible but a qualitative ranking is practical for the two members of each pair. Table 13.7 is a table of probabilities required for the sign test.

The same data[2] which were used in Table 13.2 will also be used to demonstrate the method of significance testing using the sign test. The data must be presented as in Table 13.8

From Table 13.8 for $N = 10$ it is seen that an x-value of 1 or less has a 1-tailed probability of occurrence under the null hypothesis of 0.011. If the chosen level of significance is 0.05 the null hypothesis is rejected in favour of the alternative hypothesis H_1: that treatment A is better than treatment B. If, however, the number of fewer signs, x, has been 3 then from Table 13.7 for $N = 10$ it would have been found that an x-value of 3 or less has a 1-tailed probability of occurrence of 0.172, and that hence at a chosen level of significance of 0.05, the null hypothesis would not be rejected.

What has been described are 1-tailed tests when only one alternative hypothesis, H_1, is considered. A 2-tailed test would consider not only H_1 but also H_2: that treatment B is better than treatment A. For this 2-tailed situation the probability values in Table 13.7 must be doubled and thus for the row for $N = 10$, the probabilities would become:

x	Probability
0	0.002
1	0.022
2	0.110
3	0.344
4	0.754
5	1
6	1

If the 1-tailed probability in Table 13.7 is equal to or greater than 0.5 then the corresponding 2-tailed probability is 1.

Table 13.7. For the sign test a table of probabilities associated with values as small as the observed values of x in the binomial test situation is required. In the body of the table are 1-tailed probabilities for the binomial test, when $p = q = \frac{1}{2}$. To save printing space, decimal points have been omitted for the P values. Adapted from Table IV, B, of Walker, Helen and Lev 1953 *Statistical Inference* (New York: Holt) p 458, with kind permission of the authors and publisher.

N	0	1	2	3	4	5	6	7	8	9	10	11	12	13	14	15
5	031	188	500	812	969	†										
6	016	109	344	656	891	984	†									
7	008	062	227	500	773	938	992	†								
8	004	035	145	363	637	855	965	996	†							
9	002	020	090	254	500	746	910	980	998	†						
10	001	011	055	172	377	623	828	945	989	999	†					
11		006	033	113	274	500	726	887	967	994	†	†				
12		003	019	073	194	387	613	806	927	981	997	†	†			
13		002	011	046	133	291	500	709	867	954	989	998	†	†		
14		001	006	029	090	212	395	605	788	910	971	994	999	†	†	
15			004	018	059	151	304	500	696	849	941	982	996	†	†	†
16			002	011	038	105	227	402	598	773	895	962	989	998	†	†
17			001	006	025	072	166	315	500	685	834	928	975	994	999	†
18			001	004	015	048	119	240	407	593	760	881	952	985	996	999
19				002	010	032	084	180	324	500	676	820	916	968	990	998
20				001	006	021	058	132	252	412	588	748	868	942	979	994
21				001	004	013	039	095	192	332	500	668	808	905	961	987
22					002	008	026	067	143	262	416	584	738	857	933	974
23					001	005	017	047	105	202	339	500	661	798	895	953
24					001	003	011	032	076	154	271	419	581	729	846	924
25						002	007	022	054	115	212	345	500	655	788	885

†1.0 or approximately 1.0.

Table 13.8. Test example data. x = number of fewer signs = 1; N = number of matched pairs which showed a difference = $(14 - 4) = 10$; CR, complete regression; PR. partial regression; NC, no change: P. disease progression.

Patient pair	Post-treatment follow-up		Sign
	Treatment A	Treatment B	
1	PR	CR	−
2	PR	NC	+
3	NC	P	+
4	CR	NC	+
5	CR:NR†	CR:R§	+
6	CR	P	+
7	NC	P	+
8	CR:NR‡	CR:R§	+
9	CR	NC	+
10	CR:R†	CR:R¶	+
11	P	P	0
12	CR	CR	0
13	CR	CR	0
14	CR	CR	0

†After 9 months.
‡After 7 months.
§ After 5 months.
¶ After 3 months.

SNAKE-HIPS LULU

This is Snake-Hips Lulu, a popular attraction in the Klondike gold rush in Dawson city. She only finds her way into a book of statistics to accompany the quotation by John Shaw Billings (1838–1913) which appeared in the *Medical Record* of 1889.

'Statistics are somewhat like old medical journals, or like revolvers in newly opened mining districts. Most men rarely use them, and find it troublesome so as to have them easy of access; but when they do want them, they want them badly'.

Chapter 14

Survival Rate Calculations

14.1 INTRODUCTION

Survival rate calculations make use of life tables and hence are termed a *life table method*, or alternatively an *actuarial* method. This is because it was *actuaries* who developed life tables early in the 19th century. The actuaries required such computations when working for life insurance companies, in order to estimate life expectancy of populations defined by birth year (usually the tables were constructed for a five-year period, i.e. five-year *cohort*), sex and age attained. This enabled the life insurance companies to set insurance premiums—and try and ensure that they made a profit!

The final step in the survival rate calculation is to multiply several probabilities. Thus for example to determine the survival probability to the end of the third year following initial treatment:

$$Pr\{\text{Surviving to the end of the 3rd year}\}$$
$$= Pr\{\text{Surviving in the 1st year}\}$$
$$\times Pr\{\text{Surviving in the 2nd year}\}$$
$$\times Pr\{\text{Surviving in the 3rd year}\}$$

The percentage three-year survival rate is then $100 \times Pr\{$Surviving to the end of the 3rd year$\}$. Such a survival probability is a product of separate probabilities which is limited by the time T_i at which we wish to compute the T_i-year survival probability, in this case for $T_i = 3$ years. Hence the term *product limit method*. In medicine, however, the description used for this method is usually *Kaplan–Meier* method. This is because Kaplan and Meier in 1958 were among the first to publish this method[1] applied to medicine, although they did not invent it. The terms in Table 14.1 are synonomous.

The predecessors of Kaplan and Meier included Greenwood[2] in 1926 who, for the Ministry of Health in London, published what was the first extensive

Table 14.1. Equivalent terminology.

Kaplan–Meier method
Product limit method
Life table method
Actuarial method

report on the natural duration of cancer, and included commentary on errors of sampling. This work was extended in 1961 by Ederer[3] for the calculation of standard errors. However, earlier in 1955 Merrell and Shulman[4] in the first volume of the *Journal of Chronic Diseases*, published a comprehensive description of the arithmetical calculation procedure.

Following the work of Greenwood a formerly widely quoted explanation of the actuarial/life table method for the medical profession was in 1950 by Berkson and Gage[5] and for some time, until overtaken by Kaplan and Meier, actuarial estimates of survival rates were termed as being calculated by the Berkson–Gage method. In the same year, 1950, the method of Greenwood was explained and used by Wood and Boag[6] and it can probably be said that the early 1950s first saw the real start of the now worldwide use of the actuarial/life table method in the field of estimation of cancer survival rates.

Prior to this, many papers assumed **either** that *all patients lost to follow-up* were dead from their disease **or** that *all patients lost to follow-up were alive*. Neither is a valid assumption and they lead respectively to either an underestimate or an overestimate of the survival rate. The actuarial/life table assumption for *grouped data* where w_i patients are lost to follow-up in a time interval is that on average the w_i patients were alive for half the interval. However, in the Kaplan–Meier method for non-grouped data, individual survival times are used without any grouping and there is no lost to follow-up classification, only patients dead or alive at last follow-up.

The graphical method of showing survival rate curves was not, as in the 1990s, a plot of the actuarial/life table rates (with perhaps standard error bars showing $\pm$ 1 SE or $\pm$ 2 SE) but sometimes a method which gave rise to strange presentations[7] such as in figure 14.1. Here at a cursory glance there are more patients alive at five years than at four years: a most successful treatment in terms of bringing patients back from the dead! In point of fact only the fraction of patients alive at a given time T_i have been plotted and 3/23 is larger than 3/45.

The Kaplan–Meier method is for non-grouped data, section 14.3, in which the survival time and status (i.e. dead/alive, with/without disease) is recorded for each patient and the survival times are calculated, as will be shown, at the times T_i when a death has occurred at T_i. For *disease-specific* survival rates the death at T_i will be due to the specific disease being considered, but for an

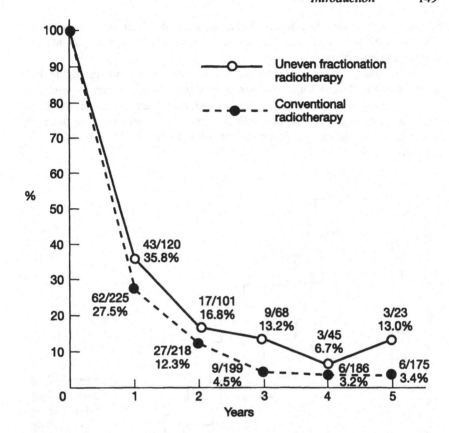

Figure 14.1. Survival rates of patients treated by two different regimes[7]. An example of how not to graph survival curves.

overall survival rate the death at T_i will be due to any cause, e.g. cancer or an intercurrent cause. For disease-specific rates when a patient has died of an *intercurrent* cause (i.e. **not** the disease in question, such as cancer) at T_i this patient survival is regarded as *censored*.

The term *censored data* means that the *endpoint* (death from a specific disease in this example) has not been reached. This is true if the patient is still alive at last follow-up and is also true for an *intercurrent cause* of death, because the patient is *lost to the risk of dying from the specific disease*, e.g. cancer.

A life table/actuarial method is also available for *grouped data*, section 14.4, where for a large series of patient data the survival times can be grouped and the arithmetic procedure made much easier. A comparison of the results of Kaplan–Meier and the grouped data method for the same series of patients is given in section 14.4.

It should also be noted that although the examples in this chapter are for cancer patient series, the life table/actuarial method is just as appropriate for other patients with a chronic disease such as kidney transplant patients and heart transplant patients.

In addition, the *endpoint* does not always have to be death and figure 14.2 is an example of four Kaplan–Meier graphs for a study where the endpoint is the time taken for a patient to request additional analgesics following an initial administration of clonidine. Four graphs are shown because there are four patient groups, each with a different initial level (0, 37.5, 75, 150 µg) of clonidine.

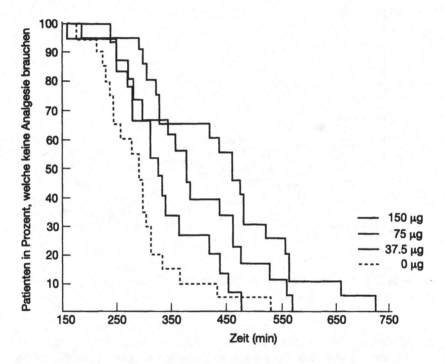

Figure 14.2. Kaplan–Meier graphs for four patient groups who receive different initial levels of intrathecal administration of clonidine. The variable on the horizontal axis is the time to request for additional pain medication. (Courtesy: Dr. Christoph H. Kindler, Department of Anästhesie, Kantonsspital, Basel).

Recently there has been an interesting argument in the *International Journal of Radiation Oncology, Biology & Physics* between Bentzen *et al*[8] who recommend that actuarial estimates be used in reporting late normal tissue effects following cancer treatment, and Caplan *et al*[9] who prefer to record crude rates such as cumulative incidence and hazard. However, Chappell[10] has pointed out that both can be correct but for different purposes.

For the biological factor of therapeutic gain and finding how bad the late complications could be, unbiased by whether the patients actually live long enough to get the complications, we need an actuarial analysis. However, from the point of view of a hospital manager, allocating resources to treatment or to re-treatment by some method which might be a different modality to that initially given, the crude figure is a better guide because the patients are automatically censored when they die.

14.2   BASIC DATA FOR THE WORKED EXAMPLES

Table 14.2 gives the basic data which will be used for the worked examples of the Kaplan–Meier method and also for the grouped data method. There are a total of 88 cases of cancer of the paranasal sinuses.

When a patient is alive at last follow-up there is obviously no further information beyond the follow-up date and therefore after that follow-up date the patient has to be removed from the analysis. This removal process is termed *censoring*, as mentioned in section 14.1.

Such censored patients may be *lost to follow-up* (e.g. emigrated) at some time prior to the date at which the analysis is being made, or at this analysis date they may be known to be alive, in which case the censoring is *withdrawal* of the patients from the analysis when still alive because obviously one cannot predict the future course of such a patient. These two classifications, lost to follow-up and withdrawn alive are combined for the life table/actuarial methods and generally denoted by w_i for *withdrawals*, when the grouped data method is used.

It is also worth pointing out that if there are many losses to follow-up then the quality of the data for analysis is poor and if there are too many withdrawn alive patients then the analysis could be being made too early before adequate information is available.

Table 14.2. Basic data for the worked examples. Of overall survival rates. Notation: A signifies that the patient was alive at the last known follow-up and D that the patient has died (cause not stated). The survival times are as they would appear in practice when a series of case notes are obtained for review: i.e. the survival times are in no particular order.

A	9	D	9	A	54	A	31
D	43	D	24	A	99	D	45
A	49	D	13	D	51	A	19
D	20	D	21	A	38	D	20
D	32	A	31	A	15	A	68
D	34	D	1	D	35	A	65
A	62	D	12	D	15	A	56
A	61	A	11	A	35	A	36
A	16	A	21	D	4	D	1
A	24	A	51	A	126	D	23
A	63	D	12	D	15	A	48
A	25	A	37	A	113	D	25
A	18	A	48	A	106	D	4
A	100	A	70	A	21	D	9
A	62	D	16	A	19	A	31
D	8	A	18	D	45	A	8
A	26	D	20	A	69	A	26
D	15	A	13	A	79	A	19
A	25	A	85	D	38	D	5
A	43	A	90	D	30	A	3
A	36	D	59	D	72	D	22
A	70	D	30	D	73	A	13

14.3   KAPLAN–MEIER LIFE TABLE METHOD FOR NON-GROUPED DATA

14.3.1   Ranking procedure for survival times

Blank forms with appropriate column headings (see section 14.3.6) are essential if one is to manually perform the Kaplan–Meier calculations with a pocket calculator, and this is the best way to understand the method! Such a form is shown in Table 14.3 where it has already been completed and where the first procedure is to rank the survival times from the *lowest to the highest*.

T_1 is the individual survival time of each of the 88 patients, *a generalised patient* is termed the ith patient, T_i in this example data of Table 14.2 is in the range 1–126 months. In Table 14.3 and Table 14.4 it is seen that there are, in certain instances, patients with the same survival time (although some are dead and some are alive). If the data had been presented with the survival times quoted to one decimal place then it would be unlikely that any patient had the same survival time as any other. In this case, the entries in the columns headed *Total deaths at T_i* [d_i] and *Alive at T_i* would be either 0 or 1.

The example in Table 14.3 therefore caters for the eventuality of multiple patients with the same survival times. It should also be noted that, in practice, the time scale for the Kaplan–Meier calculations is often in dimensions of days or weeks.

In the first column of Table 14.3 it is seen that there are 53 ranks from 1–126 months and not 88. This is because of the patients with identical survival times. If there were none with identical survival times then there would have been a total of 88 ranks for 88 patients.

14.3.2 Probability of dying at time T_i

In the third column of Table 14.3 the patients at risk at T_i are specified using the notation r_1 and this of course commmences with $r_i = 88$ and when the final rank is reached (rank= 53) when there are no more patients left, one has $r_i = 1$. If r_i does not equal 1 then this indicates that an arithmetical error has been made.

The probability of dying from all causes at T_i (we are now going to calculate the *overall survival rate* as distinct from the *cancer-specific survival rate* for which the cause of death must be specified in columns four and five for either a cancer death[CA] or a death from an intercurrent disease[ID], see section 14.3.4 and Table 14.4) is annotated as q_i where $q_i = (d_i/r_i)$.

The probability of surviving **through** T_i is therefore $(1 - q_i)$ which we will call p_i. The ninth column in Table 14.3 lists all the p_i probabilities for each T_i and their product (multiplied by 100 to obtain a percentage value) is given in the tenth column.

A patient who is alive at T_i does not contribute to the probability calculation (except in being required to compute the patients at risk in the third column). Since if a patient is alive **at** T_i then the probability of dying **at** T_i is 0 and hence the probability of surviving **through** T_i is 1 and when 1 is entered into a product calculation it does not, obviously, have any effect.

GRAVESTONE INSCRIPTION
FROM THE DEEP SOUTH OF THE USA

I told you I was sick

14.3.3   Calculation of overall survival rates

The arithmetic for the first few calculations of Table 14.3, in order to obtain the overall survival rate figures (OSR) in the tenth column are as follows for the first five percentage survival rates listed in the tenth column of this table.

OSR at 1 month   = $100 \times (0.977) = 97.7\%$

OSR at 3 months   =   97.7% since the case with $T_i = 3$ was alive and not dead at follow-up and thus the computation is $100 \times (0.977 \times 1)$

OSR at 4 months   = $100 \times (0.977 \times 0.976) = 95.4\%$ since there were 2 patients dead at $T_i = 4$ months and thus for the 4th rank $T_i = 5$ months the number of patients at risk $r_i = (85 - 2) = 83$

OSR at 5 months   = $100 \times (0.977 \times 0.976 \times 0.988) = 94.3\%$

OSR at 8 months   = $100 \times (0.977 \times 0.976 \times 0.988 \times 0.988) = 93.1\%$

Figure 14.3 graphs the results of Table 14.3 and the familiar pattern of *stairs with uneven steps* for a Kaplan–Meier calculation is demonstrated.

14.3.4   Calculation of cancer-specific survival rates

Generally this will be disease-specific for a defined disease, but our example is for cancer and hence in this instance it is cancer-specific. Table 14.4 illustrates the procedure for calculating this cancer-specific survival rate, but it is necessary that all causes of death are classified either as cancer deaths d_i^{CA} in the third column or as intercurrent deaths d_i^{ID} in the fourth column.

This can cause problems in that it is often very difficult to distinguish whether an intercurrent cause of death has occurred (i.e. when no cancer is present). Death certification is notoriously inaccurate. The accuracy of the original data (i.e. *data quality*) with respect to cause of death should therefore be taken into account before deciding if it is possible to compute a cancer-specific survival rate.

The data in Table 14.2 for cancer of the paranasal sinuses is known only in terms of dead or alive and no subdivision in terms of cancer deaths or intercurrent deaths was available. The subdivision into d_i^{CA} and d_i^{ID} in Table 14.4 is therefore *manufactured* for the sole purpose of showing the computational procedure, the results of which are given in the last three columns of Table 14.4.

The computational method for determining the q_i and p_i probabilities for the cancer-specific rates are similar to that for the overall rate as can be seen from the column headings for the two sets of q_i and p_i values.

Table 14.3. Computation procedure for the Kaplan–Meier overall survival rates at 1–126 months for the test data of 88 cancers of the paranasal sinuses, given in Table 14.2.

Rank	Survival time [T_i]	Patients at risk at T_i [r_i]	Cancer deaths at T_i [d_i^{CA}]	Intercurrent deaths at T_i [d_i^{ID}]	Total deaths at T_i [d_i]	Alive at T_i and then lost to follow-up	Pr{Dying at T_i} {All Causes} [d_i/r_i] $= q_i$	Pr {Surviving through T_i} [$1 - q_i$] $= p_i$	Overall Survival Rate at T_i [%]	Pr {Dying at T_i} {Due to Cancer} [d_i^{CA}/r_i] $= q_i$	Pr {Surviving through T_i} [$1 - q_i$] $= p_i$	Cancer Specific Survival Rate at T_i [%]
1	1	88			2	0	0.023	0.977	97.7			
2	3	86			0	1	0	1				
3	4	85			2	0	0.024	0.976	95.4			
4	5	83			1	0	0.012	0.988	94.3			
5	8	82			1	1	0.012	0.988	93.1			
6	9	80			2	1	0.025	0.975	90.8			
7	11	77			0	1	0	1				
8	12	76			2	0	0.025	0.975	88.4			
9	13	74			1	2	0.014	0.986	87.2			
10	15	71			3	1	0.042	0.958	83.5			
11	16	67			1	1	0.015	0.985	82.3			
12	18	65			0	2	0	1				
13	19	63			0	3	0	1				
14	20	60			3	0	0.050	0.950	78.2			
15	21	57			1	2	0.018	0.982	76.8			
16	22	54			1	0	0.019	0.981	75.3			
17	23	53			1	0	0.019	0.981	73.9			

Table 14.3. (continued)

Rank	Survival time [T_i]	Patients at risk at T_i [r_i]	Cancer deaths at T_i [d_i^{CA}]	Intercurrent deaths at T_i [d_i^{ID}]	Total deaths at T_i [d_i]	Alive at T_i and then lost to follow-up	Pr{Dying at T_i} [All Causes] [d_i/r_i] $= q_i$	Pr{Surviving through T_i} [$1-q_i$] $= p_i$	Overall Survival Rate at T_i [%]	Pr{Dying at T_i} [Due to Cancer] [d_i^{CA}/r_i] $= q_i$	Pr{Surviving through T_i} [$1-q_i$] $= p_i$	Cancer Specific Survival Rate at T_i [%]
18	24	52			1	1	0.019	0.981	72.5			
19	25	50			1	2	0.020	0.980	71.1			
20	26	47			0	2	0	1				
21	30	45			2	0	0.044	0.956	67.9			
22	31	43			0	3	0	1				
23	32	40			1	0	0.025	0.975	66.2			
24	34	39			1	0	0.026	0.974	64.5			
25	35	38			1	1	0.026	0.974	62.8			
26	36	36			0	2	0	1				
27	37	34			0	1	0	1				
28	38	33			1	1	0.030	0.970	60.9			
29	43	31			1	1	0.032	0.968	59.0			
30	45	29			2	0	0.069	0.931	54.9			
31	48	27			0	2	0	1				
32	49	25			0	1	0	1				
33	51	24			1	1	0.042	0.958	52.6			
34	54	22			0	1	0	1				

Table 14.3. (continued)

Rank	Survival time [T_i]	Patients at risk at T_i [r_i]	Cancer deaths at T_i [d_i^{CA}]	Intercurrent deaths at T_i [d_i^{ID}]	Total deaths at T_i [d_i]	Alive at T_i and then lost to follow-up	Pr[Dying at T_i [All Causes] $[d_i/r_i] = q_i$	Pr {Surviving through T_i} $[1 - q_i] = p_i$	Overall Survival Rate at T_i [%]	Pr [Dying at T_i [Due to Cancer] $[d_i^{CA}/r_i] = q_i$	Pr {Surviving through T_i} $[1 - q_i] = p_i$	Cancer Specific Survival Rate at T_i [%]
35	56	21			0	1	0	1				
36	59	20			1	0	0.050	0.950	50.0			
37	61	19			0	1	0	1				
38	62	18			0	2	0	1				
39	63	16			0	1	0	1				
40	65	15			0	1	0	1				
41	68	14			0	1	0	1				
42	69	13			0	1	0	1				
43	70	12			0	2	0	1				
44	72	10			1	0	0.090	0.900	45.0			
45	73	9			1	0	0.111	0.889	40.0			
46	79	8			0	1	0	1				
47	85	7			0	1	0	1				
48	90	6			0	1	0	1				
49	99	5			0	1	0	1				
50	100	4			0	1	0	1				
51	106	3			0	1	0	1				
52	113	2			0	1	0	1				
53	126	1			0	1	0	1				

Table 14.4. Computation procedure for the Kaplan–Meier cancer-specific survival rates for the test data in table 14.2, with modifications. These modifications include a classification of the cause of death, either cancer^{CA} or intercurrent cause of death^{ID}. The withdrawals are signified^{A}. The test data when ranked for 1–18 months, now becomes:

1^{CA} 1^{CA} 3^{A} 4^{CA} 4^{A} 5^{D} 5^{CA} 8^{A} 9^{CA} 9^{CA} 9^{A} 11^{A} 12^{CA}
12^{ID} 13^{CA} 13^{A} 13^{A} 15^{A} 15^{CA} 15^{ID} 16^{CA} 16^{A} 18^{A} 18^{A}

Only rows for 12 ranks are given in this table as these are a sufficient number to demonstrate the method.

Rank	Survival time $[T_i]$	Patients at risk at T_i $[r_i]$	Cancer deaths at T_i $[d_i^{CA}]$	Intercurrent deaths at T_i $[d_i^{ID}]$	Total deaths at T_i $[d_i]$	Alive at T_i and then lost to follow-up	Pr{Dying at T_i} {All Causes} $[d_i/r_i] = q_i$	Pr{Surviving through T_i} $[1-q_i] = p_i$	Overall Survival Rate at T_i [%]	Pr{Dying at T_i} {Due to Cancer} $[d_i^{CA}/r_i] = q_i$	Pr{Surviving through T_i} $[1-q_i] = p_i$	Cancer Specific Survival Rate at T_i [%]
1	1	88	2	0	2	0	0.023	0.977	97.7	0.023	0.977	97.7
2	3	86	0	0	0	1	0	1		0	1	
3	4	85	1	1	2	0	0.024	0.976	95.4	0.012	0.988	96.5
4	5	83	1	0	1	0	0.012	0.988	94.3	0.012	0.988	95.4
5	8	82	1	0	1	1	0.012	0.988	93.1	0.012	0.988	94.2
6	9	80	2	0	2	1	0.025	0.975	90.8	0.025	0.975	91.9
7	11	77	0	0	0	1	0	1		0	1	
8	12	76	1	1	2	0	0.025	0.975	88.4	0.013	0.987	90.7
9	13	74	1	0	1	2	0.014	0.986	87.2	0.014	0.986	89.4
10	15	71	2	1	3	1	0.042	0.958	83.5	0.028	0.972	86.9
11	16	67	1	0	1	1	0.015	0.985	82.3	0.015	0.985	85.6
12	18	65	0	0	0	2	0	1		0	1	

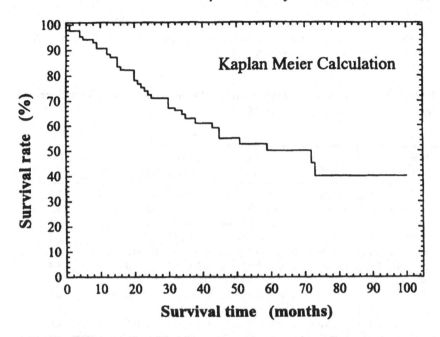

Figure 14.3. Overall survival rates for the data in Table 14.2 for 88 patients with cancer of the paranasal sinuses.

14.3.5   Standard errors

It is always important to calculate the standard errors associated with the survival rates since without the standard error there is no means of assessing the accuracy of the estimated rate. Many years ago, medical journals were actually found to quote for example a 67% five-year survival rate when in point of fact this related only to two cases out of three! In such an instance, since the standard error of a percentage P for a total sample size of N is given by

$$SE = \sqrt{(P \cdot [100 - P]/N)}$$

the associated SE with this 67% survival rate would be $\sqrt{([66.67 \times 33.33]/3)} = \sqrt{(740.7)} = 27.2\%$ and therefore the the survival rate ± 2 standard errors is 66.7% $\pm$ 54.4%—figures which speak for themselves!

Using the same notation as in Table 14.3 for T_i and r_i and d_i and q_i with additionally $Z_i = (r_i - d_i)$ and $Y_i = \sum[q_i/Z_i]$ the formula for the standard error in the overall survival rate at T_i (see column 10 in Table 14.5) is

$$SE = \sqrt{Y_i} \cdot [\text{Overall survival rate at } T_i]$$

The arithmetic calculations necessary are shown in Table 14.4 for the first 10 values of the survival rate at T_i which are quoted in Table 14.2, namely $T_i =$ 1. 4. 5. 8. 9, 12, 12, 15, 16 and 20 months.

Table 14.5. Calculation of the standard errors S_i [%] associated with the overall survival rates. P_{T_i} is the overall survival rate [%] at T_i.

Rank	T_i	r_i	d_i	$[r_i - d_i]$ $= Z_i$	q_i	q_i/Z_i	$\sum[q_i/Z_i]$ $= Y_i$	$\sqrt{Y_i}$	P_{T_i}	$S_i = P_{T_i} \cdot \sqrt{Y_i}$
1	1	88	2	86	0.023	0.000267	0.000267	0.0164	97.7	1.6
3	4	85	2	83	0.024	0.000289	0.000556	0.0236	95.4	2.2
4	5	83	1	82	0.012	0.000146	0.000702	0.0265	94.3	2.5
5	8	82	1	81	0.012	0.000148	0.000850	0.0292	93.1	2.7
6	9	80	2	78	0.025	0.000321	0.001171	0.0342	90.8	3.1
8	12	76	2	74	0.025	0.000338	0.001509	0.0390	88.4	3.4
9	13	74	1	73	0.014	0.000192	0.001701	0.0412	87.2	3.6
10	15	71	3	68	0.042	0.000618	0.002319	0.0482	83.5	4.0
11	16	67	1	66	0.015	0.000227	0.002546	0.0505	82.3	4.2
14	20	60	3	57	0.050	0.000877	0.003423	0.0585	78.2	4.6

14.3.6   Standard forms

A well designed form is essential and that shown in Tables 14.3 and 14.4 has been found to be useful in practice, particularly for teaching purposes in workshop seminars. There are a total of 13 column headings and these are sufficient for a calculation of the overall survival rate, as in Table 14.3, or of both overall and cancer-specific (Table 14.4) survival rates. The value of such forms should not be underestimated.

14.4   LIFE TABLE METHOD FOR GROUPED DATA

14.4.1   When to group data and when not to group data

I am often asked in seminars when should one decide to use grouped data in terms of the total number of patients in the series being analysed rather than use the non-grouped data Kaplan–Meier method of section 14.3.

The answer is not all that simple, because it depends in part on the length of follow-up available, on how many patients have already died and also on the pattern of deaths which are perhaps mainly grouped together in the early months post-treatment or alternatively spread out over several years with only a few deaths in any 12-month period. These are all factors which will vary from series to series.

Commonsense should be used in making an assessment. For instance when only 30 cases are being analysed then Kaplan–Meier is appropriate but when 300 are being analysed then grouping can take place. In the current example of 88 cases it will be seen later from figure 14.5 that there is indeed very little difference in the results between the two methods of computation in this particular series.

14.4.2 Censored: lost to follow-up or withdrawn alive

The basic ungrouped data used for this worked example is given in Table 14.2 where all 88 individual survival times are listed together with the patient status of either D for dead or A for alive.

The term *lost to follow-up or withdrawn alive* at the time of the last follow-up, during a given time interval $[T_{i-1} - T_i]$, is given in Table 14.6 in the the fourth column which is headed w_i. This w_i includes those patients truly lost to follow-up, i.e. cannot be located for any follow-up assessment because, for example, they might have emigrated, and also those patients who at the time of the follow-up assessment are known to be alive in interval $[T_{i-1} - T_i]$ and therefore cannot contribute any information to probability calculations for the next interval since it cannot be known at this stage whether they will survive through the next interval, or die in the next interval. They therefore have to be *censored*, i.e. *withdrawn from the analysis.* In the Kaplan–Meier method for non-grouped data there are no w_i terms in the calculations because each patient's survival time is treated individually.

14.4.3 Grouping procedure for the survival times

Six-month intervals have been chosen for the worked example, and the grouped data is given in Table 14.6. In practice, when manually processing the basic data into grouped intervals, the form design in Table 14.7 has been found over many years to be very useful.

If a patient's survival time falls on an interval boundary, such as 12 months in this example, then the *rule* is that 0.5 patient is put in the 6–12 month interval and 0.5 patient in the 12–18 month interval. The data in Table 14.2 are given in integer months, but in practice the survival times will be available to one decimal place and therefore very few will fall on an interval boundary.

Table 14.6. Grouped data for six-month intervals for the purpose of calculating overall survival rates. If a cancer-specific survival rate is required then d_i must be split into d_i^{CA} and d_i^{ID} in a similar manner to the Kaplan–Meier calculations. The addition of the total number dead and the total number alive is a useful internal consistency check to ensure that no patients have been omitted, thus $36 + 52 = 88$ in this example.

Interval number i	Time interval $T_{i-1} - T_i$ (months)	Number of deaths in interval i [d_i]	Number of cases lost to follow-up or withdrawn alive [w_i]
1	0–6	5	1
2	6–12	4	3
3	12–18	6	5
4	18–24	6.5	6.5
5	24–30	2.5	4.5
6	30–36	4	5
7	36–42	1	3
8	42–48	3	2
9	48–54	1	3.5
10	54–60	1	1.5
11	60–66	0	5
12	66–72	0.5	4
13	72–78	1.5	0
14	78–84	0	1
15	84–90	0	1.5
16	90–96	0	0.5
17	96–102	0	2
18	102–108	0	1
19	108–114	0	1
20	114–120	0	0
	> 120	0	1
Totals		36	52

14.4.4 Probability of dying at time T_i

Table 14.8 shows the calculation procedure for the overall survival rate using a series of 10 columns. w_i has already been defined and of the remaining symbols, d_i are the number of deaths in an interval and r_i is the number of patients entering an interval.

The patients at risk in an interval must take into account the withdrawals w_i and the actuarial assumption is that *a patient withdrawn in an interval has on average survived for half the interval.* The number of patients at risk n_i in

Form for Survival Rate Calculation
Grouped Data

Interval number $\{i\}$	Time interval $\{T_{i-1}-T_i\}$	Number of deaths in interval i			No. of cases lost to FU or withdrawn alive $\{w_i\}$
		From CA d_i^{CA}	Not from CA d_i^{ID}	Total d_i	
1					
2					
3					
4					
5					
6					
7					
8					
9					
10					
11					
12					
13					
14					
15					
16					
17					
18					
19					
20					
≥ 21					
	Totals				

Table 14.7 Form design for processing basic data into grouped data by time interval and patient status at last follow-up

an interval is thus given by

$$n_i = (r_i - 0.5w_i)$$

which is the seventh column in Table 14.8. The probability of dying (all causes of death) **in** an interval i is given in the eighth column as $q_i = (d_i/n_i)$. The probability of surviving **through** interval i is therefore $p_i = (1 - q_i)$.

14.4.5 Calculation of overall survival rates

The arithmetic for the first four calculations in Table 14.8 in order to obtain the overall survival rate (OSR) figures in the tenth column, are given below. The results are shown in the graph in figure 14.4 which also contains ±1 standard error bars. The individual survival rates are joined by straight lines in figure 14.4 although in practice, a smooth curve would be drawn.

OSR at 6 months = $100 \times (0.943) = 94.3\%$

OSR at 12 months = $100 \times (0.943 \times 0.950) = 89.6\%$

OSR at 18 months = $100 \times (0.943 \times 0.950 \times 0.917) = 82.1\%$

OSR at 24 months = $100 \times (0.943 \times 0.950 \times 0.917 \times 0.893) = 73.4\%$

14.4.6 Overall survival rate comparisons for grouped data and Kaplan–Meier methods

Figure 14.5 shows the comparison between the two methods for this series of 88 cases and it is quite clearly seen that they agree well to within ±1 standard error.

14.4.7 Standard errors

The standard errors are calculated in a similar manner to those for the Kaplan–Meier method which are shown in Table 14.5. For the grouped data method, the first 10 calculations, for $T_i = 6$ months to $T_i = 60$ months are shown to indicate the method of calculation of the standard error S_i [%], Table 14.9.

Table 14.8. Computation procedure for the grouped data life table method for overall survival rates for 15 six-month time intervals in the range 0–6 to 84–90 months.

Interval Number i	Survival Time T_{i-1}	Deaths in interval i d_i	Withdrawals in interval i w_i	$\frac{1}{2} w_i$	Patients entering interval i r_i	Patients at risk in interval i $n_i = (r_i - \frac{1}{2} w_i)$	Pr{Dying in interval i} (All Causes) $q_i = d_i/n_i$	Pr{surviving in interval i} $p_i = (1 - q_i)$	Overall survival rate at T_i [%]
1	0–6	5	1	0.5	88	87.5	0.057	0.943	94.3
2	6–12	4	3	1.5	82	80.5	0.050	0.950	89.6
3	12–18	6	5	2.5	75	72.5	0.083	0.917	82.1
4	18–24	6.5	6.5	3.25	64	60.75	0.107	0.893	73.4
5	24–30	2.5	4.5	2.25	51	48.75	0.051	0.949	69.7
6	30–36	4	5	2.5	44	41.5	0.096	0.904	63.0
7	36–42	1	3	1.5	35	33.5	0.030	0.970	61.1
8	42–48	3	2	1	31	30.0	0.100	0.900	55.0
9	48–54	1	3.5	1.75	26	24.25	0.041	0.959	52.7
10	54–60	1	1.5	0.75	21.5	20.75	0.048	0.952	50.2
11	60–66	0	5	2.5	19	16.5	0	1	
12	66–72	0.5	4	2	14	12.0	0.042	0.958	48.1
13	72–78	1.5	0	0	9.5	9.5	0.158	0.842	40.5
14	78–84	0	1	0.5	8	7.5	0	1	
15	84–90	0	1.5	0.75	7	6.25	0	1	

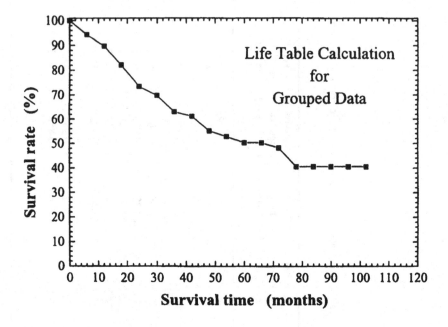

Figure 14.4. Overall survival rates for the data in Table 14.2 for 88 patients with cancer of the paranasal sinuses.

Table 14.9. Calculation of the standard errors S_i [%] associated with the overall survival rates P_{T_i}[%] at T_i. Note that $q_i = d_i/n_i$ in this table for grouped data but in Table 14.5 for the Kaplan–Meier calculations $q_i = d_i/r_i$. In the grouped data calculation $n_i = (r_i - 0.5w_i)$ and therefore if there are no withdrawals for an ith interval, i.e. $w_i = 0$, $n_i = r_i$ and thus for that i^{th} interval $q_i = d_i/r_i$.

Inter-val i	T_i	n_i	r_i	d_i	$[r_i - d_i]$ $= Z_i$	q_i	q_i/Z_i	$\sum[q_i/Z_i]$ $= Y_i$	$\sqrt{Y_i}$	P_{T_i}	$S_i =$ $P_{T_i} \cdot \sqrt{Y_i}$
1	6	87.5	88	5	83	0.057	0.000688	0.000688	0.0262	94.3	2.5
2	12	80.5	82	4	78	0.050	0.000637	0.000133	0.0364	89.6	3.3
3	18	72.5	75	6	69	0.083	0.001199	0.002525	0.0503	82.1	4.1
4	24	60.75	64	6.5	57.5	0.107	0.001861	0.004386	0.0662	73.4	4.9
5	30	48.75	51	2.5	48.5	0.051	0.001057	0.005443	0.0738	69.7	5.1
6	36	41.5	44	4	40	0.096	0.002410	0.007853	0.0886	63.0	5.6
7	42	33.5	35	1	34	0.030	0.000878	0.008731	0.0934	61.1	5.7
8	48	30.0	31	3	28	0.100	0.003571	0.012302	0.1109	55.0	6.1
9	54	24.25	26	1	25	0.041	0.001650	0.013952	0.1181	52.7	6.2
10	60	20.75	21.5	1	20.5	0.048	0.002351	0.016302	0.1277	50.2	6.4

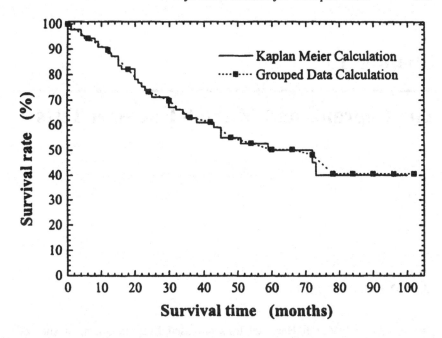

Figure 14.5. Comparison of overall survival rates calculated by two different life table/actuarial methods: Kaplan–Meier and the method for grouped data.

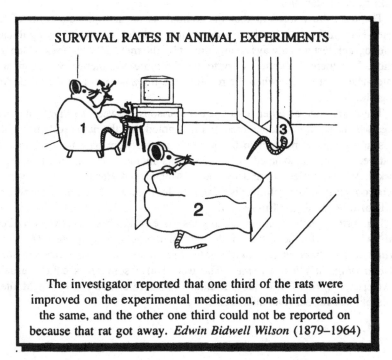

Chapter 15

The Logrank and Mantel–Haenszel Tests

15.1 INTRODUCTION

The logrank and Mantel–Haenszel tests are used for comparison of survival curves, two or more in the case of the logrank, but only two in the case of Mantel–Haenszel and the test statistic for both these tests is approximated by the chi-squared statistic.

The tests are related in that the logrank is an approximate form of the Mantel–Haenszel and it has been shown[1] that a difference which is significant by the logrank test will always be significant by the more exact Mantel–Haenszel method. It is therefore unnecessary to use the more arithmetically complicated Mantel–Haenszel except when the result of the logrank test is on the borderline of significance[1].

There are several other tests for comparisons between survival rates and curves which might be encountered and it is noted that the name for a particular test is not always consistent in the journal and software package user manual literature. Names associated with these tests include Gehan, Prentice, Peto, Kruskal, Wallis, Wilcoxon and of course Mantel and Haenszel. However, in this introductory text we limit discussion to logrank and Mantel–Haenszel; for details of other tests see Buyse *et al*[2].

For further reading on the logrank test, one of the best reviews is by Peto *et al*[3] who refer to the name of the test in the following quote: *The name logrank derives from obscure mathematical considerations which are not worth understanding: it's just a name.* The test is also sometimes called, usually by American workers who quote Mantel[4] as the reference for it, the Mantel–Haenszel[5] test for survivorship data.

15.2 THE LOGRANK TEST

15.2.1 Methodology

The methodology of the logrank test is to calculate a *score* for each patient which represents the exposure to the risk of dying if the true survival experience of all groups were identical. From the patient scores in the groups a test statistic is calculated which, when there is no difference between the survival curves, has a sampling distribution which is approximated by the χ^2 distribution. The observed value of the test statistic is then compared with the table of χ^2 values for the appropriate number of degrees of freedom, to assess whether the data are consistent with the null hypothesis of no difference in survival experience between the groups.

For any given day post-treatment when deaths occur, we argue as follows. Suppose that there are d deaths with n_A patients in group A and n_B patients in group B at risk, then if there is no real difference in survival experience between groups A and B. the expected number of deaths in group A is $d[n_A/(n_A + n_B)]$ and the expected number of deaths in group B is $d[n_B/(n_A + n_B)]$. The test uses the Kaplan–Meier method to estimate survival probabilities and the time intervals (e.g. days, weeks) are chosen, if possible, such that the event of more than one death per interval is a rare occasion. Deaths precede censoring times in the ranking.

15.2.2 Worked example

To illustrate the calculation procedure, data for cancer of the tongue have been used where the two groups differ by anatomical subsite. Group A consists of 20 cases of cancer of the anterior two-thirds tongue and group B of 28 cases of cancer of the posterior one-third tongue.

The initial procedure when using the logrank test is to rank the survival times of the combined groups of patients in order from smallest to greatest: see column (2), Table 15.1. The 48 survival times have been ranked from 13 weeks to 1105 weeks and an indication has also been given when the data are censored. In the table ^{WA} denotes a patient alive at last follow-up and the symbol ^{ID} that the patient died from another disease with cancer absent.

The latter patients can be considered censored, in as much as time to death with cancer present exceeds the recorded follow-up time, if the endpoint under review is certified death from cancer. If the endpoint is taken as death from whatever the cause, the patients annotated ^{ID} will be regarded as having exact or uncensored survival times.

Column (3), Table 15.1, tabulates the number of patients at risk prior to the corresponding death time T_i in column (2). Columns (2)–(5), Table 15.1, are similar to the second to fifth columns in Table 15.2 which shows the method of calculation for determining the T-year survival rate by the product limit method.

Columns (6) and (7), Table 15.1, are the numbers of patients at risk of dying in the two groups at time T_t and the values are calculated in a similar manner to the number of patients at risk stated in column (3). Columns (8) and (9) are for calculation of what Peto *et al* [3] term the *extent of exposure to risk of death* for the two groups. Cumulated, these extents of exposure to the risk of dying give the scores.

The summation of all the $(n_{A_i} \times q_i)$ values and all the $(n_{B_i} \times q_i)$ values in columns (8) and (9) give scores in this example of 14.44 and 23.60. These are regarded as the *expected* number of cancer deaths in each group, and knowing the *observed* number of (cancer) deaths in both groups, 13 and 25, the logrank test statistic χ^2 can be calculated with, for this example of only two groups (A and B), one degree of freedom.

The null hypothesis, that there is no difference between the survival experience of the two groups, can be tested by comparing the observed value of the test statistic $\{(E_A - O_A)^2/E_A\} + \{(E_B - O_B)^2/E_B\}$ with a table of χ^2 for one degree of freedom, part of which is reproduced below:

P-value	0.005	0.01	0.05	0.10	0.20
χ^2 for one degree of freedom	7.88	6.63	3.84	2.71	1.64

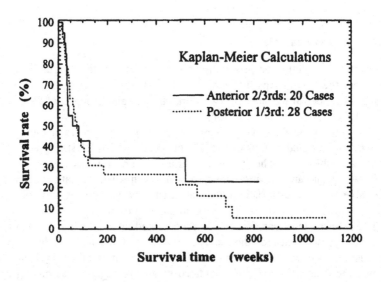

Figure 15.1. Kaplan–Meier survival curves for the two series of patients in the logrank test worked example: 20 cases of cancer of the anterior two-thirds of tongue and 28 cases of cancer of the posterior one-third of tongue.

From Table 15.1 the observed value of the logrank test statistic is 0.21, and thus the data are entirely consistent with no difference in survival between anatomical subsites anterior two-thirds and posterior one-third of tongue, $P > 0.20$. Figure 15.1 shows the Kaplan–Meier survival curves for the two patients groups A and B.

15.3 THE MANTEL–HAENSZEL TEST

15.3.1 Methodology

The null hypothesis, as in the logrank test, is H_0: there is no difference between the survival curve, with H_1: one survival curve is consistently different from the other. A word of caution, though, because if the two survival curves cross over the results of the method should be interpreted cautiously[6].

The methodology combines a series of 2×2 tables (see section 9.6) with a table formed each time T_i a death occurs in group A or group B, which could represent two treatments A and B, or an interventional treatment A and a control group B. Table 15.2 gives the format of a generalised 2×2 table for time T_i with the number of patients at risk prior to T_i stated in the last column of the table.

The expected number of deaths at T_i for group A can be shown to be as follows[6]:

$$E[d_{A_i}] = \{[(d_{A_i} + d_{B_i}) \times (d_{A_i} + s_{A_i})]/r_i\}$$

and the *variance* of the observed number of deaths in group A as[5,6] the product of the four maginal totals in Table 15.2 divided by a product term containing numbers of patients at risk, that is, $r_i^2(r_i - 1)$

$$V_{A_i} = \{[(d_{A_i} + d_{B_i})(s_{A_i} + s_{B_i})(d_{A_i} + s_{A_i})(d_{B_i} + s_{B_i})]/[r_i^2(r_i - 1)]\}$$

If K is the number of distinct events (e.g. deaths) times in the *combined* A+B group, then the Mantel–Haenszel statistic, MH, which has approximately a chi-squared distribution with one degree of freedom, is given by the formula:

$$MH = \left\{ \left(\sum_{i=1}^{K}[d_{A_i}]\right) - \left(\sum_{i=1}^{K}[E_{A_i}]\right)^2 \right\} / \left\{ \sum_{j=1}^{K} V_{A_i} \right\}$$

15.3.2 Worked example

The data in Table 15.1 for the two groups of patients A and B is rearranged as in Table 15.3 for the Mantel–Haenszel method.

The Mantel–Haenszel test statistic MH can be calculated directly from Table 15.3 for those rows which have a value greater than 0 for either d_{A_i} or d_{B_i}. There was therefore really no need to record the rows with only a 1^{WA} or

(text continued on p 174)

Table 15.1. Logrank test worked example.

Column (1), treatment group	Column (2), survival time in weeks, T_i	Column (3), no. of patients at risk at time T_i [r_i]	Column (4), no. of cancer deaths at time T_i [d_i]	Column (5), probability of dying at time T_i [$q_i = d_i/r_i$]	Column (6), no. of patients at risk prior to time T_i in group A only [n_{A_i}]	Column (7), no. of patients at risk prior to time T_i in group B only [n_{B_i}]	Column (8), extent of exposure to risk of death at T_i in group A only, [$n_{A_i} \times q_i$]	Column (9), extent of exposure to risk of death at T_i in group B only, [$n_{B_i} \times q_i$]
B	13	48	1	0.021	20†	28‡	0.42	0.58
B	17	47	1	0.021	20	27	0.43	0.57
A	18	46	1	0.022	20	26	0.43	0.57
B	22	45	1	0.022	19	26	0.42	0.58
A	25	44	1	0.023	19	25	0.43	0.57
B	26	43	1	0.023	18	25	0.42	0.58
B	27	42	1	0.024	18	24	0.43	0.57
A	30	41	1	0.024	18	23	0.44	0.56
B	31	40	1	0.025	17	23	0.43	0.58
A	34	39	1	0.026	17	22	0.44	0.56
A	35	38	1	0.026	16	22	0.42	0.58
A	36	37	1	0.027	15	22	0.41	0.59
B	37[WA]	36	0	0	14	22	0	0
A	38	35	1	0.029	14	21	0.40	0.60
A	39	34	1	0.029	13	21	0.38	0.62
B	40	33	1	0.030	12	21	0.36	0.64
A	43	32	1	0.031	12	20	0.38	0.63
B	44	31	1	0.032	11	20	0.35	0.65
B	47	30	1	0.033	11	19	0.37	0.63
B	48	29	1	0.034	11	18	0.38	0.62
A	60	28	1	0.036	11	17	0.39	0.61
B	61	27	1	0.037	10	17	0.37	0.63
B	65	26	1	0.038	10	16	0.38	0.62

Table 15.1. (continued)

A	68[WA]	25	0	0	10	15	0	0
B	69[WA]	24	0	0	9	15	0	0
B	71	23	1	0.043	9	14	0.39	0.61
A	77[WA]	22	0	0	9	13	0	0
B	78	21	1	0.048	8	13	0.38	0.62
A	82[WA]	20	0	0	8	12	0	0
B	83	19	1	0.053	7	12	0.37	0.63
A	86	18	1	0.056	7	11	0.39	0.61
B	88[ID]	17	0	0	6	11	0	0
B	95	16	1	0.063	6	10	0.38	0.63
B	108	15	1	0.067	6	9	0.40	0.60
A	121[ID]	14	0	0	5	8	0	0
A	123	13	1	0.077	5	8	0.38	0.62
B	130	12	1	0.083	5	7	0.42	0.58
B	186	11	1	0.091	4	7	0.36	0.64
B	213[WA]	10	0	0	4	6	0	0
A	324[WA]	9	0	0	4	5	0	0
B	481	8	1	0.125	3	5	0.38	0.63
A	520	7	1	0.143	3	4	0.43	0.57
A	546[WA]	6	0	0	2	4	0	0
A	568	5	1	0.200	1	4	0.20	0.80
B	685	4	1	0.250	1	3	0.25	0.75
B	711	3	1	0.333	1	2	0.33	0.67
A	819	2	1	0.500	1	1	0.50	0.50
B	1105	1	1	1.000	0	1	0	1.00

Totals $= 14.44(E_A)$ $= 23.60(E_B)$

†Total in group A
‡Total in group B
WA Alive at last follow-up.
ID Intercurrent deaths (regarded as censored observations in this example).

O_A = no. of cancer deaths in treatment group $A = 13$.
O_B = no. of cancer deaths in treatment group $B = 25$.
$O_A + O_B = 38$.

$E_A + E_B = 38.04$

Logrank test statistic $= \dfrac{(O_A - E_A)^2}{E_A} + \dfrac{(O_B - E_B)^2}{E_B}$

$= 0.144 + 0.083$

$= 0.21.$

Table 15.2. Format of a 2×2 table for deaths and survivors for when at least one death occurs at time T_i, that is, at least d_{A_i} or d_{B_i} must be non-zero.

Treatment group	Deaths at T_i	Survivors at T_i	No. of patients at risk prior to T_i
A	d_{A_i}	s_{A_i}	$d_{A_i} + s_{A_i}$
B	d_{B_i}	s_{B_i}	$d_{B_i} + s_{B_i}$
A + B	$d_{A_i} + d_{B_i}$	$s_{A_i} + s_{B_i}$	r_i

a 1^{ID} and no deaths, in order to calculate the MH-statistic. However, by including them it is a good internal consistency check on the arithmetic, as the ranks are seen to total 48 as there are no ties in this example for any T_i value.

In the MH-statistic formula the term $\sum_{i=1}^{38}(d_{A_i}) = 13$ because there are only 13 deaths in treatment group A. $K = 38$ because the total cancer deaths, $\sum(d_i)$ in Table 15.1, in this example are 38, whereas there are 48 patients. It is also reiterated that group B would always be the *controls*.

Evaluation of the term $\sum_{i=1}^{38}\{E[d_{A_i}]\}$ is for this example the sum of 38 ($= K$) terms because there are 38 deaths at different survival times T_i in the combined group A+B. Each term is obtained from the numbers in the following columns of Table 15.3; compare with the formula for $E[d_{A_i}]$ given earlier:

$$E[d_{A_i}] = \{[\text{9th column}] \times [\text{3rd column}]/[\text{9th} + \text{10th columns}]\}$$

and in this example [9th column] $= 1$ for each T_i. Thus

$$
\begin{aligned}
\sum_{i=1}^{38} E[d_{A_i}] = {}& \{20/48\} + \{20/47\} + \{20/46\} + \{19/45\} + \{19/44\} \\
& + \{18/43\} + \{18/42\} + \{18/41\} + \{17/40\} + \{17/39\} \\
& + \{16/38\} + \{15/36\} + \{14/35\} + \{13/34\} + \{12/33\} \\
& + \{12/32\} + \{11/31\} + \{11/30\} + \{11/29\} + \{11/28\} \\
& + \{10/27\} + \{10/26\} + \{9/23\} + \{8/21\} + \{7/19\} \\
& + \{7/18\} + \{6/16\} + \{6/15\} + \{5/13\} + \{5/12\} + \{4/11\} \\
& + \{3/8\} + \{3/7\} + \{1/5\} + \{1/4\} + \{1/3\} + \{1/2\} + \{0/1\} \\
= {}& 14.445
\end{aligned}
$$

Table 15.3. Data for the Mantel–Haenszel test worked example.

Rank i	Time of death* T_i weeks	Treatment A §$d_{A_i} + s_{A_i}$	d_{A_i}	w_{A_i}	Treatment B §$d_{B_i} + s_{B_i}$	d_{B_i}	w_{B_i}	Totals (A+B) §$d_{A_i} + d_{B_i}$	§$s_{A_i} + s_{B_i}$
1	13	20	0	0	28	1	0	1	47
2	17	20	0	0	27	1	0	1	46
3	18	20	1	0	26	0	0	1	45
4	22	19	0	0	26	1	0	1	44
5	25	19	1	0	25	0	0	1	43
6	26	18	0	0	25	1	0	1	42
7	27	18	0	0	24	1	0	1	41
8	30	18	1	0	23	0	0	1	40
9	31	17	0	0	23	1	0	1	39
10	34	17	1	0	22	1	0	1	38
11	35	16	1	0	21	0	0	1	37
12	36	15	1	0	21	0	0	1	36
13	37	14	0	0	21	0	1[WA]	0	35
14	38	14	1	0	20	0	0	1	34
15	39	13	1	0	20	0	0	1	33
16	40	12	0	0	20	1	0	1	32
17	43	12	1	0	19	0	0	1	31
18	44	11	0	0	19	1	0	1	30
19	47	11	0	0	18	1	0	1	29
20	48	11	0	0	17	1	0	1	28
21	60	11	1	0	16	0	0	1	27
22	61	10	0	0	16	1	0	1	26
23	65	10	0	0	15	1	0	1	25
24	68	10	0	1[WA]	14	0	0	1	24
25	69	9	0	0	14	0	1[WA]	0	23
26	71	9	0	0	13	1	0	1	22
27	77	9	0	1[WA]	12	0	0	0	21
28	78	8	0	0	12	1	0	1	20
29	82	8	0	1[WA]	11	0	0	0	19
30	83	7	0	0	11	1	0	1	18
31	86	7	1	0	10	0	0	1	17
32	88	6	0	0	10	0	1[ID]	0	16
33	95	6	0	0	9	1	0	1	15
34	108	6	0	0	8	1	0	1	14
35	121	6	0	1[ID]	7	0	0	0	13
36	123	5	0	0	7	1	0	1	12
37	130	5	1	0	6	0	0	1	11
38	186	4	0	0	6	1	0	1	10
39	213	4	0	0	5	0	1[WA]	0	9
40	324	4	0	1[WA]	4	0	0	0	8
41	481	3	0	0	4	1	0	1	7
42	520	3	1	0	3	0	0	1	6
43	546	2	0	1[WA]	3	0	0	0	5
44	568	1	0	0	3	1	0	1	4
45	685	1	0	0	2	1	0	1	3
46	711	1	0	0	1	1	0	1	2
47	819	1	1	0	1	0	0	1	1
48	1105	0	0	0	1	1	0	1	0

* Generally this will be termed *event time* because the date of death will not be the endpoint for all analyses.

w_{A_i} = no. of patients lost to follow-up or withdrawn alive between T_i and T_{i+1} in group A

$d_{A_i} + s_{A_i}$ = no. of patients *at risk prior to time* T_i in group A

d_{A_i} = no. of deaths at time T_i in group A

s_{A_i} = no. of survivors at time T_i in group A

§ These are the four marginal totals in the 2 × 2 table in Table 15.2

NOTE: No. of cancer deaths in group A = 13 and in group B = 25. Intercurrent deaths[ID] are regarded as censored in this example as well as those withdrawn alive[WA] from the analysis; see also footnotes at bottom of Table 15.1.

To calculate the final denominator in the formula for the MH-statistic we require $\sum_{i=1}^{38} V_{A_i}$, which again is a summation of 38 values of V_{A_i}; where, referring to Table 15.3,

$$V_{A_i} = \frac{\{\text{Product of the 4 marginal totals labelled } \S \text{ in column headings of Table 15.3}\}}{[\text{9th} + \text{10th columns}]^2 \times [\text{9th} + \text{10th columns} - 1]}$$

Without making reference to Table 15.3, one can calculate the 38 V_{A_i} values from the 38 (2 × 2) tables corresponding to the *event times* as V_{A_i} is computed from the marginal totals in these 2 × 2 tables. Nevertheless, whatever computational method is used, even for a relatively small A+B group of 48 patients, the arithmetic involved is very lengthy and Mantel–Haenszel tests should be made using a statistical software package: there are too many opportunities for arithmetical errors if one attempts the test using only a pocket calculator.

The calculation of ΣV_{A_i} from the data in the marginal total columns[§] in Table 15.3 will therefore be as follows, with the last term always being 0. This 38th term corresponds to the last row in Table 15.3:

$$\sum_{i=1}^{38} V_{A_i} = [20 \times 28 \times 1 \times 47]/\{[48]^2 \times 47\}$$
$$+ 20 \times 27 \times 1 \times 46/\{[47]^2 \times 46\}$$
$$+ 20 \times 26 \times 1 \times 45/\{[46]^2 \times 45\}$$
$$+ 19 \times 26 \times 1 \times 44/\{[45]^2 \times 44\}$$
$$+ \ldots + 1 \times 1 \times 1 \times 1/\{[2]^2 \times 1\}$$
$$+ 0 \times 1 \times 1 \times 0/\{[1]^2/0\}$$
$$= 8.017$$

The value of the MH-statistic for this example is therefore 0.26 and as with the logrank test worked example, $P > 0.20$, and the data are thus entirely consistent with no difference in survival between anatomical subsites anterior two-thirds and posterior one-third of cancer of the tongue:

$$\text{MH} = (13 - 14.445)^2/8.017 = 0.260$$

Examples of some of the 2 × 2 tables are given in Table 15.4 and the marginal totals can be seen to correspond to those in Table 15.3 but, on balance, the quickest method for the computations is to set up a table like 15.3 rather than draw up all the 2 × 2 tables, a total of 38 tables in this example.

Finally, a reminder from the second paragraph of section 15.1 where it was stated *difference by the logrank test will always be significant by the more exact Mantel–Haenszel method* and therefore we would expect the MH-statistic to be larger than the logrank statistic, which it is, because 0.26 > 0.21.

Table 15.4. Examples of two 2 × 2 tables for the worked example: see rows ranked $i = 1$ and $i = 15$ in Table 15.3. A and B are the two patient groups. d is the number of patients who died at T_i and s is the number of patients who are alive between T_i and T_{i+1} and $r = [d + s]$ is the number of patients who were at risk before death at time T_i.

2 × 2 table for $T_i = 13$			2 × 2 table for $T_i = 15$		
	d	s		d	s
A	0	20	20	A	1	12	13
B	1	27	28	B	0	20	20
	1	47	48		1	32	33

THE 'DECEASED' LEFT THE FUNERAL SWEARING

An event with an extremely small probability of occurrence was reported under the above heading by the *Daily Telegraph* in the early 1970s, from Caracas. 'When grave diggers shovelled the first spadefuls of earth into a grave in the village at Pecaya, Venezuela, the 'dead' man, Roberto Rodriguez, who had collapsed after a heart attack, burst open the lid of the coffin, scrambled out of the grave and ran home shouting and swearing. His mother-in-law, who was standing at the graveside, dropped dead from shock. She will be buried in the grave prepared for her son-in-law, after doctors have made absolutely certain there is no mistake this time'.

Chapter 16

Regression and Correlation

16.1 INTRODUCTION

A common weather prediction saying is *red sky at night, shepherd's delight*. In effect this is a statement concerning an association between two events, namely the colour of the sky and the weather. A less trivial example could be the consideration of the possible relationship between two variables X and Y, each of which can be measured numerically. For example:

Dose and response.

Habit and disease incidence.

In the former, the study could be of a drug treatment and tumour response and in the latter, study of the correlation between smoking and lung cancer.

If an association is suspected, then a useful initial step would be to draw a scatter diagram to give an indication of any possible correlation between X and Y. An example is in Figure 16.1 where even without the straight line in this figure you can see that there is a linear trend between $Y =$ relative risk of lung cancer and $X =$ asbestos exposure. The straight line is called the *regression line* and its equation is:

$$Y = a + bX$$

where $a = -908$ (with a standard error of ± 171) and $b = +861$ (with a standard error of ± 75). The parameter b is given the special name *regression coefficient*, which should not be mixed up with *correlation coefficient*, usually denoted by r, which is different. In Figure 16.1 $r = +0.98$.

This regression line could be used to estimate the value of Y for any X in the range 1000–3000 or by extrapolation the value of Y for say $X = 3500$, just beyond the last measured value of (X, Y) which is for X almost 3000. This is the use of a *regression line*, that is, for prediction, whereas the use of a *correlation coefficient r* is to measure the strength of the association between X and Y. The value of r can range between $+1$ and -1 and the nearer the value

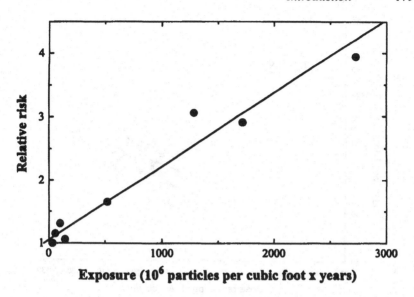

Figure 16.1. Relationship between asbestos exposure (particle-years) and the relative risk of lung cancer. The correlation coefficient is $r = +0.98$ and the regression line is $Y = -908 + 861X$.

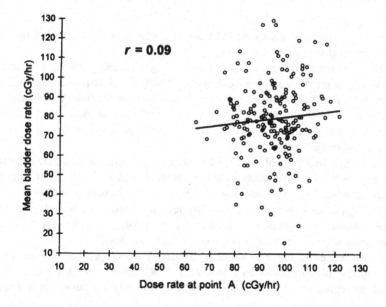

Figure 16.2. An example of when a regression line should not be drawn on a scatter plot.

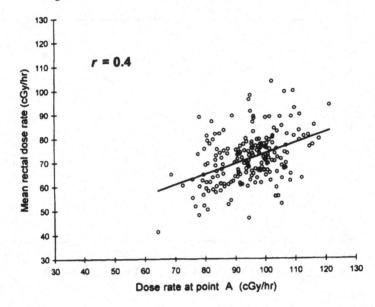

Figure 16.3. A further example of when a regression line should not be drawn on a scatter plot. This indicates that a value of $r = 0.4$ is not close enough to 1 to indicate that a regression line would be useful for purposes of prediction.

of r to 1 the better the association. Figure 16.1 can therefore be said to indicate a good correlation as $r = +0.98$.

In the treatment of cancer of the cervix using radioactive sources a reference dose point, point A, has often been used, although this is now somewhat out of date with the availability of computer dose distributions in three dimensions. Figure 16.2 relates the point A dose rate to the mean bladder dose rate and Figure 16.3 to the mean rectal dose rate[1]. These are good examples of when **not** to construct a regression line.

In Figure 16.2 the value of the correlation coefficient $r = +0.09$ indicates quite clearly that there is no association between X and Y and therefore drawing a regression line, which is only useful for prediction, is totally useless.

In Figure 16.3 the points are hardly less scattered even though the value of the correlation coefficient is higher at $r = +0.4$, but again, for the same reasons, the drawing of a regression line is again useless.

However, even worse is to come. There are also regression *curves*, using polynomial expressions such as $Y = a + bX + cX^2$ and $Y = a + bX + cX^2 + dX^3$ as well as regression *lines* $Y = a + bX$, but **not** regression curves as in Figure 16.4.

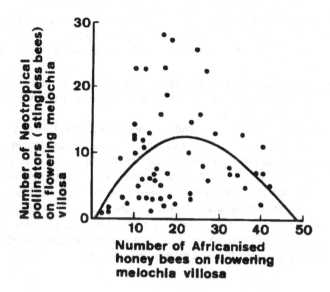

Figure 16.4. These scatter diagram data were published by D W Roubik in 1978 issue of *Science*, with the following abstract.

'The Africanized honey bee, a hybrid of European and African honey bees, is thought to displace native pollinators. After experimental introduction of Africanized honey bee hives near flowers, stingless bees became less abundant or harvested less resource as visitations by Africanized bees increased. Shifts in resource use caused by colonising Africanised honey bees may lead to population decline of Neotropical pollinators.'

The full, almost semi-circular, curve is a polynomial

$$Y = -0.516 + 1.08X - 0.023X^2$$

and was seriously drawn by Roubik to demonstrate non-linear regression. A response in *Science*, 1978 volume 202, page 823, by R M Hazen drew attention to the *rather fanciful curve fitting* of Roubik and with tongue in cheek drew a path through virtually all the points on the scatter diagram in a series of whirls and loops suggesting that a better regression would be the *flight of the bumble bee*! Roubik's reply included the comment:

'It seems to me that biologists are often obliged to take a different view of quantitative data from that of physical scientists. They have more or less set rules, while we must often try to discover nature's meanings. And there is a lot of slop in nature.'

Another word of warning, because the subject of regression and correlation is one in which there are many mistakes in the literature. Even if there is good correlation between X and Y it **must** also make sense and not be like the story of Figure 16.5.

Many other examples of the misuse of correlation and regression can be found in the literature and it is possibly the topic in statistics which is most frequently incorrectly applied.

The objective of proving a correlation between X and Y is to show that a relationship exists between these two variables, so that having demonstrated the existence of this relationship, it can be used within some theoretical framework. Blind use of regression formulae, just because they exist, can be very misleading. If $Y = a$ *cause* and $X = an$ *effect*, one must be careful not to draw too many conclusions if there may be several other possible causes. Cause and effect in medicine are seldom so simple as to be explained by a single straight line. It should also be remembered that it is the size of the correlation coefficient, r, which is important in deciding if a relationship is linear.

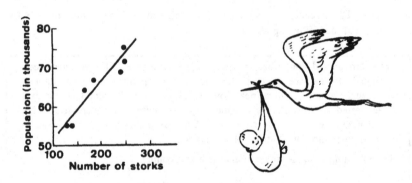

Figure 16.5. A regression line for real[2] data for the population of Oldenburg in Germany at the end of each year versus the number of storks observed in that year: 1930–1936. Confusing correlation, just because it exists, can lead to peculiar implications. Thus one should not conclude[3] from this data that *Anyone who draws the incorrect conclusion that storks bring babies and proceeds to shoot storks in the hopes of reducing the population will be disappointed!*

16.2 METHOD OF LEAST SQUARES FOR ESTIMATION OF THE SLOPE AND INTERCEPT OF A STRAIGHT LINE

A method is required to determine the slope, a, and intercept, b, of a straight line (see Figure 1.8) but if, for example, six different people draw the line through a scatter diagram of points by eye, then there will probably be six different *best fit* lines to the data. To avoid this, several methods of calculation are available, of which the most commonly used is the method of least squares. To illustrate the method see the first two columns of Table 16.1 (on page 185) which are headed X and Y and Figure 16.6 which plots all 10 (X, Y) points. The formulae required for the calculations are

$$\text{Intercept } b = \frac{\sum X \sum XY - \sum X^2 \sum Y}{(\sum X)^2 - N(\sum X^2)}$$

$$\text{Slope } a = \frac{\sum X \sum Y - N \sum XY}{(\sum X)^2 - N(\sum X^2)}$$

where N is the number of points on the scatter diagram, $N = 10$ for Figure 16.6 and the straight line is

$$Y = (0.93)X + 0.60$$

The additional notation in this figure is relevant to the next section, section 16.3.

As an example of a real data situation in which a straight line is of interest for prediction purposes, consider that[4] in the *British Medical Journal*, entitled *The victims of Chernobyl in Greece: induced abortions after the accident*. To calculate the slope and intercept of the straight line let

$$X = (\text{Birth year}) = 1980$$

$$Y = \frac{(\text{No of live births in Greece during the month of January})}{1000}$$

then a table (similar to Table 16.1) is prepared (Table 16.2 on page 186) and the values of a and b compute to

$$a = -0.23$$
$$b = +10.6$$

The values of X from 1–6 correspond to the birth years 1981–1986 and therefore to predict the number of live births in Greece in January 1987. If the 1981–1986 trend continues, we calculate

$$Y_{1987} = (-0.23 \times 7) + 10.6 = 9.0$$

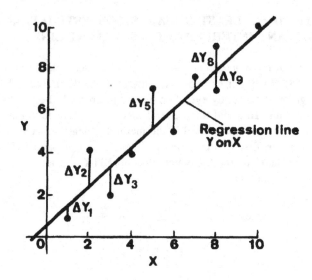

Figure 16.6. Regression line of Y on X, showing the deviation ΔY of the points from the line.

and therefore·the live birth statistic for January 1987 is predicted to be 9000. The number of births that month was actually only some 7000, showing a most unexpected reduction of more than 20% of the expected figure, see Figure 16.7 (page 186). This was considered by the authors to be due to the Chernobyl accident in that:

'The Chernobyl accident took place on 26 April 1986, but the extent of the catastrophe became apparent in Greece a few days later. During May there was panic because of conflicting data and false rumours. By June more reliable information became available.'

and

'Many obstetricians initially thought it prudent to interrupt otherwise wanted pregnancies or were unable to resist requests from worried pregnant women and their husbands. Within a few weeks misconceptions in the medical profession were largely cleared.'

The striking reductions which occurred in the January 1987 figures were not repeated in February and March 1987.

Table 16.1. Calculation of the slope and intercept of a straight line by the method of least squares.

X	Y	X^2	XY
1	1	1	1
3	2	9	6
2	4	4	8
4	4	16	16
6	5	36	30
5	7	25	35
8	7	64	56
7	7.5	49	52.5
8	9	64	72
10	10	100	100
$\sum X = 54$	$\sum Y = 56.5$	$\sum X^2 = 368$	$\sum XY = 376.5$

$$a = \frac{\sum X \sum Y - N \sum XY}{(\sum X)^2 - N(\sum X^2)}$$
$$= \frac{(54 \times 56.5) - (10 \times 376.5)}{(54)^2 - (10 \times 368)}$$
$$= \frac{714}{764} = 0.93$$

$$b = \frac{\sum X \sum XY - \sum X^2 \sum Y}{(\sum X)^2 - N(\sum X^2)}$$
$$= \frac{(54 \times 376.5) - (368 \times 56.5)}{(54)^2 - (10 \times 368)}$$
$$= \frac{461}{764} = 0.60$$

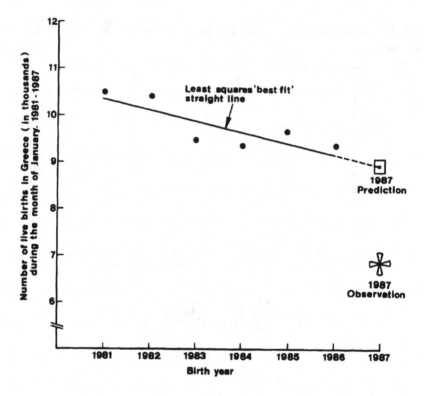

Figure 16.7. Data for live births in Greece in the month of January for 1981–1987.

Table 16.2. Calculation of the slope and intercept of a straight line by the method of least squares: see Figure 16.5[4].

X	Y	X^2	XY
1	10.5	1	10.5
2	10.3	4	20.6
3	9.5	9	28.5
4	9.4	16	37.6
5	9.6	25	48.0
6	9.3	36	55.8
$\sum X = 21$	$\sum Y = 58.6$	$\sum X^2 = 91$	$\sum XY = 201.0$

16.3 REGRESSION LINES

From the formulae in section 16.2 it is seen that the following four summations are required: $\sum X$, $\sum Y$, $\sum XY$, $\sum(X^2)$, where $\sum$ means *sum of all the ...
values*. From Table 16.1 the *least squares best fit straight line* was calculated as

$$Y = (0.93)X + 0.60$$

and this is drawn on the scatter diagram, in Figure 16.6. This straight line is called the *regression line of Y on X* and may be used for predicting values of Y for given values of X.

This assumes that Y is *dependent* upon X, and an example would be when $Y =$ tumour growth and $X =$ time. Tumour growth can obviously be dependent on time; however, the reverse *cannot* be true, since time is *not* dependent upon tumour growth. Situations arise when it is not clear whether X depends on Y or Y depends on X. In that case, two regression lines are calculated Y *on* X as in Figure 16.6, and also X *on* Y.

The choice of independent variable is then usually the one that will be used to predict values of the other. The regression equation can then be regarded as a prediction formula and if the correlation is good, then the two regression lines will be close together. The point at which they cross is $(\bar{X}, \bar{Y})$ the sample mean, and this provides a good check on the arithmetic.

The method of least squares used to calculate the *best fit* straight line minimises the squares of the deviations, ΔY, from the line. This is shown in Figure 16.6 where the vertical bars drawn from points to the line represent ΔY. The best fit straight line has been chosen such that the sum of all ΔY^2 values is a minimum.

In Table 16.1 the *slope a* and the *intercept b* were calculated and the formulae for the standard errors in a and b are

$$\sigma_a = \sqrt{\frac{N[\sum(\Delta Y^2)]}{(N-2)[N\sum X^2 - (\sum X)^2]}}$$

$$\sigma_b = \sqrt{\frac{[\sum(\Delta Y)^2(\sum X)^2]}{(N-2)[N\sum X^2 - (\sum X)^2]}}$$

where $\sum \Delta Y^2$ is the sum of the squares of the deviations of each point from the line.

$$\Delta Y = \begin{pmatrix} Y\text{-value of the} \\ \text{point on the} \\ \text{graph:} \\ observation \end{pmatrix} - \begin{pmatrix} (aX + b) \text{ which is} \\ \text{the } Y\text{-value of the} \\ \text{point on the line} \\ calculated \text{ using} \\ \text{the straight line} \\ \text{formula} \end{pmatrix}$$

Using the formulae for σ_a and σ_b, and when quoting two standard errors, we have $a = 0.93 \pm 0.28$ and $b = 0.60 \pm 1.64$.

16.4 PEARSON'S CORRELATION COEFFICIENT

The distribution of points scattered on Figure 16.6 makes it easy to visualise a linear relationship between Y and X, with the straight line passing through the central section of the band of points. In statistical terminology, it can be said that *there is good correlation between Y and X* and this can be shown numerically by calculating a *correlation coefficient* to describe the position of the straight line $Y = aX + b$ relative to the observations. The formula for the Pearson correlation coefficient, r (named after its originator), is

$$r = \frac{\sum[(X - \bar{X})(Y - \bar{Y})]}{\sqrt{[\sum(X - \bar{X})^2][\sum(Y - \bar{Y})^2]}}$$

An alternative formula does not contain $\bar{X}$ and $\bar{Y}$, but although it looks more complicated, many of the summation ($\sum$) terms have already been calculated to determine a and b and it is therefore more convenient to use in practice (see Table 16.3).

$$r = \frac{N[\sum XY] - [\sum X \sum Y]}{\sqrt{[N[\sum X^2] - [\sum X]^2][N[\sum Y^2] - [\sum Y]^2]}}$$

For straight lines such as those in Figure 16.1 and Figure 16.6 we say that there is *positive correlation* and this is indicated by the positive sign for r whereas in Figures 16.7 and 16.8 there is *negative correlation* and in Figures 16.2 and 16.3 there is *no correlation*.

16.5 TESTING FOR A SIGNIFICANT CORRELATION: AN APPLICATION OF THE t-TEST

In Table 16.3 it is seen that $r = +0.92$. A perfect positive correlation would be $r = +1$ and a perfect negative correlation would be $r = -1$. No correlation would be $r = 0$. A correlation coefficient of $r = +0.92$ is therefore good correlation.

r can be calculated for any scatter diagram but whether the results show a *significant correlation* between X and Y should then be tested before values of Y are predicted using the straight-line relationship. The significance test is made using the t-test, where

$$t = \frac{r\sqrt{N - 2}}{\sqrt{1 - r^2}}$$

Table 16.3. Calculation schedule to determine the Pearson correlation coefficient, r.

$$\Sigma X = 54, \ \bar{X} = 54/10 = 5.4$$
$$\Sigma Y = 56.5, \ \bar{Y} = 56.5/10 = 5.65$$

X	Y	Y^2 (used in method 2)	$X - \bar{X}$	$Y - \bar{Y}$	$(X - \bar{X})^2$	$(Y - \bar{Y})^2$	$(X - \bar{X})(Y - \bar{Y})$
1	1	1	−4.4	−4.65	19.36	21.62	20.46
3	2	4	−2.4	−3.65	5.76	13.32	8.76
2	4	16	−3.4	−1.65	11.56	2.72	5.61
4	4	16	−1.4	−1.65	1.96	2.72	2.31
6	5	25	0.6	−0.65	0.36	0.42	−0.39
5	7	49	−0.4	1.35	0.16	1.82	−0.54
8	7	49	2.6	1.35	6.76	1.82	3.51
7	7.5	56.25	1.6	1.85	2.56	3.42	2.96
8	9	81	2.6	3.35	6.76	11.22	8.71
10	10	100	4.6	4.35	21.16	18.92	20.01
		ΣY^2 = 397.25			$\Sigma(X - \bar{X})^2$ = 76.4	$\Sigma(Y - \bar{Y})^2$ = 78.0	$\Sigma(X - \bar{X})(Y - \bar{Y})$ = 71.4

Method 1

$$r = \frac{\Sigma[(X - \bar{X})(Y - \bar{Y})]}{\sqrt{[\Sigma(X - \bar{X})^2][\Sigma(Y - \bar{Y})^2]}}$$

$$a = \frac{71.4}{\sqrt{(76.4 \times 78)}} = 0.92$$

Method 2

$$\Sigma X = 54 \qquad \Sigma Y = 56.5$$
$$\Sigma XY = 376.5 \text{ from Table 16.1}$$
$$(\Sigma X)^2 = (54)^2 = 2916$$
$$(\Sigma Y)^2 = (56.5)^2 = 3192.25$$
$$\Sigma X^2 = 368 \text{ from Table 16.1}$$
$$\Sigma Y^2 = 397.25$$

$$r = \frac{N[\Sigma XY] - [\Sigma X \Sigma Y]}{\sqrt{[N[\Sigma X^2] - [\Sigma X]^2][N[\Sigma Y^2] - [\Sigma Y]^2]}}$$

$$r = \frac{(10 \times 376.5) - (54 \times 56.5)}{\sqrt{[10 \times 368 - 2916][10 \times 397.25 - 3192.25]}}$$

$$= \frac{714}{\sqrt{(764 \times 780.25)}}$$

$$= 0.92$$

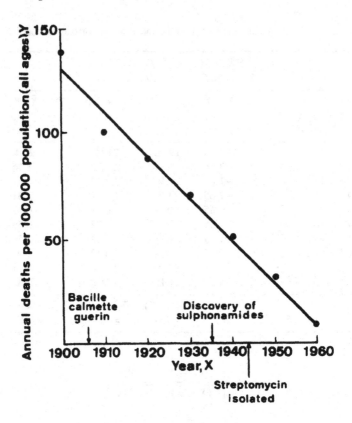

Figure 16.8. Correlation of deaths due to tuberculosis with time from 1900 to 1960. The regression line is $Y = 130 - 2X$ and the correlation coefficient is $r = -0.99$.

which is derived from $t = r/(\text{Standard deviation of } r)$. The number of degrees of freedom is given by

$$DF = N - 2$$

where N is the number of points on the scatter diagram.

The null hypothesis is

H_0: there is no linear association between Y and X.

Since $r = +0.92$ and $N = 10$, the derived t-statistic is

$$t = (0.92 \times 2.83)/\sqrt{1 - 0.846} = 6.6.$$

From Table 11.2 for $DF = N - 2 = 8$, $P = 0.05$, the t-statistic is 2.31. Since 6.6 is greater than 2.31 the null hypothesis, H_0, is rejected at the 0.05

level of significance and we can say that a significant correlation ($P < 0.05$) exists between X and Y. This test is a two-tailed test example, since we have tested for a critical value of t greater than $+2.31$ or less than -2.31. If there is no linear association and H_0 is true, then the derived t-statistic would have been greater than $+2.31$ or less than -2.31 in only 5% of any series of identical trials with $N = 10$.

16.6 SPEARMAN'S RANK CORRELATION COEFFICIENT

The Spearman correlation coefficient, usually given the symbol ρ or r_s is, as its name suggests, used when the calculations are made on the ranks of the observations, rather than, as with Pearson's correlation coefficient, r, on the observations themselves. ρ is interpreted in the same way as r but it should be realised that since one has less information when only the rankings are given, then ρ is not as informative as it would be if calculated from the *measured variables*, as with r.

Considering the formula for r in Method 2 within Table 16.3, the Spearman rank correlation coefficient is obtained using the same formula but inserting the ranks rather than the measured values. However, because the difference between the adjacent ranks is always 1 this formula can be simplified to become

$$\rho = 1 - \left[6 \sum (d^2)\right] / [N\{N^2 - 1\}]$$

where d is the difference between the rankings of the same item in each series. Since the difference term in the formula is d^2 (summed over all N pairs of ranks) it does not matter whether d is positive or negative. Table 16.4 gives an example of the calculation schedule to determine ρ using the above formula.

If two items in one or other of the series of measurements tie, that is, have the same rank, such as third equal, then the ranks are added and the sum divided by the number of items sharing the same rank. Thus for example with two tied ranks which would for different measurement values be equal to 3 and 4, for tied (equal) measurement values, each will be given a rank of $(3 + 4)/2 = 3.5$.

Table 16.5 gives the significance levels of ρ for small samples. For 11 or more samples Table 16.6 (page 193) should be used with the following formula for the number of degrees of freedom DF

$$\text{DF} = [\text{Number of pairs} - 2]$$

In the example in Table 16.4 $\rho = 0.714$ and the sample size $= 7$. The correlation is therefore (just) not significant at the 5% level since the critical value in Table 16.5 is $\rho = 0.750$.

Table 16.4. Calculation schedule to determine a Spearman rank correlation coefficient ρ. Rank r_X is the rank in the series of measurements X and rank r_Y is the rank in the series of measurements Y. In this example there are seven pairs of measurements which have been ranked and thus $N = 7$.

Rank r_X	Rank r_Y	d	d^2
3	1	2	4
2	4	2	4
1	2	1	1
4	3	1	1
6	5	1	1
5	7	2	4
7	6	1	1
		$\sum(d^2) = 16$	

$$\rho = 1 - [(6 \times 16)]/[7 \times \{7^2 - 1\}]$$
$$\rho = 0.714$$

Table 16.5. Significance levels of ρ for small samples[6]. The sample size equals the number of pairs.

Sample size	5% level	1% level
4 or less	None	None
5	1.000	None
6	0.886	1.000
7	0.750	0.893
8	0.714	0.857
9	0.683	0.833
10	0.648	0.794

16.7 KENDALL'S τ RANK CORRELATION COEFFICIENT

Kendall's (tau) τ is also a rank correlation coefficient and is closely related to Spearman's ρ. It takes longer to calculate than ρ but τ can be extended to study partial correlation[5]. However, the power of the test for the same level of significance[7] is smaller for τ than for ρ. Table 16.7 gives an example of the calculation schedule to determine τ.

For further reading on rank correlation (ρ and τ) see the textbooks by Snedecor and Cochran[6] and by Sachs[7].

Table 16.6. Correlation coefficients at the 5% and 1% level of significance. n = degrees of freedom where $n = (n' - 2)$ where n' is the number of pairs on which the correlation is based. This table gives an exact test of significance for correlation coefficient r and is equivalent to the t-test of the regression. $t = \{r[\sqrt{(n'-2)}]\}/\{\sqrt{(1-r^2)}\}$. From Fisher and Yates *Statistical Tables for Biological, Agricultural and Medical Research* (6th edn, 1974, table VII, p 63). Courtesy Longman Group UK Limited.

n	.1	.05	.02	.01	.001
1	98769	99692	999507	999877	9999988
2	90000	95000	98000	990000	99900
3	8054	8783	93433	95873	99116
4	7293	8114	8822	91720	97406
5	6694	7545	8329	8745	95074
6	6215	7067	7887	8343	92493
7	5822	6664	7498	7977	8982
8	5494	6319	7155	7646	8721
9	5214	6021	6851	7348	8471
10	4973	5760	6851	7079	8233
11	4762	5529	6339	6835	8010
12	4575	5324	6120	6614	7800
13	4409	5139	5923	6411	7603
14	4259	4973	5742	6226	7420
15	4214	4821	5577	6055	7246

n	.1	.05	.02	.01	.001
16	4000	4683	5425	5897	7084
17	3887	4555	5285	5751	6932
18	3783	4438	5155	5614	6787
19	3687	4329	5034	5487	6652
20	3598	4227	4921	5368	6524
25	3233	3809	4451	4869	5794
30	2960	3494	4093	4487	5541
35	2746	3246	3810	4182	5189
40	2573	3044	3578	3932	4896
45	2428	2875	3384	3721	4648
50	2306	2732	3218	3541	4433
60	2108	2500	2948	3248	4078
70	1954	2319	2737	3017	3799
80	1829	2172	2565	2830	3568
90	1726	2050	2422	2673	3375
100	1638	1946	2301	2540	3211

Table 16.7. Calculation schedule to determine a Kendall τ rank correlation coefficient. The same observations are used in this table as in Table 16.4.

Step 1
Rearrange the two rankings in Table 16.4 so that one of them r_X is in the order $1, 2, 3, \ldots, N$

Rank r_X	1	2	3	4	5	6	7
Rank r_Y	2	4	1	3	7	5	6

Step 2
• Take each rank r_Y in turn, count how many of the ranks to the *right* of it are smaller than it, and add these counts

• Thus for $r_Y = 2$ the count is 1 since there is only one value of r_Y smaller than 2 (i.e. $r_Y = 1$) to the right of $r_Y = 2$.

• The six counts are 1, 2, 0, 0, 2, 0, there being no need to count the extreme right rank

• The total $Q = 5$

Step 3
Calculate τ using the formula

$$\tau = 1 - [4Q/(N\{N - 1\})]$$

Thus $\tau = 1 - [20/(7 \times 6)] = 0.524$

LES RELIQUES AUTHENTIQUES

TETE A ORBITELLO
1 TETE A MONTPELLIER

1 TETE A NAPLES
TETE A SMAXIMIN EN PROVENCE

1 BRAS A ROME

1 BRAS A MILAN

1 BRAS A MARSEILLE

1 BRAS A CAPOUE

S¹ BLAISE

1 BRAS A NOTRE-DAME DE PARIS

1 CORPS A S¹MARIE DE ROME

1 BRAS A BASSE-FONTAINE

FRAGMENTS A BRINDES A MALINES A LISBONNE

FRAGMENTS A RAGUSE A VOLTERRE A ANVERS

Reconstitution de St-Blaise d'après ses Reliques
(en cas d'oubli prière d'en faire part)

Always check your arithmetic and if for example the total
number of observations is N, make sure that when you add up
the values of all observations in the individual cells they
actually total N: otherwise you might end up like St. Blaise
and have to repeat all your arithmetic.

Chapter 17

Analysis of Variance

17.1 INTRODUCTION

Analysis of variance, generally termed ANOVA, is an extension of the t-test. Suppose that the objective of a study is is to discover whether there are differences in the means of several independent groups. The problem is therefore how to measure the extent of the differences among the means. If we had only two groups then we would measure the difference between the sample means and use the two-sample t-test for unpaired data: see sections 11.5 and 11.6. With more than two means it is of course technically possible to make multiple t-tests on all possible pairs of means, but *making multiple tests increases the probability of making a type I error*. In section 8.4 this is defined as the error of rejecting the null hypothesis H_0 when H_0 is true. The risk associated with this error is the alpha risk which is specified by using a P-value.

What is required in this situation when a single t-test is not appropriate, is a single measure that summarises the differences between several means and a method of simultaneously comparing these means in one step. This problem is solved by the use of ANOVA and the statistical significance test used is the F-test. Table 17.1 lists some simple examples of null hypotheses H_0 which can be tested using ANOVA, a statistical technique which has applications in many fields of interest, not only in medicine.

One-way (sometimes called one-factor) ANOVA is a generalisation of the *unpaired t-test* and is appropriate for any number of groups. A table for a one-way ANOVA could be as in Table 17.2 where there are two treatment levels and two measurements of the response are made for each treatment level.

Two-way (or two-factor) ANOVA is an extension of the *paired t-test* and is ANOVA in which there is a more complicated structure than the one-way classification because the data is classified in two ways and not in one way. An example is data on birth weights[1] in which the null hypothesis H_0 is that differences in birth weights are independent of birth rank (i.e. whether the child is first, second, third.... in the family) or of maternal age.

Table 17.1. Null hypotheses H_0 which can be studied by one-way ANOVA.

Description of n sample groups and the factor of interest	Parameter measured by a numerical value Means: $\bar{X}_1, \bar{X}_2, \bar{X}_3, \ldots, \bar{X}_n$	Null hypothesis
Teaching methods A, B. C, D	Score assessment	H_0: no difference in results from the 4 teaching methods, as measured by $\bar{X}_1, \bar{X}_2, \bar{X}_3, \bar{X}_4$
Petrol costs in cities, towns and country areas	Cost/litre	H_0: no difference in cost of petrol in the 3 geographical locations, as measured by $\bar{X}_1, \bar{X}_2, \bar{X}_3$
3 groups of newborn infants with similar population parameters for birth weight, sex, race and any other factor which might influence weight gain, are given diets A, B and C	Weight at end of 3 months	H_0: no difference in the effectiveness of the 3 diets as measured by $\bar{X}_1, \bar{X}_2, \bar{X}_3$
On an industrial assembly line there are 5 groups each of 8 experienced operators who assemble a particular computer part: groups A, B. C, D, E	Time taken for assembly	H_0: no difference in the efficiency of the 5 groups as measured by $\bar{X}_1, \bar{X}_2, \bar{X}_3, \bar{X}_4, \bar{X}_5$

Table 17.2. Table for one-way ANOVA for two groups (a and c, and b and d) of measurement.

		Treatment level	
		T_1	T_2
Response	R_1	a	b
	R_2	c	d

H_0 is tested by estimating the share of the total variation in birth weight which is attributable to differences in birth rank, and the share which is attributable to maternal age, leaving a *residual variation* which is not attributable to either factor.

Explanations are given below of a few of the more frequently used terms in Chapters 17 and 18. Some have been modified and extended from entries in the dictionary[2] sponsored by the International Epidemiological Association.

Covariate. A variable that is possibly predictive of the outcome under study. A covariate may be of direct interest to the study or may be a confounding variable or effect modifier[2]. The terms variate and variable are synonymous.

Dependent and Independent Variables. A *dependent* variable has a value which is dependent on the effect of other *independent* variable(s) in the relationship under study. In an outcome of a study with *dependent* variable variation, we seek to explain such variation by the influence of *independent* variables[2]. In the example of Figure 16.1 the *dependent* variable is the risk of lung cancer and the *independent* variable, only one in this case, is exposure to asbestos. The *dependent* variable (Y) is the one predicted by a regression equation, e.g. $Y = a + bX$.

Confounding Variable. Confounding is a situation in which a measure of the effect of an exposure on risk is distorted because of the association of exposure with other factor(s) that influence the outcome under study[2]. If confounding is present then it can completely destroy the entire study. A variable that appears to be protective may, after control of confounding, be found to be harmful[3]. Age and social class are often confounding variables in epidemiological studies. Confounding may be the explanation[3] for the relationship between coffee consumption and the risk of coronary heart disease, since it is known that coffee consumption is associated with cigarette smoking: people who drink coffee are more likely to smoke than people who do not drink coffee. It is also well known that cigarette smoking is a cause of coronary heart disease. It is thus possible that the relationship between coffee consumption and coronary heart disease merely reflects the known causal association of smoking with the disease. In this situation, smoking counfounds the apparent relationship between coffee consumption and coronary heart disease.

Multivariate Analysis. A set of techniques used when the variation in several variables has to be studied simultaneously. It is an analytical method that allows the simultaneous study of two or more dependent variables[2]. It differs from univariate or bivariate analysis (one-way or two-way ANOVA) in that it directs attention away from the analysis of the mean and variance of a single variable, or from the pair-wise relationship between two variables. Attention is therefore directed towards the analysis of *covariances* or correlations which reflect the extent of the relationship among three or more variables.

Sum of Squares and Mean Square. One problem often encountered with ANOVA is that there are two equivalent terminologies for *sum of squares* SS, and *mean square* MS (which is in fact a *variance*). These are:

$$SS_{RESIDUAL} \text{ is the same as } SS_{WITHIN}$$

$$SS_{REGRESSION} \text{ is the same as } SS_{BETWEEN}$$

The *within* and *between* refer to groups of measurement, and SS_{WITHIN} relates to variability arising from the sampling technique and $SS_{BETWEEN}$ to the variability between the groups of measurements.

17.2 THE *F*-TEST

The *F*-test is an integral part of analysis of variance. It is used for testing for a significant difference between two variances S_1^2 and S_2^2 where the *F*-statistic is

$$F = S_1^2/S_2^2$$

and where S_1^2 must always be the larger variance. Values of the *F*-statistic are given in Tables 17.3(*a*, *b*) where DF_1 is the number of degrees of freedom for the larger variance and DF_2 the number for the smaller variance. Some of these data are presented graphically in Figure 17.1.

It must be remembered that the *F*-test is a test for the comparison of two independent estimates of variance, and this condition fails if the observations in the two samples are paired. Also, there is the underlying assumption for the *F*-test that the populations from which the samples are drawn are normally distributed.

As a simple example to illustrate the use of Tables 17.3(*a*, *b*), consider 10 patients placed in two different hospital wards, whose systolic blood pressures are:

Ward A patients: 170, 180, 160, 200, 180 mm Hg
Ward B patients: 210, 140, 160, 230, 150 mm Hg.

The mean blood pressure of both groups is the same, equal to 178 mm Hg and the problem is to study whether the variability of the blood pressures is also the same for both groups, so that they can be assumed to be from the same population. If they do all come from the same population, then they may be paired with each other to assess, for example, the effect of an anti-hypertensive drug. The derived F-statistic is $1570/220 = 7.16$ and from Table 17.3 for $DF_1 = DF_2 = 4$ degrees of freedom and a chosen level of significance of 0.05, F is 6.4. Since 7.16 is greater than 6.4 we reject the null hypothesis at the 0.05 level of significance, which is

H_0: Both groups of patients are from the same population.

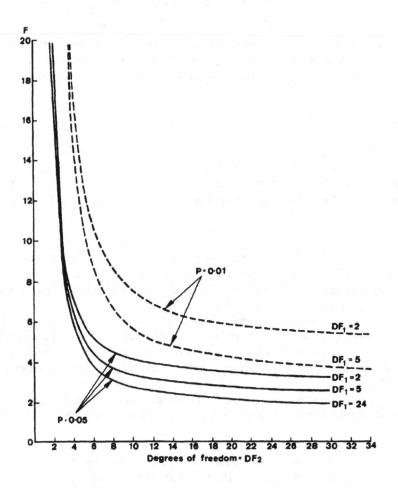

Figure 17.1. The F-statistic as a function of degrees of freedom and probability levels, see data in Tables 17.3(a, b).

Table 17.3. (*a*) Values of the *F*-statistic for selected degrees of freedom and a probability level of *P* = **0.05**. DF₁ must always correspond to the greater variance. From Fisher and Yates, *Statistical Tables For Biological, Agricultural and Medical Research* (6th edn, 1974, table V, p 53). Courtesy Longman Group UK Limited.

DF₂ \ DF₁	1	2	3	4	5	6	8	12	24	∞
1	161.4	199.5	215.7	224.6	230.2	234.0	238.9	243.9	249.0	254.3
2	18.51	19.00	19.16	19.25	19.30	19.33	19.37	19.41	19.45	19.50
3	10.13	9.55	9.28	9.12	9.01	8.94	8.84	8.74	8.64	8.53
4	7.71	6.94	6.59	6.39	6.26	6.16	6.04	5.91	5.77	5.63
5	6.61	5.79	5.41	5.19	5.05	4.95	4.82	4.68	4.53	4.36
6	5.99	5.14	4.76	4.53	4.39	4.28	4.15	4.00	3.84	3.67
7	5.59	4,74	4.35	4.12	3.97	3.87	3.73	3.57	3.41	3.23
8	5.32	4.46	4.07	3.84	3.69	3.58	3.44	3.28	3.12	2.93
9	5.12	4.26	3.86	3.63	3.48	3.37	3.23	3.07	2.90	2.71
10	4.96	4.10	3.71	3.48	3.33	3.22	3.07	2.91	2.74	2.54
11	4.84	3.98	3.59	3.36	3.20	3.09	2.95	2.79	2.61	2.40
12	4.75	3.88	3.49	3.26	3.11	3.00	2.85	2.69	2.50	2.30
13	4.67	3.80	3.41	3.18	3.02	2.92	2.77	2.60	2.42	2.21
14	4.60	3.74	3.34	3.11	2.96	2.85	2.70	2.53	2.35	2.13
15	4.54	3.68	3.29	3.06	2.90	2.79	2.64	2.48	2.29	2.07
16	4.49	3.63	3.24	3.01	2.85	2.74	2.59	2.42	2.24	2.01
17	4.45	3.59	3.20	2.96	2.81	2.70	2.55	2.38	2.19	1.96
18	4.41	3.55	3.16	2.93	2.77	2.66	2.51	2.34	2.15	1.92
19	4.38	3.52	3.13	2.90	2.74	2.63	2.48	2.31	2.11	1.88
20	4.35	3.49	3.10	2.87	2.71	2.60	2.45	2.28	2.08	1.84
21	4.32	3.47	3.07	2.84	2.68	2.57	2.42	2.25	2.05	1.81
22	4.30	3.44	3.05	2.82	2.66	2.55	2.40	2.23	2.03	1.78
23	4.28	3.42	3.03	2.80	2.64	2.53	2.38	2.20	2.00	1.76
24	4.26	3.40	3.01	2.78	2.62	2.51	2.36	2.18	1.98	1.73
25	4.24	3.38	2.99	2.76	2.60	2.49	2.34	2.16	1.96	1.71
26	4.22	3.37	2.98	2.74	2.59	2.47	2.32	2.15	1.95	1.69
27	4.21	3.35	2.96	2.73	2.57	2.46	2.30	2.13	1.93	1.67
28	4.20	3.34	2.95	2.71	2.56	2.44	2.29	2.12	1.91	1.65
29	4.18	3.33	2.93	2.70	2.54	2.43	2.28	2.10	1.90	1.64
30	4.17	3.32	2.92	2.69	2.53	2.42	2.27	2.09	1.89	1.62
40	4.08	3.23	2.84	2.61	2.45	2.34	2.18	2.00	1.79	1.51
60	4.00	3.15	2.76	2.52	2.37	2.25	2.10	1.92	1.70	1.39
120	3.92	3.07	2.68	2.45	2.29	2.17	2.02	1.83	1.61	1.25
∞	3.84	2.99	2.60	2.37	2.21	2.10	1.94	1.75	1.52	1.00

Table 17.3. (*b*) Values of the *F*-statistic for selected degrees of freedom and a probability level of *P*= **0.01**. DF_1 must always correspond to the greater variance. From Fisher and Yates, *Statistical Tables For Biological, Agricultural and Medical Research* (6th edn, 1974, table V, p 55). Courtesy Longman Group UK Limited.

DF_2 \ DF_1	1	2	3	4	5	6	8	12	24	∞
1	4052	4999	5403	5625	5764	5859	5982	6106	6234	6366
2	98.50	99.00	99.17	99.25	99.30	99.33	99.37	99.42	99.46	99.50
3	34.12	30.82	29.46	28.71	28.24	27.91	27.49	27.05	26.60	26.12
4	21.20	18.00	16.69	15.98	15.52	15.21	14.80	14.37	13.93	13.46
5	16.26	13.27	12.06	11.39	10.97	10.67	10.29	9.89	9.47	9.02
6	13.74	10.92	9.78	9.15	8.75	8.47	8.10	7.72	7.31	6.88
7	12.25	9.55	8.45	7.85	7.46	7.19	6.84	6.47	6.07	5.65
8	11.26	8.65	7.59	7.01	6.63	6.37	6.03	5.67	5.28	4.86
9	10.56	8.02	6.99	6.42	6.06	5.80	5.47	5.11	4.73	4.31
10	10.04	7.56	6.55	5.99	5.64	5.39	5.06	4.71	4.33	3.91
11	9.65	7.20	6.22	5.67	5.32	5.07	4.74	4.40	4.02	3.60
12	9.33	6.93	5.95	5.41	5.06	4.82	4.50	4.16	3.78	3.36
13	9.07	6.70	5.74	5.20	4.86	4.62	4.30	3.96	3.59	3.16
14	8.86	6.51	5.56	5.03	4.69	4.46	4.14	3.80	3.43	3.00
15	8.68	6.36	5.42	4.89	4.56	4.32	4.00	3.67	3.29	2.87
16	8.53	6.23	5.29	4.77	4.44	4.20	3.89	3.55	3.18	2.75
17	8.40	6.11	5.18	4.67	4.34	4.10	3.79	3.45	3.08	2.65
18	8.28	6.01	5.09	4.58	4.25	4.01	3.71	3.37	3.00	2.57
19	8.18	5.93	5.01	4.50	4.17	3.94	3.63	3.30	2.92	2.49
20	8.10	5.85	4.94	4.43	4.10	3.87	3.56	3.23	2.86	2.42
21	8.02	5.78	4.87	4.37	4.04	3.81	3.51	3.17	2.80	2.36
22	7.94	5.72	4.82	4.31	3.99	3.76	3.45	3.12	2.75	2.31
23	7.88	5.66	4.76	4.26	3.94	3.71	3.41	3.07	2.70	2.26
24	7.82	5.61	4.72	4.22	3.90	3.67	3.36	3.03	2.66	2.21
25	7.77	5.57	4.68	4.18	3.86	3.63	3.32	2.99	2.62	2.17
26	7.72	5.53	4.64	4.14	3.82	3.59	3.29	2.96	2.58	2.13
27	7.68	5.49	4.60	4.11	3.78	3.56	3.26	2.93	2.55	2.10
28	7.64	5.45	4.57	4.07	3.75	3.53	3.23	2.90	2.52	2.06
29	7.60	5.42	4.54	4.04	3.73	3.50	3.20	2.87	2.49	2.03
30	7.56	5.39	4.51	4.02	3.70	3.47	3.17	2.84	2.47	2.01
40	7.31	5.18	4.31	3.83	3.51	3.29	2.99	2.66	2.29	1.80
60	7.08	4.98	4.13	3.65	3.34	3.12	2.82	2.50	2.12	1.60
120	6.85	4.79	3.95	3.48	3.17	2.96	2.66	2.34	1.95	1.38
∞	6.64	4.60	3.78	3.32	3.02	2.80	2.51	2.18	1.79	1.00

17.3 ONE-WAY ANOVA: WORKED EXAMPLE

To illustrate the construction of a one-way analysis of variance table for a one-factor hypothetical experiment, the effects of 3 drugs A, B and C, the results of which are measured by the parameter X are shown in Table 17.4. In practice, sample sizes of only $n = 4$ would imply a very poorly designed experiment. Nevertheless, when an experiment involves small animals such as rats or mice. the investigator will often ask the statistician at the start of the experiment 'Can I use only a very few animals because they are so expensive?'. Expense has nothing to do with statistical significance!

In analysis of variance the variability of the observations is split up into two components.

• A component that depends on the differences *between* the means of the 3 treatment group populations. Note that the word **between** is synonomous with the word **regression**: see section 17.1.

• A component that measures the variability *within* the 3 groups of 4 measurements each, which arises from the sampling technique used. Note that the word **within** is synonomous with the word **residual**: see section 17.1.

To examine the data in more detail the responses in Table 17.4 can be re-written, including the values of the means $\bar{X}_j$ as shown in Table 17.5, and by doing so, provide the *within* group variations.

The *grand mean* (i.e. mean of means) $\bar{X}$ of all the 12 X_{ji} responses is calculated from the row $\sum X_{ji}$ in Table 17.4 and is $\bar{X} = (140 + 160 + 144)/12 = 37$. Using this grand mean $\bar{X}$ the responses in Table 17.5 can be re-written as shown in Table 17.6, and by doing so, provide the *between* group variations.

From Table 17.6 it is seen that each response X_{ji} consists of three components which can be written as:

X_{ji} = Grand mean $\bar{X}$ +   Between (regression)   +   Within (residual)
                                          groups   groups
                                          variations   variations

$$X_{ji} = \bar{X} + (\bar{X}_j - \bar{X}) + (X_{ji} - \bar{X}_j)$$

		A treatment group		
	The	component		A residual
$X_{ji} =$	grand	+ which is a	+	variability of
	mean	deviation from		individual
		the grand mean		observations

Thus:

$$(X_{ji} - \bar{X}) = (\bar{X}_j - \bar{X}) + (X_{ji} - \bar{X}_j)$$

Table 17.4. One-way ANOVA table. The 12 values (of X) in the body of the table are the experimental results (i.e. responses): 4 measurements for each of the 3 treatments. The table layout is similar to that of Table 17.2. Note that in practice the number of results per treatment do not have to be equal. They have only been made equal with $n = 4$ in this hypothetical example to make the arithmetic easier. If there are j treatment groups and i measurements per group then the individual responses are termed X_{ji} where in this example $j = 1, 2, 3$ and $i = 1, 2, 3, 4$. Each value of the response X_{ji} is thus classified in terms of a column number and a row number in the table. Thus for example $X_{23} = 39$.

| | Treatment group j | | | |
	Drug A $[j=1]$	Drug B $[j=2]$	Drug C $[j=3]$	
	36	44	35	Row $i=1$
	37	37	40	Row $i=2$
Response X_{ji} $[i=1,2,3,4]$ $[j=1,2,3]$	34	39	33	Row $i=3$
	33	40	36	Row $i=4$
$\sum X_{ji}$	140	160	144	
Mean $\bar{X}_j$	35	40	36	

Table 17.5. Rearrangement of the response data in Table 17.4 with the X_{ji} values tabulated as the value of the mean $\bar{X}_j$ and the variation from this mean. The variations are the *within (residual)* group variations.

| Treatment group j | | |
Drug A $[j=1]$	Drug B $[j=2]$	Drug C $[j=3]$
$35+1$	$40+4$	$36-1$
$35+2$	$40-3$	$36+4$
$35-1$	$40-1$	$36-3$
$35-2$	$40+0$	$36+0$

Table 17.6. Rearrangement of the response data in Table 17.5 with the variation from the grand mean $\bar{X}$ tabulated. This table shows both the *within* (*residual*) group variations and the *between* (*regression*) group variations.

	Treatment group j	
Drug A $[j = 1]$	Drug B $[j = 2]$	Drug C $[j = 3]$
$37 - 2 + 1$	$37 + 3 + 4$	$37 - 1 - 1$
$37 - 2 + 2$	$37 + 3 - 3$	$37 - 1 + 4$
$37 - 2 - 1$	$37 + 3 - 1$	$37 - 1 - 3$
$37 - 2 - 2$	$37 + 3 + 0$	$37 - 1 + 0$

Table 17.7. ANOVA table for sum of squares, where n_i is the number of measurements per treatment group. For the data of Tables 17.4–17.6 we have $n_i = 4$ for $i = 1, 2, 3$. $SS_{TOTAL} = SS_{BETWEEN} + SS_{WITHIN}$.

Source of variation	Sum of squares of deviations (SS)
Between treatment group variation	$SS_{BETWEEN}(SS_{REGRESSION}) = n_i \sum_j (\bar{X}_j - \bar{X})^2$
Within measurement group variation	$SS_{WITHIN}(SS_{RESIDUAL}) = \sum_j \sum_i (X_{ji} - \bar{X}_j)^2$
Total sum of squares	$SS_{TOTAL} = \sum_j \sum_i (X_{ji} - \bar{X})^2$

An appropriate overall measure of variation which can be split up into useful parts is the *sum of squares of deviations* and a table can be constructed for this type of decomposition: see Table 17.7.

The one-way ANOVA table format as in Table 17.7 is constructed for the response data of Tables 17.4–17.6 in Table 17.9. However, since this is an introductory text the various steps in the arithmetical procedure are given first in Table 17.8 to aid understanding.

The F-statistic is defined as:

$$F = [MS_{BETWEEN}/MS_{WITHIN}]$$

and if there is no difference between the means of all the 3 treatment groups the value of the F-statistic should be close to 1. As the differences between the sample means become larger, the numerator of the F-statistic also becomes larger (i.e. $MS_{BETWEEN}$ becomes larger) whereas the variation within the measurements SS_{WITHIN} remains unchanged.

Table 17.8. Arithmetic for the construction of the one-way ANOVA Table 17.9 which is for the response data of Tables 17.4–17.6. The formulae in Table 17.7 are used for the calculations.

$SS_{BETWEEN} = 4(4 + 9 + 1) = \mathbf{56}$

Looking at the 12 sets of numbers in Table 17.6 it is seen that the *second number* in each of the three $[j = 1, 2, 3]$ columns is 2 [for $j = 1$], 3 [for $j = 2$] and 1 [for $j = 3$] and corresponds to $(\bar{X}_j - \bar{X})$.

Thus $SS_{BETWEEN} = 4(2^2 + 3^2 + 1^2) = \mathbf{56}$

$SS_{WITHIN} = (1 + 4 + 1 + 4$
$\qquad\qquad +16 + 9 + 1 + 0$
$\qquad\qquad +1 + 16 + 9 + 0) = \mathbf{62}$

These numbers are also obtained from Table 17.6 and are the *third number* in each of the three $[j = 1, 2, 3]$ columns before they are squared.

Thus $SS_{WITHIN} = (1^2 + 2^2 + 1^2 + 2^2$
$\qquad\qquad +4^2 + 3^2 + 1^2 + 0^2$
$\qquad\qquad +1^2 + 4^2 + 3^2 + 0^2) = \mathbf{62}$

It is also noted that these above numbers which are squared also appear in Table 17.5.

$SS_{TOTAL} = 56 + 62 = \mathbf{118}$

and this can be checked using the formula in Table 17.7. SS_{TOTAL} is the sum of the squares of the combined *second and third numbers* in each of the three $[j = 1, 2, 3]$ columns

$= (-2 + 1)^2 + (-2 + 2)^2 + (-2 - 1)^2 + (-2 - 2)^2$
$\quad +(+3 + 4)^2 + (+3 - 3)^2 + (+3 - 1)^2 + (+3 + 0)^2$
$\quad +(-1 - 1)^2 + (-1 + 4)^2 + (-1 - 3)^2 + (-1 + 0)^2$
$= (1 + 0 + 9 + 16 + 49 + 0 + 4 + 9 + 4 + 9 + 4 + 9 + 16 + 1) = \mathbf{118}$

Thus for large F-values we reject the null hypothesis H_0 of equal means, whereas for values of the F-statistic close to 1 we do not reject H_0 and must conclude that we cannot show a difference in the treatments studied. The alternative hypothesis H_1 is that *at least one of the pairs of means is not equal*. However, it does **not** specify which pairs are not equal: **only** that one or more pairs are unequal.

For the example in Table 17.9, $F = 28.00/6.89 = 4.06$ and for $DF_1 = 2$ and $DF_2 = 9$ the critical value of the F-statistic from Tables 17.3(*a*, *b*) for $P = 0.05$ is $F = 4.26$. The derived F-statistic is 4.06 and this is less than 4.26. Thus the null hypothesis H_0 would not be rejected at the 0.05 level of

Table 17.9. One-way ANOVA table for the treatment response data in Tables 17.4–17.6. Mean square (MS) = SS/DF. DF_{TOTAL} = (No. of responses −1), $DF_{BETWEEN}$ = (No. of treatment groups −1) and DF_{WITHIN} = (DF_{TOTAL} − $DF_{BETWEEN}$) but note that $MS_{TOTAL} \neq MS_{BETWEEN} + MS_{WITHIN}$.

Source of variation	Sum of squares	Degrees of freedom (DF)	Mean square
Between groups	$SS_{BETWEEN} = 56$	2	$MS_{BETWEEN} = 28.00$
Within groups	$SS_{WITHIN} = 62$	9	$MS_{WITHIN} = 6.89$
Total	$SS_{TOTAL} = 118$	11	$MS_{TOTAL} = 10.73$

significance (just!).

The example just described has been worked out from first principles, but there is also a *quicker method* to calculate $SS_{BETWEEN}$ and SS_{TOTAL} and then from the equality

$$SS_{TOTAL} = SS_{BETWEEN} + SS_{WITHIN}$$

to calculate SS_{WITHIN} and then proceed as before to derive the F-statistic. The formulae used are given below where from Table 17.4 we have the following for W, Z and Y, where the total number of responses is $3 \times [n_i = 4] = N = 12$

$$W = \left\{ \sum_{\substack{j=1 \\ (ALL\ j)}}^{j=3} X_{ji} \right\}^2 / N = \{140 + 160 + 144\}^2 / 12 = 197136/12 = 16428$$

$$Z = \left\{ \sum_{\substack{j,i=1,1 \\ (ALL\ j,i)}}^{j,i=3,4} X_{ji}^2 \right\} = \{1296 + 1369 + 1156 + 1089$$
$$+ 1936 + 1369 + 1521 + 1600$$
$$+ 1225 + 1600 + 1089 + 1296\} = 16546$$

$$Y = \left\{ \sum_{\substack{j=1 \\ (ALL\ j)}}^{j=3} \left(\left[\sum X_{ji} \right]^2 / n_i \right) \right\} = \{140^2/4 + 160^2/4 + 144^2/4\}$$
$$= \{4900 + 6400 + 5184\} = 16484$$

$$SS_{TOTAL} = [Z - W] = \mathbf{118}$$

which agrees with the value we calculated from first principles,

$$SS_{BETWEEN} = [Y - W] = \mathbf{56}$$

which also agrees with what we calculated from first principles.

17.4   TWO-WAY ANOVA

Two-way (or two-factor) ANOVA has already been referred to in section 17.1 and Table 17.10 continues the example previously mentioned[1] and is the two-way extension of a one-way ANOVA table such as given in Table 17.9.

17.4.1   Example 1

Table 17.10. Two-way ANOVA table for a study of birth weight analysed by birth rank and maternal age[1,4]. In this instance there are two sources of variation between groups: birth rank and maternal age.

Source of variation	Sum of squares	Degrees of freedom (DF)	Mean square
Birth rank	$SS_{BIRTH\ RANK} = 56.81$	3	$MS_{BIRTH\ RANK} = 18.94$
Age	$SS_{AGE} = 7.43$	4	$MS_{AGE} = 1.86$
Birth rank × Age	$SS_{INT} = 22.72$	12	$MS_{INT} = 1.89$
Within groups	$SS_{WITHIN} = 6022.68$	4690	$MS_{WITHIN} = 1.28$

Table 17.10 gives the share of the total variation in birth weight which is attributable to differences in birth rank, and the share which is attributable to differences in maternal age, leaving a residual (within groups) which is not attributable to either factor but as mentioned earlier, is due to variations arising from the sampling technique. $SS_{BIRTH\ RANK}$ and SS_{AGE} are therefore in effect $SS_{BETWEEN1}$ and $SS_{BETWEEN2}$.

In addition to the independent direct effects of birth rank and maternal age, there may also be an *interaction* effect, which is why the terms SS_{INT} and MS_{INT} are required. The derived F-statistics are given below.

$$F = 18.94/1.28 = 14.80$$

This is significant, $P < 0.001$, showing that if H_0 were true, then such a difference in mean squares would occur due to chance in less than 1 in 1000 times. H_0 is rejected and it is therefore concluded that birth weights are not independent of birth rank.

$$F = 1.86/1.28 = 1.45 \textbf{ and } F = 1.89/1.28 = 1.48$$

This result is not significant since $P > 0.05$ and it is therefore concluded that birth weight is not independent of maternal age and of the joint effect of maternal age and birth rank. Hence if birth weights are classified by maternal age, we should not expect significant differences in mean birth weights between age groups, once the effects of birth rank have been taken into account[4].

The hypotheses which can be tested by two-way ANOVA techniques are the null hypothesis

H_0: There is no interaction between factor-A and factor-B

the alternative hypothesis being

H_1: There is an interaction present.

The null hypothesis

H_0: All of the factor-A means are equal

the alternative hypothesis being

H_2: At least one factor-A population mean does not equal another factor-A population mean.

The null hypothesis

H_0: All of the factor-B means are equal

the alternative hypothesis being

H_3: At least one factor-B population mean does not equal another factor-B population mean.

17.4.2 Example 2

Table 17.11. Hypothetical data for two-way ANOVA. Suppose that the measurements are subjective competence scores and that factor-A is country (countries A1 and A2) and that factor-B is education (high level B1 and low level B2).

		Country		
		A1	A2	Mean
Education	B1	5	7	
	B1	4	5	
Mean		{4.50}	{6.00}	5.25
	B2	4	6	
	B2	4	6	
	B2	3	6	
Mean		{3.67}	{6.00}	4.83
Mean		4.00	6.00	5.00

$$SS_{EDUCATION} = 4(5.25 - 5.00)^2 + 6(4.83 - 5.00)^2 = 0.42 \text{ and } DF_{EDUCATION} = 1$$

$$SS_{COUNTRY} = 5(4.00 - 5.00)^2 + 5(6.00 - 5.00)^2 = 10.00 \text{ and } DF_{COUNTRY} = 1$$

The sum of squares for the combined Country+Education variable is obtained using the means in the { } brackets as follows:

$$SS_{COUNTRY+EDUCATION} = 2(4.50 - 5.00)^2 + 2(6.00 - 5.00)^2$$
$$+ 3(3.667 - 5.00)^2 + 3(6.00 - 5.00)^2$$
$$= 0.50 + 2.00 + 5.33 + 3.00 = 10.83$$

The sum of squares for the interaction is obtained using the formula

$$SS_{INTERACTION} = SS_{COUNTRY+EDUCATION} - SS_{EDUCATION} - SS_{COUNTRY}$$

$$SS_{INTERACTION} = (10.83 - 0.42 - 10.00) = 0.41$$

and where r = no. of education types and c = no. of countries the associated degrees of freedom are $DF_{INTERACTION} = (rc - 1) - (r - 1) - (c - 1) = [r - 1][c - 1] = 1$.

SS_{TOTAL} is found by subtracting the grand mean (5.00) from each observation (there are 10), squaring these differences and adding them up:

$$SS_{TOTAL} = (0 + 1 + 1 + 1 + 4 + 4 + 0 + 1 + 1 + 1) = 14.00$$

$$SS_{WITHIN} = SS_{TOTAL} \; SS_{COUNTRY+EDUCATION} = (14.00 - 10.83) = 3.17$$

The two-way ANOVA table analogous to the one-way Table 17.9 is Table 17.12.

Table 17.12. Two-way ANOVA table for the data and calculations in Table 17.11.

Source of variation	Sum of squares (SS)	Degrees of freedom (DF)	Mean square (MS)
Country	10.00	1	10.00
Education	0.42	1	0.42
Interaction	0.41	1	0.41
Within groups	3.17	6	0.53
Total	14.00	9	

The derived F-statistics for education level ($F = 0.42/0.53 = 0.79$) and for the interaction between level of education and country ($F = 0.42/0.53 = 0.79$) are less than 1 and are not significant, whereas for the country factor $F = 10.00/0.53 = 18.87$ with from Table 17.3(a, b) $DF_1 = 1$ and $DF_2 = 6$, giving a critical value of the F-statistic of $F = 5.99$ for $P = 0.05$ and of $F = 13.74$ for $P = 0.01$. This is a significant result, $P < 0.01$, since $18.87 > 13.74$, and we conclude that the null hypothesis H_0 can be rejected and the competence scores are **not** independent of country.

17.5 RELATIONSHIP BETWEEN CORRELATION COEFFICIENT AND A ONE-WAY ANOVA TABLE

Using as an example[5] one-way ANOVA for birth weight and mother's height we have the information in Table 17.13.

Table 17.13. ANOVA table for measurements of birth weight (Y) as a function of height of mother (X): regression line is $Y = a + bX$.

Source of variation	Sum of squares (SS)	Degrees of freedom (DF)	Mean square (MS)
Between groups	$SS_{BETWEEN}(SS_{REGRESSION}) = 1.48$	1	$MS_{BETWEEN} = 1.4800$
Within groups	$SS_{WITHIN}(SS_{RESIDUAL}) = 20.39$	98	$MS_{WITHIN} = 0.2081$

If there were no real association between Y and X then $MS_{BETWEEN}$ and MS_{WITHIN} would be about the same size, and therefore the derived F-statistic should be about 1. Whereas if Y and X are associated then $MS_{BETWEEN}$ is larger than MS_{WITHIN} and F is greater than 1. For the data in Table 17.12 the derived F-statistic is $F = (1.4800/0.2081) = 7.11$. From Table 17.3($a$, b) the critical values of the F-statistic for $P = 0.01$ are $F = 7.08$ for $DF_{1,60}$ and $F = 6.85$ for $DF_{1,120}$. No values are tabulated for a $DF_{1,98}$ but we can say that the critical value of the F-statistic for $DF_{1,98}$ is between 7.08 and 6.85 and therefore $F = 7.11$ is significant, $P < 0.01$.

The correlation coefficient, r, was discussed in the previous chapter in section 16.4 and we can use the ANOVA table to calculate r in a different way since

$$r = SS_{BETWEEN}/SS_{TOTAL}$$

which for the data in Table 17.13 gives $r^2 = (1.48/21.87) = 0.0676$ and therefore $r = 0.26$. Such a value of r, which is much less than 1, will not indicate a good correlation, as can be seen from the scatter diagrams in Figures 16.2 and 16.3 for $r = 0.09$ and $r = 0.40$.

PLACE THE DECIMAL POINT CORRECTLY

Popeye's superhuman strength for deeds of derring-do comes from consuming a can of spinach. The discovery that spinach was as valuable a source of iron as red meat was made in the 1890s, and it proved a useful propaganda weapon for the meatless days of the second world war. A statue of Popeye in Crystal City, Texas, commemorates the fact that single-handedly he raised the consumption of spinach by 33%. America was *strong to finish 'cos they ate their spinach* and duly defeated the Hun. Unfortunately, the propaganda was fraudulent; German chemists reinvestigating the iron content of spinach had shown in the 1930s that the orginal workers had put the decimal point in the wrong place and made a tenfold overestimate of its value. Spinach is no better for you than cabbage, Brussels sprouts or broccoli. For a source of iron Popeye would have been better off chewing cans.

Chapter 18

Multivariate Analysis: The Cox Proportional Hazards Model

18.1 INTRODUCTION

A simple regression model assumes that an outcome or response variable (e.g. survival from cancer following treatment) can be explained mainly in terms of one explanatory or predictor variable. On the basis on this assumption it is accepted that the many other factors which may influence the response variable are individually only of minor importance and their collective effect is zero. This will rarely be a valid assumption for the various prognostic variables which may affect cancer survival and therefore a multiple regression model is neccessary.

Cox[1] developed such a model for life tables (see also the paper by Breslow[2]) which enables an assessment to be made of the extent to which prognostic variables are associated with survival. He illustrated the use of the model for leukaemia but it has also been used for other forms of survival data such as those for cardiac studies, including studies of heart transplant survivals. Another cardiac example is the analysis by the Norwegian Multicentre Study Group[3] who applied the model to study the reduction in mortality and reinfarction in patients surviving acute myocardial infarction.

Multivariate analysis gives a series of coefficients or hazard ratios, each one of which indicates the magnitude of the effect on outcome of that particular variable, if all the other variables are assumed to be held constant.

The Cox model is now very frequently used in radiation oncology and, for example, from 1996 onwards there is seldom an issue of the *International Journal of Radiation Oncology Biology & Physics* which includes any discussion of survival and prognostic factors which does not use the Cox model.

It must, however, be emphasised that the Cox model is not the only model for multivariate analysis, although it is the commonest in radiation oncology, and several alternative models exist such as the *log-logistic* model[4,5] and the *linear logistic* model[6]. The latter has been used for analyses of case-control studies[7].

The statistical formulation of such models, including that of Cox, is extremely complicated and not totally appropriate for an introductory text. Thus much of what is included in this chapter should be regarded as advanced, not introductory material, and be considered as optional reading for those commencing a study of medical statistics

Also, the use of the Cox model will always require a computer programme, and will never be used manually with a pocket calculator. The computations are far too extensive for this to take place. However, care must be taken in reading computer software user manuals for instructions on data entry.

Finally, it is noted that a regression model provides a means of utilising the available information, but when the number of possible prognostic covariates is large, the number of iterations before convergence is reached may be high or the method may fail because the covariates are closely related to one another. It is therefore advisable to undertake an exploratory univariate analysis data analysis (see Chapter 17) to avoid including too many prognostic factors.

18.2 HAZARD FUNCTION, CUMULATIVE DISTRIBUTION FUNCTION AND SURVIVOR FUNCTION

The Cox regression model is often referred to as a *proportional hazards model* and since the term *hazard* is relatively unknown in the medical literature it is now discussed briefly before explaining the assumptions underlying the proportional hazards model.

The *hazard function* is also known as the *force of mortality* or the instantaneous death rate or the age-specific (or time-specific) failure rate. This latter term relates to renewal theory[8] when failure time data in industry, such as failure times of generator field windings, are studied. Expressed in terms of probability functions, the hazard function, denoted by $\lambda(t)$, is the ratio of the probability density function $f(t)$ to the survivor function $(1 - F(t))$ at time t, where $F(t)$ is the distribution function.

$$\lambda(t) = f(t)/(1 - F(t))$$

The *distribution function* or *cumulative distribution function*, $F(t)$, increases with t and equals 1 at the maximum value of t. It is the probability that the time variable takes a value less than or equal to t. The *survivor function*, $1 - F(t)$, is the probability that the time variable takes a value greater than t. The *probability density function*, $f(t)$, is a function whose integral over the range t_1 to t_2, for a continuous variable t, is equal to the probability that the time variable T takes a value in that range.

Figure 18.1(a,b,c) illustrates $f(t)$, $F(t)$ and $1 - F(t)$ schematically for the exponential distribution. It is seen that for this particular distribution the hazard function is constant, equal to λ. An exponential model for the survival experience of patients treated for cancer therefore would imply an instantaneous death rate which is constant irrespective of elapsed time since diagnosis.

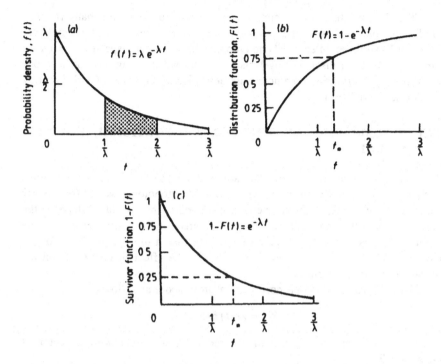

Figure 18.1. Since the **hazard function** $\lambda(t)$ is the ratio of the probability density function to the survivor function, $\lambda(t) = \lambda$ for the exponential distribution. There is a one-to-one correspondence between hazard function and the probability density function. Thus the hazard function is not a constant for any other distribution. For example, for the two parameter (c, b) Weibull distribution which has a positively skewed density function $f(t)$ equal to $(ct^{c-1}/b^c) \exp{-(t/b)^c}$ and a survivor function equal to $\exp{-(t/b)^c}$, the hazard function is ct^{c-1}/b^c which decreases with t if $c < 1$ and increases with t if $c > 1$.

(a) **Probability density function.** Probability that t is in the range $1/\lambda$ to $2/\lambda$ equals the area beneath the curve between these limits (the shaded area). Expressed as an integral this area is $\int_{t_1}^{t_2} f(t)\mathrm{d}t$ where $t_1 = 1/\lambda$ and $t_2 = 2/\lambda$. Total probability equals unity and thus the total area beneath the curve is $\int_0^\infty f(t)\mathrm{d}t = 1$.

(b) **Distribution function.** The probability density function $F(t)$ is the first derivative of the distribution function,

$$f(t) = \frac{\mathrm{d}}{\mathrm{d}t}[F(t)] \quad \text{and} \quad \frac{\mathrm{d}}{\mathrm{d}t}[1 - \mathrm{e}^{-\lambda t}] = \lambda \mathrm{e}^{-\lambda t}$$

The probability that t lies between 0 and t_* is 0.75.

(c) **Survivor function.** The probability that t exceeds t_* is 0.25. If the survivor function is expressed as a percentage survival rate, the above survival curve indicates a 25% t_*-year survival rate if t is in yearly units.

The density, distribution, survivor and hazard functions for many families of distributions including the binomial, lognormal, Poisson and Weibull are given by Hastings and Peacock[9] and graphical methods for inspecting the hazard function have been discussed by Nelson[10]. Survival distributions and methods of estimating distribution parameters are described by Gross and Clark[11] and by Kalbfleisch and Prentice[4].

18.3 ASSUMPTIONS OF THE COX PROPORTIONAL HAZARDS MODEL

The underlying assumption is that the ratio of the risks of dying in two subgroups is constant over time (*proportional hazards*). The monotonic weight (or the *risk ratio*) associated with one parameter is assumed to be constant whatever the values of the other parameters, except if interactions are introduced into the model. (This is a common problem for all regression models, not only for the Cox model.) Another basic assumption is that the effect of covariates on the hazard function is log-linear.

Cox[1,12] proposed the proportional hazards model as follows:

$$\lambda(t, Z) = \lambda_0(t) \exp(Z \cdot \beta)$$

where $\lambda(t, Z)$ denotes the hazard at time t for a patient with a vector of covariates Z.

That is,

$$\lambda(t, Z) = \lambda_0(t) \exp(\beta_1 Z_1 + \beta_2 Z_2 + \ldots + \beta_p Z_p)$$

where β is a $p \times 1$ vector of unknown parameters (*weights*) and $\lambda_0(t)$ is the unknown hazard function for a reference patient with covariates Z identically equal to zero. It is the underlying form of the hazard, and is not specified in terms of a distribution such as the Weibull, and is determined by the form of the data under study.

For a given Z, $\exp(\beta_1 Z_1 + \beta_2 Z_2 + \ldots + \beta_p Z_p)$ is a constant so that the hazard function $\lambda(t, Z)$ is a constant multiple of the underlying hazard $\lambda_0(t)$. Consequently, Cox introduced the term *proportional hazards model*.

18.4 HAZARD EQUATION FOR THREE PROGNOSTIC FACTORS

To illustrate this hypothetically, suppose that there are three prognostic factors, $Z_1 =$ (patient's age-50) in years, $Z_2 =$ sex and $Z_3 =$ disease stage at presentation. Z_2 may take a value 1(= male) or 0(= female) and Z_3 entails equal spaced ordering of stages. Then $\lambda_0(t)$ gives the hazard for a reference patient who is aged 50 years ($Z_1 = 0$), female ($Z_2 = 0$) and stage code 0 ($Z_3 = 0$). $\lambda_0(t)$ describes the *force of mortality* for a patient all of whose covariates are zero.

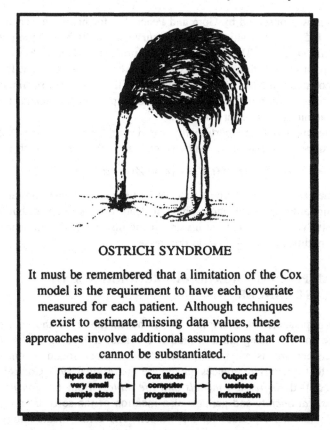

OSTRICH SYNDROME

It must be remembered that a limitation of the Cox model is the requirement to have each covariate measured for each patient. Although techniques exist to estimate missing data values, these approaches involve additional assumptions that often cannot be substantiated.

Input data for very small sample sizes	→	Cox Model computer programme	→	Output of useless information

If, however, the patient had been 55 years old instead of 50, male instead of female, and stage code 1 instead of 0; and if the weights β given to the individual covariates had been estimated as $\beta_1 = 0.1$, $\beta_2 = 0.3$ and $\beta_3 = 0.3$, the hazard would be

$$\lambda_0(t)\exp(0.1 \times 5 + 0.3 \times 1 + 0.3 \times 1) = \lambda_0(t)\exp(1.1) = 3.0\,\lambda_0(t)$$

That is, the corresponding hazard would always be 3.0 times greater than that for the reference patient, and this ratio of risks would remain constant with time.

18.5 RELAXATION OF THE ASSUMPTION OF PROPORTIONALITY OF HAZARDS

The prognosis for leukaemia (used by Cox[1] as an illustration) and, for example, inoperable stage III breast cancer[13] are not particularly good but the assumption of the proportional hazards model that the ratio of hazards is constant with time is acceptable. However, in cancer series with a significant number of long-term survivors it may not be reasonable to assume that the hazard ratio is independent of time. This was observed by Gore and Peacock[14] for breast cancer data from Edinburgh, 1954–64, for which long-term survival results have been published by Langlands *et al* [15]. They found that the proportional hazards model was unsuitable for use with this data and Gore has also commented that for breast cancer, a covariate such as stage which is initially of prognostic importance, fades into relative unimportance at later follow-up, with the annual mortality rate independent of stage after 10 years.

The assumption of proportionality can be relaxed in various ways, for example prognostic variables Z may also change with time in which case

$$\lambda(t; Z(t)) = \lambda_0 \exp(Z(t) \cdot \beta')$$

The multiple $\exp(Z(t) \cdot \beta')$ being a function of time is no longer constant and so the proportionality factor changes. This will lead to a less elegant, but perhaps more realistic view of the natural history of the disease as the influence of the covariates modifies as the follow-up progresses.

18.6   EXAMPLE: LOCOREGIONAL RECURRENCE OF BREAST CANCER

This section presents an example[16] from the literature of the results of a multivariate analysis using the Cox model. It relates to locoregional recurrence of breast cancer and is a typical presentation of those found in oncological journals in that the only references quoted are the original 1972 Cox reference[1] (which is entitled *Regression models and life tables*) from the *Journal of the Royal Statistical Society* and perhaps also mention of the software programme used to calculate the P values for this model. The actual methodology of the Cox model is seldom, if ever, discussed in oncological journals and one has to look elsewhere to rather esoteric statistical papers.

In this example, as with all such multivariate analyses, a univariate analysis is first performed and then the data relating to those prognostic factors found to be statistically significant (at a chosen probability level P. which does not necessarily have to be $P = 0.05$) are entered into the multivariate analysis programme.   Exceptions do exist, however, when even if there is a non-significant univariate result the data for the factor are still entered into the multivariate analysis: but this is not standard procedure.

Many examples of the use of the Cox model are to be found in the literature, but studies of prognostic factors for the subsequent outcome after treatment of locoregional recurrence of breast cancer are of major interest because of the high incidence worldwide of breast cancer. Also, because the outcome of patients with locoregional recurrence after mastectomy has often been described as fatal, the identification of even a small subgroup of patients with a favourable prognosis is very important.

Table 18.1 from Willner *et al*[16] clearly shows that some prognostic factors which are considered significant when using only univariate analysis are not

Example 219

significant when using multivariate analysis: this is a typical situation and clearly shows when multiple prognostic factors are being studied, a series of univariate analyses is not adequate.

If significance which was apparent in univariate analysis disappears in the multivariate stage it means (1) it should be considered non-significant and (2) this is probably because it is a surrogate for (i.e. highly correlated with) one of the other variables.

It can also be true that a factor which is not significant in univariate analysis can be significant in multivariate analysis, such as age at RD (< / > 50 years) in Table 18.1, although most studies would have discarded those prognostic factors not found to be significant at the univariate stage, This situation is not ideal, and it could be proposed that factors associated with a $P < 0.20$ in the univariate stage be entered into the multivariate model.

Two major papers previously published on locoregional recurrent of breast cancer, before the 1997 analysis of Willner et al[16] of the University of Wurzburg patient data, are those of Halverson et al[17] in 1992 from the Mallinckrody Institute of Radiology, St. Louis, and of Schwaibold et al[18] in 1991 from the Fox Chase Cancer Center, Philadelphia. All give examples of multivariate analysis but that of Willner is the most exhaustive.

It is noted that all authors: Willner et al, Halverson et al and Schwaibold et al, although finding several factors significant in univariate analysis, excluded them from multivariate analysis owing to the small number of patients with all data available for multivariate analysis (the *Ostrich syndrome* of page 217). This is a very important point to note when undertaking multivariate analysis: exclude poor quality data.

Finally, I end this section with the following quotation from Willner et al[16], which places in perspective the justification for the use of multivariate analyses in oncology.

'The complex inter-relationships of prognostic factors underlines the importance of multivariate survival analysis of all available prognostic factors and treatments'.

It is stressed that multivariate analysis should, with care (because of any situation with small numbers), take precedence over any of the univariate analyses. However, differences between univariate and multivariate analyses do give some information about correlations (*confounding*) between different variables and also sometimes unexpected correlations.

Table 18.1. Results of Willner *et al*[16] following univariate and multivariate analyses for prognostic factors influencing post-recurrence survival rate. NS = not significant; pT = primary tumour status; PD = primary diagnosis; RD = recurrence diagnosis and * indicates that data for these factors were excluded from the multivariate analysis because of a large number of missing values. The conclusions of Willner *et al* were 'A highly favourable subgroup is those with a solitary chest wall or axillary recurrent nodule (in a patient aged > 50 years) with a disease-free interval ≥ 1 year, with a pT1-2pN0 primary tumour, without tumour necrosis, and whose recurrence is locally controlled'.

Prognostic factors	Univariate analysis	Multivariate analysis
pT status (T1,2/3,4)	$P < 0.001$	$P < 0.01$
Grading (G 1,2/3,4)	$P < 0.01$	*
Lymphatic vessel invasion	$P < 0.001$	NS
Blood vessel invasion	$P < 0.01$	NS
Tumor necrosis	$P < 0.001$	$P < 0.01$
Hormonal receptor status	$P < 0.01$	*
Age at PD (< / > 50 years)	NS	NS
Axillary node status at PD	$P < 0.001$	$P < 0.05$
Postmastectomy irradiation	NS	NS
Postmastectomy chemotherapy	$P < 0.01$	NS
Postmastectomy hormonal therapy	$P < 0.05$	NS
Site of recurrence (chest, axilla, supra, combined)	$P < 0.01$	$P < 0.001$
Time to recurrence (< / > 1 year)	$P < 0.001$	$P < 0.01$
Scar recurrence	$P < 0.001$	NS
Size of largest recurr. nodes	$P < 0.05$	NS
No. of recurr. nodes	$P < 0.001$	$P < 0.001$
Age at RD (< / > 50 years)	NS	$P < 0.05$
Surgical removal of recurrence	$P < 0.001$	NS
Irradiation of recurrence	NS	NS
Irradiation dose (< / > 50 Gy)	NS	NS
Target volume of irradiation (small/total site/total locoregional)	$P < 0.05$	$P = 0.05$
Irradiation planning procedure	NS	NS
Hormonal therapy for recurrence	NS	NS
Chemotherapy for recurrence	$P < 0.01$	NS
Local control in recurr. site	$P < 0.001$	$P < 0.05$

18.7 METHODOLOGY

The paper by Christensent[19] in the journal *Hepatology* has to date one of the fullest descriptions of Cox model methodology and for this reason is recommended reading. The Christensen patient group is theoretically constructed using three variables and totals 30 with:

death (categorised as 1) the endpoint in 18/30

and with 12/30 censored (categorised as 0).

Serum albumin (gm/litre) is on a continuous scale of measurement (21–36),

as is bilirubin (micromoles/litre) also on a continuous scale of

measurement (19–332): and is entered into the model as $\log_{10}$bilirubin.

Alcoholism is categorised as either present (1) or absent (0).

18.7.1 Relationship Between Survival Probability and Hazard

Figure 18.2 shows a Kaplan–Meier survival graph for what Christensen terms the *cumulative survival probability* (the adjective cumulative is not standard terminology for life table survival graphs) and it is seen how this survival curve is related to the *cumulative hazard*.

18.7.2 Likelihood Ratio Test

To fit the Cox model to Christensen's data, which has three variables (i.e. covariates) albumin, $\log_{10}$bilirubin and alcoholism, which in the equation for the hazard in section 18.3, will be Z_1, Z_2 and Z_3, a total of seven Cox analyses can be studied, Table 18.2. There will be three models involving only one covariate, three including two covariates and one including all three covariates.

Estimation and significance testing of a given model involves the concept of *likelihood*: which means the probability of the observed data being *explained* by a certain model. The overall significance of a model is based on the ratio between the likelihood $L(0)$ of a model in which the covariates show no covariation with the survival time (the regression coefficients b (termed β in the equation in section 18.3) all being zero: in this example $b_1 = b_2 = b_3 = 0$ because there are only three covariates) and the likelihood $L(b)$ of the model with the b coefficient(s). The b coefficients are estimated in such a way that $L(b)$ is maximised. Thus the estimated parameters (such as the b coefficients) of a Cox model are called *maximum likelihood estimates*. The underlying hazard is not estimated.

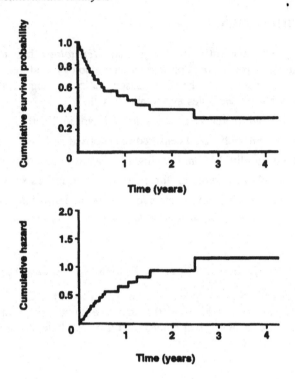

Figure 18.2. If the survival probability as a function of time (t) is denoted by $S(t)$ and the estimated cumulative hazard by Λ the one can be estimated from the other by using the relationship $\Lambda(t) = -\log_e S(t)$ and $S(t) = e^{-\Lambda(t)}$ The survivor function in Figure 18.1, $(1 - F(t))$, is the same as $S(t)$ in this figure. One of the problems in understanding the Cox model is the multiplicity of symbols, not all of which are consistent in the different publications. However, for this section I have retained the Christensen[19] notation.

The greater the value of $L(b)$ or the less the value of the likelihood ratio $L(0)/L(b)$, the better the model actually *explains* or fits the observed data[4]. The significance of each model can be tested statistically using the relationship

$$\chi^2 = -2 \times \log_e(L(0)/L(b))$$
$$= -2 \times (\log_e L(0) - \log_e L(b))$$
$$= 2 \times (\log_e L(b) - \log_e L(0))$$

where the degrees of freedom equals the number of coefficients estimated in the model[4]. Christensen's results[19] of seven Cox regression analyses are given in Table 18.2 which is typical of the spectrum of data obtained in computer printouts for the Cox model, although it will not necessarily be in the same layout as in Table 18.2.

18.7.3 Table of Results

Table 18.2. Results of seven Cox regression analyses, after Christensen[19].
χ^2 values are computed from the formulae given above. This should really be called a χ^2 *model* but for simplicity the word *model* has been omitted.

$\log_e L(0)$ is the same for all the models, i.e. -52.319 **DF** = degrees of freedom.
$L(b)$ and hence χ^2 depend on the variable(s) included. For example for model 1, $\log_e L(b)$ is -36.825 and thus $\chi^2 = 2(-36.825 - (-52.319)) = 30.99$ as seen in the second column of the table.

P is the level of statistical significance for χ^2 and P^{***} the level of statistical significance of the regression coefficient b.

Variables are: ALB for albumin, LGB for $\log_{10}$bilirubin and ALC for alcoholism.

b is the regression coefficient. $SE(b)$ is the standard error of b.

Z is the normal deviate which equals $b/SE(b)$. The significance of each regression coefficient (b) can be estimated by comparing Z^2 with the χ^2 squared distribution with DF= 1. This is sometimes termed the *Wald test*[19]. If $Z > 1.96$, (see Chapter 3 for discussion of the normal distribution and the standard normal deviate) then b is significantly different from zero at the $P = 0.05$ level of significance: two-tailed test. However, if $Z < 1.96$ for certain coefficients, b, this should not be taken to mean that these variables have no effect on prognosis, but only that the effect is too small to be shown up at the $P = 0.05$ level of significance with the number of patients in the study.

The relative importance of the variables is given by the numerical value of Z. The greater the value of Z the more significant it is in the model. It is seen in the table for model 7, for example, that the variables decrease in the following order: ALB, LGB, ALC with ALC being insignificant.

Model number	χ^2	DF	P	Variables included	b	$SE(b)$	Z	P^{***}
1	30.99	1	< 0.0001	ALB	−0.42	0.089	−4.71	< 0.0001
2	21.24	1	< 0.0001	LGB	4.44	1.06	4.17	< 0.0001
3	8.79	1	0.003	ALC	1.55	0.55	2.82	0.005
4	35.89	2	< 0.0001	ALB	−0.35	0.10	−3.43	0.0006
				LGB	2.36	1.11	2.12	0.03
5	32.50	2	< 0.0001	ALB	−0.39	0.094	−4.16	< 0.0001
				ALC	0.79	0.64	1.23	0.22
6	25.13	2	< 0.0001	LGB	3.88	1.06	3.66	0.0002
				ALC	1.14	0.59	1.93	0.056
7	37.04	3	< 0.0001	ALB	−0.32	0.11	−3.07	0.002
				LGB	2.25	1.11	2.03	0.04
				ALC	0.71	0.66	1.08	0.28

18.7.4   Relative Risk: Ratio Between Hazards

Using the values of the regression coefficients, b, it is possible to estimate relative risk (defined in section 22.2 for cancer with reference to irradiated populations such as the atomic bomb survivors in Hiroshima and Nagasaki, as the *number of cancer cases in the irradiated population to the number of cases expected in the unexposed population*. A relative risk of 1.1 indicates a 10% increase in cancer due to radiation, compared with the normal incidence of the baseline/reference group).

In the Cox model analyses the relative risks are the ratios between hazards attributable to various levels of a particular variable, when all other variables are unchanged. In Table 18.2 considering model 3 and the variable ALC, this model predicts the relative risk of ALC to non-ALC by the equation

$$RR = e^{1.55 \times 1}/e^{1.55 \times 0} = 4.7/1 = 4.7$$

since for this single variable (ALC only) model 3, $b = 1.55$.

In model 4, considering ALB, a 1 gm lower concentration of serum albumin (ALB), e.g. 29 gm/litre compared to 30 gm/litre, is associated with an increase in relative risk of

$$RR = e^{-0.35 \times 29}/e^{-0.35 \times (-1)} = 1.42$$

since $b = -0.35$ in Table 18.2. The relative risk associated with a 1 gm/litre lowering in ALB is independent of the *absolute* level of ALB.

An example of relative risks for actual clinical data (as distinct from Christensen's constructed data[19] is given in Figure 18.3 for the Oslo and Stockholm post-mastectomy megavoltage radiotherapy trials[20] where the results were analyses using the Cox model.

One of the conclusions was that the effect of the radiation was significantly related to the size of the primary tumour. The test for a trend indicated that the relative risk of distant metastasis was significantly lower for irradiated patients with small than with large tumours: $\chi^2 = 6.74$, $P < 0.01$. A similar trend was observed for the relative risk of death: $\chi^2 = 5.66$, $P < 0.02$.

A second clinical example[21], in this instance for osteogenic sarcoma, is given in Table 18.3 where three factors were found to have individual significant prognostic value for improved survival, namely the accomplishment of complete metastasectomy, the presence of a solitary metastasis, and the administration of adequate salvage chemotherapy. The relative risk of death in the absence of these factors is respectively 5.2, 3.5 and 2.2.

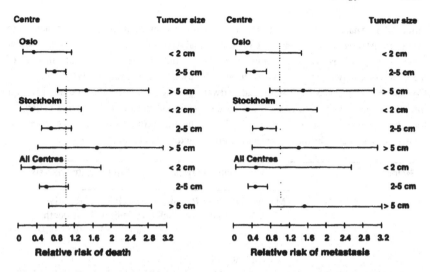

Figure 18.3. (Left) Relative risk of death for irradiated node-positive patients *versus* non-irradiated node-positive patients by tumour size and by treatment centre. (Right) Relative risk of distant metastasis for irradiated node-positive patients *versus* non-irradiated node-positive patients by tumour size and treatment centre[20].

18.7.5 Stepwise Selection of Variables

Stepwise selection of variables can be either by forward or backward selection[19]. With the *forward stepwise method* the model is built up stepwise by including at each step the variable giving the largest reduction in the likelihood ratio or equivalently the largest increase in the χ^2 model (see legend to Table 18.2) Thus in the first step, ALB (see model 1) would be included because this variable gives the highest significant χ^2 of all possible models with one variable (i.e. models 1–3).

In the next step, LGB would be added (model 4) because this variable increases χ^2 significantly

$$(35.89 - 30.99 = 4.90 \text{ with DF } = 1 \text{ and thus } P < 0.05)$$

in contrast to ALC (model 5) which only gives an insignificant increase in χ^2

$$(32.50 - 30.99 = 1.51 \text{ with DF } = 1 \text{ and thus } P > 0.20).$$

The values of DF are calculated as the difference between the number of estimated coefficients in the models being compared. Inclusion of ALC in a model comprising ALB and LGB (model 7) does not lead to a significant increase in χ^2

$$(37.05 - 35.89 \geqslant 1.15 \text{ with DF } = 1 \text{ and thus } P > 0.20).$$

Table 18.3. Analysis of potential prognostic factors for survival from the first metastatic event of osteogenic sarcoma[21]. Univariate analysis was made using Kaplan–Meier survival calculations and the logrank test to determine the level of significance P of the difference between two survival curves. Stepwise multivariate analysis was by the Cox model using the BMDP statistical software package. The significance level, P, for the multivariate analysis was obtained by the likelihood ratio test (sometime called the *log likelihood* ratio test, see section 18.7.2). The final column in the table gives the relative risk of death in the *absence* of a given factor. CI=confidence interval.

Factor	Univariate analysis			Multivariate analysis	
	Median survival time (months)		P value (Logrank)	P value (Likelihood ratio test)	Relative risk of death (95% CI)
	Factor present	Factor absent			
Complete metastasectomy	40	10	< 0.001	< 0.001	5.2 (2.4–11.6)
Solitary metastasis	150	12	0.001	0.005	3.5 (1.5–8.5)
Adequate salvage chemo.	24	10	0.026	0.022	2.2 (1.1–4.2)
Primary site other than femur & humerus	35	12	0.073	0.097	
Relapse free interval > 21 months	25	12	0.030	0.152	
Primary protocol T10 or SSG8	22	12	0.150	0.269	
Responder to adequate salvage chemotherapy	68	12	0.024	0.462	

Therefore model 4 would be the final model if the forward stepwise selection method was used.

When using the *backward stepwise method* one starts with a model which includes all the variables, and then non-significant variables are removed stepwise from the model by excluding the most non-significant variable at each step until each remaining variable contributes significantly to the model.

Thus one would start with model 7 and then remove ALC because this variable is non-significant. This would lead to model 4 which would be the fmal model because both variables, ALB and LGB are statistically significant.

When the analyses are not too complicated, such as the Christensen[19] data for 30 patients and three covariates, forward selection and backward selection lead to the same final model. However, with more complex analyses which include many variables, the two different methods of selection may lead to slightly different models.

18.8   EXAMPLES: BREAST CANCER

An interesting use of the Cox model has been with the Nottingham Breast Cancer Study[22] in which a prognostic index was obtained:

$$\text{Index } I = (0.17 \times \text{ Tumour Size})$$
$$+ (0.76 \times \text{ Lymph Node Stage})$$
$$+ (0.82 \times \text{ Tumour Grade})$$

A group of 387 patients were available for analysis and values of the regression coefficients b and of the normal variate Z are listed in Table 18.4 for each possible prognostic factor.

Summarising previous commentary, the Cox model allows each variable to be evaluated independently, taking into account the effects of all other variables and the regression coefficients, b, show how much each factor contributes to the hazard which is inversely related to survival. A positive value of b therefore indicates a poorer survival time as the given variable increases.

Table 18.4.   Values of b and Z from the Cox analysis[22]. Notation for statistical significance is * for $P < 0.01$ and *** for $P < 0.001$. The coding for these significant variables is as follows. Size (in cm): Lymph node stage: (A= 1: tumour absent from all 3 nodes sampled, B= 2: tumour in low axillary node only, C= 3: tumour in apical and/or internal mammary gland). Tumour grade: I, II and III.

Variable	b	Z
Age	−0.0162	1.02
Menopausal state	0.524	1.50
Size	0.172	2.92*
Lymph node stage	0.763	5.29***
Tumour grade	0.822	4.56***
Cell reaction	0.091	0.62
Sinus histiocytosis	−0.204	1.26
Oestrogen receptor content	−0.340	1.72
Adjuvant therapy	−0.332	0.83

The larger the value of index I the worse the prognosis for the patient and Figure 18.4 shows survival curves[22] according to lymph node biopsy stage and according to index I.

This index $I = (0.2 \times \text{ Size} + \text{Stage} + \text{Grade})$ has also been used for more extensive data[23] with a 15-year survival and clearly separated patient groups prognoses using the recommended 3.4 and 5.4 values of I. The 15-year survival rates for the three groups were 80%, 42% and 13% for a population of 1629 cases of operable breast cancer.

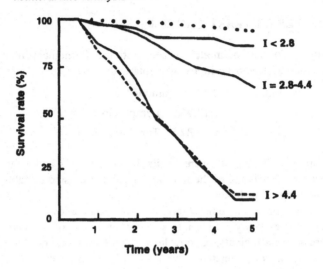

Figure 18.4. Survival curves of primary breast cancer patients according to lymph node stage (A, B, C) and index I (> 2.8 (64 cases), 2.8–4.4 (169 cases) and > 4.4 (65 cases)) The solid lines are those for the values of I. The top dotted line is for the survival of an age-matched population free of breast cancer and the lower dashed line is for 25 patients identified as having a poor prognosis by a previous index. These authors[22] also proposed for simpler practical use rounding the coefficients so that $I = [0.2 \times$ Size + Stage + Grade]. They noted that the results in this figure were reproduced almost exactly if the index values I are changed to 3.4 and 5.4 from 2.8 and 4.4.

Univariate analysis of clinical and histological data is undertaken using a χ^2 test using contingency tables whereas testing for a significant difference in survival between two patient groups is undertaken using the logrank test. What is not done is to use a χ^2 test and a 2×2 contingency table for, say five-year survivals, where the classes are Survival > 5 years and Survival < 5 years for the two patient groups.

This would use only a part of the available information, discarding much which can be useful. The logrank test involves a life table calculation: for a single survival curve a Kaplan–Meier life table calculation would be used whereas with the logrank test one is in effect comparing two Kaplan–Meier curves.

As an example, in a French multicentre study of the prognostic value of steroid receptors after long-term follow-up of 2257 operable breast cancers[24], population characteristics such as hormonal status (pre-menopausal and post-menopausal) and TNM stage were studied in univariate analysis using the χ^2 test.

Kaplan–Meier survival rates were determined for oestrogen receptors (ER) and progesterone receptors (PR) with the ER and PR characteristics classified as negative, positive or not determined. *Univariate analysis* was performed for disease-free intervals and overall survivals. At five years the overall survival was 90.1% for ER+ tumours, 91.5% for PR+ tumours and 91.4% when both receptors were positive. All differences were significant, $P < 0.01$ between positive and negative results.

For *multivariate analysis* using the Cox model in this study[24] the following covariates were used: age (with a cut-off chosen to be 35 years), menopausal status, clinical stage, tumour size, histological type and grade of tumour, and steroid receptors. It was found that PR status had no statistical significance regarding either the disease-free interval or mortality risk through breast cancer. Also no significant differences were found in the relative risks associated with either ER or PR status among premenopausal versus post-menopausal patients. Some of the results are given in Table 18.5 as an example of relative risks obtained using Cox modelling.

DON'T JUMP TO CONCLUSIONS

From a 1996 issue of the *International Journal of Radiation Oncolgy Biology & Physics* (**34** 745–747, Dr M.S. Anscher writing an Editorial on adjuvant theory for stage C prostate cancer and PSA) the following experiment was recorded of 'the scientist who obtained a grant to study how much each leg contributed to a frog's ability to jump. After telling his frog to jump he noted that a frog with 4 legs jumped 4 feet, one with 3 legs jumped 3 feet, one with 2 legs jumped 2 feet and a frog with 1 leg jumped 1 foot. Finally, he removed the frog's last leg. No matter how loudly he yelled at the frog it did not move. From these data, the scientist concluded that a frog with no legs cannot hear.'

Table 18.5. Study of oestrogen and progesterone receptors and breast cancer prognosis, in which multivariate analysis with the Cox model was used[24]. ESN: entry step number into the Cox model. The relative risks (RR) are referred to the best prognosis group (RR = 1). 95% confidence intervals (CI) for the RRs are given in brackets. ***: not significant with the upper group. NS: not significant. SBR: Scarff, Bloom and Richardson method of grading. The main conclusion of this study was that by multivariate analysis, steroid receptor status has a relative limited predictive value when compared with the well established prognostic criteria of early breast cancer.

Variable	ESN	RR (& 95% CI) Disease-free survival	ESN	RR (& 95% CI) Metastasis-free survival	ESN	RR (& 95% CI) Overall survival
Histology grade						
SBR						
I	1	—	1	—	1	—
II		2.09 (1.61–2.72)		2.32 (1.69–3.18)		2.45 (1.71–3.50)
III		2.50 (1.57–3.98)***		2.86 (1.67–4.88)***		2.80 (1.54–5.07)***
Axillary nodes						
pN						
None	2	—	2	—	2	—
1–3		1.56 (1.28–1.89)		1.85 (1.50–2.31)		1.65 (1.30–2.08)
> 3		2.29 (1.36–3.86)		2.99 (1.77–5.17)		3.23 (1.82–5.69)
Tumour size						
< 20 mm	3	—	3	—	3	—
20–50 mm		1.46 (1.22–1.74)		1.56 (1.27–1.97)		1.37 (1.06–1 76)
> 50 mm		2.05 (1.31–3.22)		2.06 (1.26–3.38)***		1.90 (1.07–3.34)

Table 18.5. (continued)

Variable	ESN	RR (& 95% CI) Disease-free survival	ESN	RR (& 95% CI) Metastasis-free survival	ESN	RR (& 95% CI) Overall survival
Age (years)						
> 35	4	1	4	1		NS
< 35		2.78 (1.91–4.03)		2.78 (1.85–4.17)		
ER status						
+ve	5	1	5	1	4	1
−ve		1.35 (1.13–1.61)		1.45 (1.19–1.75)		1.85 (1.55–1.95)
TNM stage		NS		NS	5	
I						1
II						1.38 (1.05–1.82)
III						1.49 (1.01–2.70)***
Hormonal status		NS		NS	6	
Pre-menopausal						1
Postmenopausal.						1.27 (1.02–1.58)

Chapter 19

Sensitivity and Specificity

19.1 DEFINITIONS

In *epidemiology* (which can be defined as the study of the distribution and determinants of health related states of events in defined populations), the terms *sensitivity* and *specificity* and the related *positive predictive value* and *negative predictive value* are used in the assessment of the predictive values of a set of symptoms or procedures as they relate to a certain disease.

These terms are defined below, as they would relate to a screening test in a population of individuals some of whom have the disease and some of whom do not, Table 19.1.

Table 19.1. Notation: a = true positives; b = false positives; c = false negatives; d = true negatives.

Test result	Disease status		Total
	Present	Absent	
Positive	a	b	$(a + b)$
Negative	c	d	$(c + d)$
Total	$(a + c)$	$(b + d)$	

Sensitivity is the probability of a positive test in people with the disease. Using Table 19.1, Sensitivity $= a/(a + c)$.

Specificity is the probability of a negative test in people without the disease. Using Table 19.1, Specificity $= d/(b+d)$. Thus $(1-Specificity)$ is the probability of a positive test in people without the disease: this is the value of X for a *receiver operating characteristic (ROC) curve* which graphs Y = Sensitivity *versus* $X = (1-$ Specificity$)$.

232

Positive predictive value (PPV) is the probability of the person having the disease when the test is positive. Using Table 19.1, PPV $= a/(a + b)$.

Negative predictive value (NPV) is the probability of the person not having the disease when the test is negative. Using Table 19.1, NPV $= d/(c + d)$.

19.2 PROBLEMS OF DEFINING NORMALITY AND ABNORMALITY

The meaning of the words normality and abnormality in the English language preceded by many years the Gaussian or normal distribution which was only discovered in 1733, see Chapter 3. *Statistical normality* and *common usage normality* do not always have the same meaning (just as statistical significance and clinical significance are not always the same) unless the distribution of test results for the normal and abnormal populations follow two different Gaussian probability distributions.

Thus one inappropriate model for normality/abnormality is a single Gaussian curve in which 5% of the population are taken to be abnormal just because of test results lying in one of two 2.5% tails, Figure 19.1 (top left). Kramer[1] has pointed out that such *statistogenic abnormality* often results in further testing and occasionally in unnecessary therapy, terming this phenomenon the *Ulysses syndrome*![2]. The analogy is with Ulysses' two-year odyssey between the end of the Trojan war and his return home, during which time he experienced a number of needless and dangerous adventures.

Figure 19.1 (top right) shows a model with two non-overlapping independent probability distributions, not necessarily Gaussian. However, in practice, the division between normality and abnormality is not so clearly defined and is more like the situation in Figure 19.1 (bottom left) or Figure 19.1 (bottom right). Defining abnormality for Figure 19.1 (bottom left) is obviously extremely difficult and for Figure 19.1 (bottom right) can also be problematical as a cut-off point has to be defined between a and b which will inevitably result in some misclassifications.

19.3 RECEIVER OPERATING CHARACTERISTIC (ROC) CURVE

The type of graph most often encountered for assessing the ability of a screening test to discriminate between healthy and diseased persons is a *receiver operating characteristic (ROC) curve*, which plots Sensitivity *versus* (1−Specificity), and for which an example is shown in Figure 19.2 for PSA (prostate specific antigen)[4].

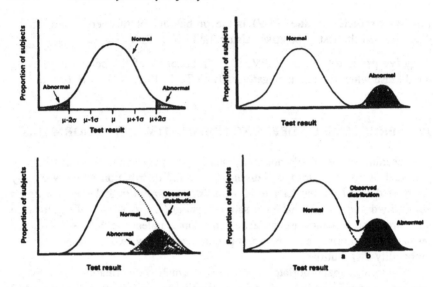

Figure 19.1. Normality/abnormality modelling. (Top left) A single Gaussian distribution. This innapropriate statistical model of normality/abnormality is sometimes referred to as the *ghost of Gauss*[1,3]. (Top right) Model using two non-overlapping independent probability distributions. (Bottom left) Model using two distributions which are merged within the overall observed distribution. (Bottom right) A bimodal distribution model.

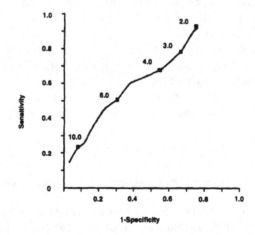

Figure 19.2. ROC curve of prostate-specific antigen (PSA): modified from Brawer[4]. Different points on this ROC curve represent different choices for a cut-off point. That for PSA= 10.0 ng/ml represents a point on the curve that results in a specificity of 0.9 but only 0.2 sensitivity, whereas for PSA= 2.0 ng/ml the point on the ROC curve represents a specificity of 0.3 and a 0.95 sensitivity.

Example   235

19.4   EXAMPLE: PROSTATE CANCER EARLY DETECTION

The importance of early diagnosis of prostate cancer can be seen from the cancer incidence and mortality figures[5] in the USA for 1996 where prostatic cancer is the leading cancer in males with 317,100 cases: greater than lung cancer with 98.900 cases. That effective treatment can be given is seen in the fact that prostatic cancer is the second leading cancer cause of death with 41,400 cases compared to 94,400 lung cancer deaths in males.

The diagnostic value of PSA is in the immunohistochemical identification of prostate carcinomas. The American Cancer Society[6] recommends that for over 50 years of age, annual digital rectal examination and PSA should be performed for the early detection of prostate cancer in asymptomatic males: and if either is abnormal then further evaluation should be considered.

The PSA test is an example of a test for which higher values of the test result reflect greater degrees of abnormality (i.e. a greater probability of the existence of prostatic cancer). To use such a test the choice of a *cut-off value* must be made. A low cut-off will result in greater sensitivity (i.e. of the probability of a positive test in people with the disease) whereas a high cut-off value will result in greater specificity (i.e. of the probability of a negative test in people without the disease).

This *inverse relationship* always exists when the test result is measured on a continuous scale as in Figure 19.2 and the relationship between specificity and sensitivity with PSA cut-off is demonstrated in Figure 19.3. The choice of cut-off point depends on the particular situation being considered and for the example under consideration, early diagnosis of prostatic cancer, Brawer[4] states the following.

'For prostate cancer most efforts are directed towards increasing specificity (probability of a negative test in males without prostatic cancer). This stems from the likelihood that men are not going to be tested only once in their lifetime, but will undergo serial tests, perhaps annually as suggested by the ACS. Thus a *false negative* test is likely to be of less significance. The test result may become positive while the malignancy is still curable.

'In contrast, false positive tests result in a large burden in terms of increased expenditures for subsequent uneccessary medical procedures and increased anxiety for misdiagnosed patients ... A strategy to reduce false positives necessitates tests that have increased specificity.'

Figure 19.4 is an example[4] of the variation of positive predictive value (PPV: probability of the person having the disease when the test is positive) as a function of PSA cut-off and age group. It is seen that there is an age-specific pattern indicating that a single PSA cut-off may be innappropriate.

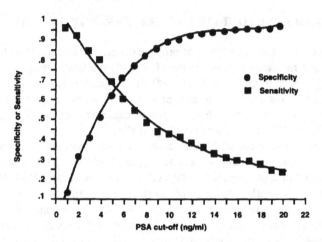

Figure 19.3. Inverse relationship of PSA sensitivity and specificity for males undergoing ultrasound guided prostate needle biopsy[4].

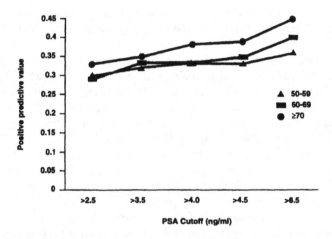

Figure 19.4. Positive predictive value for PSA cut-offs of 2.5, 3.5, 4.0, 4.5, and 6.5 ng/ml and different age groups in men undergoing ultrasound guided prostate needle biopsy[4]. Note that PPV tends to increase with advancing age and with higher PSA cut-off levels.

Example   237

19.5   EXAMPLE: NUCLEAR MEDICINE IMAGING INTERLABORATORY COMPARISON STUDIES

In the example in section 19.4 the aim was to increase specificity, which brings with it a reduction in sensitivity. This example is chosen to illustrate a situation where sensitivity is more important than specificity.

The WHO and IAEA have organised[7] blind studies for nuclear medicine gamma camera imaging devices for a wide spectrum of manufacturer's equipment and for some 16 countries. The test object, Figure 19.5, was a *liver phantom* (i.e. an object which when imaged mimics the image of a human liver) containing 10 targets simulating liver tumours, with the same diameter of 2 cm but with different target-to-background radioactivity ratios.

The laboratories in the survey did not know the position or number of targets (i.e. it was a blind study) and were requested to report how many, and in which positions (grid reference axes, $x =$A, H and $y =$1, 7 are seen in Figure 19.5) the target images were seen. The analysis was made by country and by gamma camera type using sensitivity, specificity and ROC curves.

Table 19.2. Interlaboratory comparison study.

Laboratory report	Actual phantom	
	Tumour present	Tumour absent
Tumour present	No. of true +s	No. of False +s
Tumour absent	No. of False −s	No. of True −s

The results were initially assessed using the data classification (analogous to that in Table 19.1) in the format of Table 19.2 and typical group (gamma camera group, laboratory country group) ROC curves[7] are shown in Figure 19.6. What is termed the *Guess line* in Figure 19.6 is the line for which (Sensitivity + Specificity)= 1 and represents the results which would be obtained purely by chance. Any diagnostic test with such an ROC, where Sensitivity always equals (1−Specificity) would provide totally equivocal diagnostic information.

The area beneath the ROC curve is always greater than 0.5 and the accuracy of the evaluation increases as the area under the ROC increases. This area was calculated for the WHO–IAEA survey using an algorithm[8] developed by the College of American Pathologists and, for example, for the participating countries, the *ROC areas* were in the range 0.77–0.95: the worst performance country being indicated by the area 0.77.

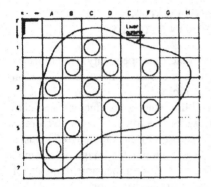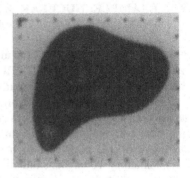

Figure 19.5. (Left) Schematic diagram of the IAEA–WHO liver phantom indicating the position of the 10 targets. (Right) Gamma camera image of the liver phantom showing the grid points outside the liver to assist in identifying the location of the targets.

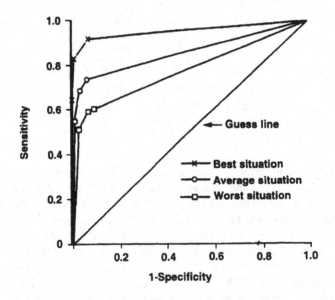

Figure 19.6. Typical ROC curves for the interlaboratory comparison survey.

19.6 KAPPA TEST

Altman[9] describes well how to measure inter-observer agreement, using as data the assessments of 85 xeromammograms by two radiologists (A and B) where the xeromammogram reports are given as one of four results: Normal; Benign disease; Suspected cancer; Cancer.

A measure of agreement is required (this is not a case of hypothesis testing as in the example in Figure 13.1 when 10 radiologists assessed whether a large film or small film was preferable for intravenous urogram imaging and a Wilcoxon test was used) between radiologist A and radiologist B rather than a test of association such as might be undertaken using the χ^2 test.

Table 19.3. Inter-observer basic data for assessment of 85 xeromammograms by two radiologists, after Altman[9] and taken from a larger study by Boyd et al[10].

	Radiologist B				
Radiologist A	Normal	Benign	Suspected ca.	Cancer	Total
Normal	21	12	0	0	33
Benign	4	17	1	0	22
Suspected cancer	3	9	15	2	29
Cancer	0	0	0	1	1
Total	28	38	16	2	85

As Altman points out, the simplest approach is to count how many exact agreements were observed between A and B, which from Table 19.3 is $54/85 = 0.64$. However, the disadvantages with this method of merely quoting a 64% measure of agreement that it does not take into account of where the agreements occurred and also the fact that one would expect a certain amount of agreement between radiologist A and radiologist B purely by chance, even if they were guessing their assessments.

The complete theory underpinning the kappa (κ) test, including the calculation of confidence intervals and including a weighted kappa test where all disagreements are not treated equally, has been given by Altman[9].

The expected frequencies along the diagonal of Table 19.3 are given in Table 19.4 from which it is seen for these data that the number of agreements expected by chance is 26.2 which is 31.% of the total, i.e. 26.2/85. What the kappa test gives is the answer to the question of how much better the radiologists were than 0.31.

The maximum agreement is 1.00 and the kappa statistic gives the radiologists' agreement as a proportion of the possible scope for performing better than chance, which is $1.00 - 0.31$

$$\kappa = (0.64 - 0.31)/(1.00 - 0.31) = 0.47$$

There are no absolute definitions for interpreting κ but it has been suggested[9,11] that the guidelines in Table 19.5 can be followed, which in the example considered here means that there was moderate agreement between radiologist A and radiologist B.

Table 19.4. Calculation of the expected frequencies for the kappa test, after Altman[9].

Assessment	Expected frequency
Normal	$33 \times (28/85) = 10.87$
Benign	$22 \times (38/85) = 9.84$
Suspected cancer	$29 \times (16/85) = 5.46$
Cancer	$1 \times (3/85) = 0.04$

Table 19.5. Guidelines for the interpretation of the κ statistic[9,11].

κ values	Strength of agreement
< 20	Poor
0.21–0.40	Fair
0.41–0.60	Moderate
0.61–0.80	Good
0.81–1.00	Very Good

Another example of the use of the kappa test has been in the interpretation of contrast enhanced magnetic resonance imaging (MRI) of the breast[12]. In this case there were three observers. A series of κ values were calculated so that all observers were tested against each other for three different *imaging sequences*: T_1 weighted pre-contrast; T_2 weighted pre-contrast; T_1 weighted post-contrast. For each imaging sequence three *lesion features* were assessed: conspicuity; signal intensity; contour. In addition, for observer A, a first reading was tested against a second reading.

The resultant series of 40 values of κ led the authors[12] to conclude that 'there is a significant observer variability and a substantial learning curve in the interpretation of breast MRI, as well as variability in the analysis of dynamic data'. Figure 19.7 is a frequency diagram of the total number of lesions reported by each radiologist per examination.

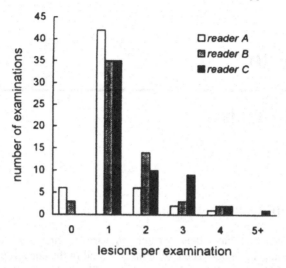

Figure 19.7. Frequency histogram of the total number of lesions reported by each radiologist per examination[12]. There was significant disagreement between radiologists in the number of reported lesions per examination with radiologist C (the least experienced observer) reporting the highest number of lesions.

Did you ever have the measles, and if so, how many? Charles Farrar Browne (1834–1867) in *The Census*, using the pseudonym Artemus Ward.

Chapter 20

Clinical Trials

20.1 INTRODUCTION

Entire books can and have been written on various aspects of clinical trial design[1-6], journals have published guidelines[7-9] and in the encyclopaedic multi-author textbooks on oncology, chapters are usually included on clinical trials design[10-13].

This chapter can only therefore review the major topics relating to clinical trials but it will emphasise the possible problems and pitfalls which can arise, both in design and analysis stages, since it is extremely useful in practice to be aware of *what not to do* in order to avoid making the clinical trial results so biased that they would be useless.

20.2 TRIAL AIMS AND OBJECTIVES

One fault which sometimes occurs is that an attempt is made to answer too many questions at the same time in a single trial. At best there should be one major question asked of the trial and a maximum of two subsidiary questions and it is essential that these aims and objectives are clearly specified in the protocol.

The five main questions to be considered in the design stage are as follows.

1. The clinical question, i.e. what treatment methods are being investigated.
2. The clinical material, i.e. what population is being studied.
3. The design of the study, e.g. phase I, II or III; randomised or non-randomized.
4. Statistical analyses and quality assurance considerations.
5. Endpoints, i.e. what measure(s) of patient welfare.

With regard to 1 the two main pitfalls when considering the clinical questions are as follows.

1. A choice of a clinically inappropriate question within the current clinical environment.

2. A choice of an unacceptable question in terms of patient and/or physician acceptance.

Question **4** will have many parts to consider, not least the following, and some will be inter-related with question **5**.

1. How is the criterion of success to be defined and measured and for what improvement in success is it considered worthwhile organising a clinical trial?
2. What level of statistical significance are we prepared to accept when analysing the results?
3. Given the number of patients available for entry into the trial, what is likely to be the duration of the trial?
4. Can historical controls be used?

20.3 TRIAL DESCRIPTION BY PHASES I–III

Trials are in general described as phase I, II, III or IV, where phase IV studies are post-marketing surveillance instituted by the pharmaceutical industry. It is the earlier phases I–III which concern us in this chapter and Table 20.1 defines these phases. However, it must be noted that alternative terminology is also sometimes used and one can encounter for example, phase IIA and IIB trials. The adjectives *early*, *pilot* and *preliminary* are less clearly defined and can refer to phase I or II trials depending on the physician undertaking the trial.

There is no internationally agreed wording for definitions of phases I–III but the essence of these phases is encapsulated in Table 20.1 and the summary of typical parameter values is given in Table 20.2 after Buyse[13]. As an example of the use of the term IIA and IIB, Table 20.3 is taken from Burdette and Gehan[1].

Table 20.1. Descriptions of phase I–III terminology.

Phase I
Dose escalation with a toxicity endpoint to determine the tolerance dose schedule
Phase II
Efficacy studies at defined dose levels where the trial screens for clinical activity and the endpoint is response. These are trials for a patient group for which treatment appears promising
Phase III
Randomised studies of new treatment *versus* current standard therapy. Endpoint is survival or time to disease progression

Table 20.2. Typical phase I–III parameter values: after Buyse[13].

Parameter	Phase I	Phase II	Phase III
No. of patients	5–15+	15–50+	50–5000+
Duration of trial	Weeks	Months	Years
Randomised ?	Never	Sometimes	Always
Multicentre ?	Never	Sometimes	Often
Site-specific ?	Sometimes	Often	Always

Table 20.3. Terminology of Burdette and Gehan[1].

Early trial: (Phase I)
   Several regimes of management are applied in order to choose the most suitable for further study.
Preliminary trial: (Phase IIA)
   If regime is not effective in early trial, this trial is planned to determine whether agent and/or operation deserves additional study.
Follow-up trial: (Phase IIB)
   This trial should provide estimate of effectiveness of a regime passing the preliminary trial.
Comparative trial: (Phase III)
   Comparison of effectiveness of new regime with standard regime or with another new method of management.

20.4   PHASE I TRIALS

Phase I trials are usually conducted with previously treated patients and the statement by Burdette and Gehan[1] some 30 years ago is still valid. 'No single formal design for a phase I study can be recommended. Proposals for such plans have been made, but the knowledge and intuition of the investigator is much more important than a specific experimental design'.

However, as an example of treatment allocation for phase I studies, Simon[12] describes a starting dose of one-tenth the LD_{10} mg.m^{-2} of body surface in the most sensitive patients and that the dose is increased for subsequent patients in a series of preplanned steps. Cohorts of 3–6 patients are treated at each dose level after the observation time for acute toxicity effects has ended for the patients treated at the lower dose level: see Table 20.4[14] for definitions of toxic levels in man.

Then if no dose limiting toxicity (DLT) is found, Table 20.4[14], the dose is escalated for the next cohort. However, it can be that if the incidence of DLT is one-third then three more patients are treated at the same dose level:

but if no further observation of DLT is noted then the dose level escalates to the next level. The stopping rule for the closure of the phase I study and the passage to a phase II investigation is when the incidence of DLT is greater than one-third in the initial cohort of patients treated at a dose level. The phase II recommended dose is often taken as the highest dose for which the incidence of DLT is less than one-third and usually more than six patients are treated at the recommended dose[12]. A commonly used dose level scheme is a modified Fibonacci series[5,12].

2nd. level = Double the 1st. level
3rd. level = 67% greater than the 2nd. level
4th. level = 50% greater than the 3rd. level
5th. level = 40% greater than the 4th. level
6th. level = 33% greater than the 5th. level and this 33% figure is used
      for all subsequent levels.

Table 20.4. Definitions of toxic levels in man, after Carter *et al* [14].

Term	Definition
Subtoxic dose	A dose that causes consistent changes in haematological or biochemical parameters and might thus herald toxicity at the next higher dose level or with prolonged administration.
Minimal toxic dose	The smallest dose at which one or more of three patients show consistent, readily reversible drug toxicity.
Recommended dose for a phase II trial	The dose that causes moderate, reversible toxicity in most patients.
Maximum tolerated dose	The highest safely tolerable dose.

Phase I trials reported in oncological journals are relatively few compared with phase II and III trials and they are mainly drug trials and not a combination of drug and radiotherapy.

This is illustrated in 1996 in *Radiotherapy & Oncology* where there was only one phase I trial reported[15] although there were also two termed *pilot* studies. In this phase I study of a hypoxic cell sensitiser in combination with conventional radiotherapy, as an example of a recent phase I investigation, 14 patients were entered, side effects were vomiting (2/14), arthralgia (1/14) and cramping (1/14): with none in 10/14. There was a range of tumour sites, of number of doses (1–35) and of total doses (1–43.5 g). The authors also stated[15] that this was 'an abbreviated phase I study and it was realised in the beginning that a formal phase I study should be repeated in the future'. In 1995 in *Radiotherapy & Oncology* there were no papers on phase I trials.

Nevertheless, many phase I trials are in fact sponsored, by for example the National Cancer Institute[16] and these include the following.

1. Evaluation of adults with relatively normal organ function.
2. Those with abnormal function.
3. Young children.
4. The elderly.

20.5 RANDOMISATION

Phase III trials are always randomised and phase II trials are sometimes randomised, Table 20.2. The objective of randomisation is to ensure that there is no bias which will make the results invalid. Thus for example the allocation of patients into two treatment groups A and B must be by a method which eliminates any preconceived opinions that a particular patient might be more suitable for A than for B. There are three types of randomisation, single, stratified and balanced: sections 20.5.2–20.5.4.

20.5.1 Methods of Randomisation

Methods of randomisation usually involve a series of random numbers, Table 20.5, where the method is to allocate treatment to one arm (A) of the trial when a patient receives an odd number and to the other treatment arm of the trial (B) when the patient receives an even number: or vice-versa. In a *double-blind trial* where only the trial medical coordinator, pharmacist and statistician have a knowledge of which patient is treated by A or B. and where neither the patient's physician nor the patient has this knowledge, allocation to A or B can be organised by handing to the patient at trial registration the information in a sealed envelope which is then passed to the pharmacist so that either the drug or the placebo can be dispensed.

Another method of randomisation, no longer so often used, is to depend on birth dates: but this is an approximation since the number of odd dates in a year does not equal the number of even dates.

20.5.2   Single Randomisation

Single randomisation is the allocation of the patient intake into the two arms of the the trial A and B without any subdivision by any patient characteristic such as age, sex or histology, Figure 20.1. If 20 random numbers are as follows:

20, 17, 42, 28, 23, 17, 59, 66, 38, 61, 02, 10, 86, 10, 51, 55, 92, 52, 44, 25.

There are 20 pairs of 2-digit numbers. If treatment A = odd number and treatment B = even number and we have 20 patients entering the trial, then the patient allocation for treatment would be as given below:

Patient	Treatment	Patient	Treatment
1	B	11	B
2	A	12	B
3	B	13	B
4	B	14	B
5	A	15	A
6	A	16	A
7	A	17	B
8	B	18	B
9	B	19	B
10	A	20	A

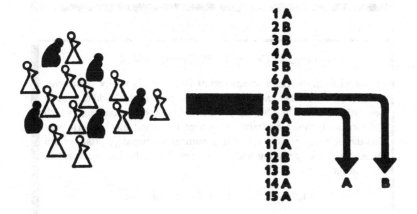

Figure 20.1. Single (sometimes called *simple*) randomisation where the allocation of patients amongst subgroups (i.e. class) is left to chance.

RANDOM NUMBER MACHINE

The cartoon was first published[3] under the legend 'Random numbers can be produced by elotronic devices or by simple machines'.

However, nothing to do with randomisation can be quite as strange as the report in the early 1970s in the *London Evening Standard* under the title *'A Bucket of Pills—and it's Help Yourself'*.

'The doctor who piled all the drugs in his surgery into a bucket in the waiting room, attaching a label telling his patients to help themselves and not to bother him.'

and

'Even when the General Medical Council did act, it could get the oddest answers—the doctor with the pills in the plastic bucket simply retorted that his treatment was no more random than that of other doctors.'

FIBONACCI SERIES OF NUMBERS

Fibonacci from the Latin means son of Bonacci, his family's surname, but he was more often known during his lifetime in the Middle Ages as Leonardo of Pisa. In 1202 he published *'A Book on Counting'* which included such topics as fractions, square roots and algebra. He was also the discoverer of an interesting number relationship. You build a Fibonacci series by starting with 1 and then adding the last two numbers to build the next.

1, 1, 2, 3, 5, 8, 13, 21, 34, 55 and so on

The same sort of relationship often appears in nature. The bumps on the outside of a pineapple are arranged in spirals, some going clockwise and some anticlockwise, and the ratio is 8:13, i.e. 8 in one direction and 13 in the other. Similarly, the ratio of spirals you see in a pine cone is 5:8 and in the heart of a sunflower is 34:55.

Table 20.5. Random numbers. These can be obtained from standard books of tables (as below) or can be generated by computer software as and when necessary. From Fisher and Yates, *Statistical tables for Biological, Agricultural and Medical Research* (6th edn, 1974, table XXXIII, p 134). Courtesy Longman Group UK Limited.

03	47	43	73	86	36	96	47	36	61	46	98	63	71	62
97	74	24	67	62	42	81	14	57	20	42	53	32	37	32
16	76	62	27	66	56	50	26	71	07	32	90	79	78	53
12	56	85	99	26	96	96	68	27	31	05	03	72	93	15
55	59	56	35	64	38	54	82	46	22	31	62	43	09	90
16	22	77	94	39	49	54	43	54	82	17	37	93	23	78
84	42	17	53	31	57	24	55	06	88	77	04	74	47	67
63	01	63	78	59	16	95	55	67	19	98	10	50	71	75
33	21	12	34	29	78	64	56	07	82	52	42	07	44	38
57	60	86	32	44	09	47	27	96	54	49	17	46	09	62
18	18	07	92	46	44	17	16	58	09	79	83	86	19	62
26	62	38	97	75	84	16	07	44	99	83	11	46	32	24
23	42	40	64	74	82	97	77	77	81	07	45	32	14	08
52	36	28	19	95	50	92	26	11	97	00	56	76	31	38
37	85	94	35	12	83	39	50	08	30	42	34	07	96	88
70	29	17	12	13	40	33	20	38	26	13	89	51	03	74
56	62	18	37	35	96	83	50	87	75	97	12	25	93	47
99	49	57	22	77	88	42	95	45	72	16	64	36	16	00
16	08	15	04	72	33	27	14	34	09	45	59	34	68	49
31	16	93	32	43	50	27	89	87	19	20	15	37	00	49
68	34	30	13	70	55	74	30	77	40	44	22	78	84	26
74	57	25	65	76	59	29	97	68	60	71	91	38	67	54
27	42	37	86	53	48	55	90	65	72	96	57	69	36	10
00	39	68	29	61	66	37	32	20	30	77	84	57	03	29
29	94	98	94	24	68	49	69	10	82	53	75	91	93	30
16	90	82	66	59	83	62	64	11	12	67	19	00	71	74
11	27	94	75	06	06	09	19	74	66	02	94	37	34	02
35	24	10	16	20	33	32	51	26	38	79	78	45	04	91
38	23	16	86	38	42	38	97	01	50	87	75	66	81	41
31	96	25	91	47	96	44	33	49	13	34	86	82	53	91
56	67	40	67	14	64	05	71	95	86	11	05	65	09	68
14	90	84	45	11	75	73	88	05	90	52	27	41	14	86
68	05	51	18	00	33	96	02	75	19	07	60	62	93	55
20	46	78	73	90	97	51	40	14	02	04	02	33	31	08
64	19	58	97	79	15	06	15	93	20	01	90	10	75	06

20.5.3   Stratified Randomisation

Another type of randomisation is stratified randomisation when the patient intake consists of two distinct classes (the black and white symbols in Figures 20.1 and 20.2), for example, X and Y, which may for instance be male and female. The aim is to ensure that no matter how many of the intake are in class X or class Y, one half of each class receive treatment A and the other half receive treatment B. Figure 20.2 shows this schematically. However, a balancing process should also be incorporated into any stratified randomisation.

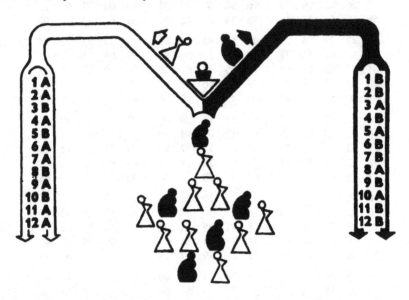

Figure 20.2. Stratified randomisation in which treatment arms are allocated equally (though randomly) to each subgroup (i.e. class) within the population.

20.5.4   Balanced Randomisation

From the patient allocation by single randomisation it is seen that there is an unequal final division between the two treatment groups. If the approximate number of patients who will enter the trial is known, then it is possible to have balanced randomisation, in which a certain block of patients will be balanced; e.g. 25 for treatment A and 25 for treatment B.

The balancing will be made for the last two or three patients, thus if after 47 patients have been allocated there are 25 assigned to A and 22 to B. then the final intake of three patients will be allocated to B to make up the total to 25, rather than continue with the randomisation process, Figure 20.3.

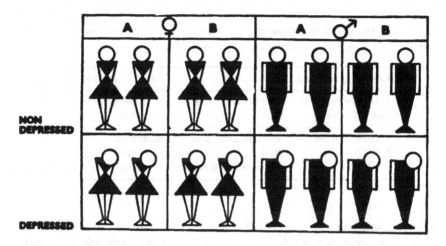

Figure 20.3. Balanced (sometimes termed *factorial*) randomisation ensures that specific numbers of patients of each characteristic are allocated equally within each subgroup (i.e. class) of arms A and B.

20.6   HISTORICAL CONTROLS

The use of historical controls is not to be recommended as it can obviously be misleading and bias the results because, for instance, of even minor changes in treatment and in recording of data. Randomised studies are always the better option.

   An illustration by Byar *et al* [17] demonstrates rather well the problem of using historical controls, Table 20.6, where in this example their use would have provided an incorrect conclusion since there is no significant difference between oestrogen therapy and placebo for equivalent time periods (i.e. the last 2.5 years of the seven year trial) but if the placebo data is obtained from the first 2.5 years (and thus by definition are historical controls when compared with oestrogen therapy results for the final 2.5 year time period of the trial) there is a significant result $P < 0.01$.

20.7   PHASE II TRIALS

Phase II trials have already been referred to in Tables 20.1–20.3 where their purpose and trial endpoints have been defined and typical parameters stated. These are recapitulated in Table 20.7.

   The most serious error in a phase II trial is to reject an effective drug from further study since a false negative error may mean that a drug of potential

Table 20.6. An example of a situation when historical controls did not work[17]. The trial extended over seven years and accrued a total of 2313 prostatic carcinoma patients.

Oestrogen trial arm	Placebo trial arm	Result of an analysis
Patients entered in last 2.5 years	Patients entered in last 2.5 years	No significant difference
Patients entered in last 2.5 years	Patients entered in first 2.5 years (**Historical controls**)	Significant difference $P < 0.01$

interest is lost forever. Hence most phase II trial designs try to minimise this error[18]. One example is the design of Gehan[19] which is a frequently used design and has therefore been used in this section as a representative design example.

However, it should be noted that there are also designs by Fleming[6] and by Simon[12] and that the Fleming design with two or three stages is now widely used.

Table 20.7. Summary of commentary on phase II trials in Tables 20.1–20.3.

Efficacy studies at defined dose levels where the trial screens for clinical activity and the endpoint is response. These are trials for a patient group for which treatment appears promising.

or

If the regime is not effective in a phase I study, a phase II trial is planned to determine whether the agent and/or operation deserves additional study. (Phase IIA)

or

It should provide an estimate of the effectiveness of a regime passing a phase I study. (Phase IIB)

and

Number of patients:	15–50+
Duration of phase II trial:	months
Randomised:	sometimes
Multicentre:	sometimes
Site-specific:	often

20.7.1 Gehan's Design

Gehan's design of a phase II study[19] has had its methodology described in several publications on trial design[1,6,12,13] with the most exhaustive reference on patient numbers required being given by Machin and Campbell[16].

The plan controls the probability of a false negative result (i.e. a type II error with an associated risk of this type of error given by the probability β, see Tables 8.1 and 8.4, and therefore looks to optimise the power $(1 - \beta)$, which is the probability of obtaining a correct result, i.e. of a treatment benefit when a treatment benefit actually exists.

The Gehan plan controls the probability of a false negative result, i.e. the β risk, by calculating the probability that the first r patients do not respond to the treatment for a pre-specified response π to the drug. The initial sample size is determined as the smallest value of r such that the probability of r consecutive failures is less than a given error rate β:

$$\text{Pr}(r \text{ successive patients failing on the drug}) = \beta = (1 - \pi)^r$$

where Pr(of an individual patient responding to a particular treatment)= π and where π is constant for all patients and can be termed the *minimum efficacy*.

If β is specified together with π then the above equation $\beta = (1 - \pi)^r$ can be solved[6] to give

$$n_1 = \log(\beta)/\log(1 - \pi)$$

where n_1 is the number of patients to be recruited to the first stage of the phase II trial (i.e. phase IIA[1]), assuming that r_1 responses will be observed in these n_1 patients. If none of the n_1 patients respond, then the drug is rejected for any further study (i.e. for a phase IIB trial).

However, if at least one patient responds, i.e. $r_1 > 0$, then an additional n_2 patients are treated in order to attain a given precision (ϵ, see Table 20.10 legend) of the response rate π. Table 20.8 gives values of r for given values of π and Table 20.9 the number of patients (n_1) required for a phase IIA trial for a given efficacy (i.e. therapeutic effectiveness) π and a given power $(1 - \beta)$. Table 20.10 gives the additional number of patients (n_2) required for the phase IIB trial.

Table 20.8. Probability of a given number of successive failures (r) for given probabilities (π) of patient response (i.e. treatment efficacy $= \pi)^6$.

Number of consecutive patients (r)	$\pi = 0.10$	0.20	0.30	0.40	0.50	0.60	0.90
1	0.9000	0.8000	0.7000	0.6000	0.5000	0.4000	0.1000
2	0.8100	0.6400	0.4900	0.3600	0.2500	0.1600	0.0100
3	0.7290	0.5120	0.3430	0.2160	0.1250	0.0640	
4	0.6561	0.4096	0.2401	0.1296	0.0625	0.0256	
5	0.5905	0.3277	0.1681	0.0503	0.0313	0.0102	
6	0.5314	0.2621	0.1176	0.0467	0.0156	0.0041	
7	0.4783	0.2097	0.0824	0.0280	0.0078		
8	0.4305	0.1678	0.0576	0.0168			
9	0.3874	0.1342	0.0404	0.0101			
10	0.3487	0.1074	0.0282	0.0060			
20	0.1216	0.0115					
30	0.0424						

20.7.1.1 *Number of Patients Required for a Phase IIA Trial*

Table 20.9. Number of patients $(n_1$ required for a phase IIA trial for a given treatment efficacy (π) and given power $(1 - \beta)^{1.6}$. $(1 - \beta)$ can be thought of as 'the chance of a conclusive result' and β as 'the rejection rate permitted by the investigator'.

Therapeutic effectiveness (π)	Power $(1 - \beta)$			
	$(1 - \beta) = 0.80$	0.90	0.95	0.99
0.05	32	45	59	90
0.10	16	22	29	44
0.15	10	15	19	29
0.20	8	11	14	21
0.30	5	7	9	13
0.50	3	4	5	7
0.60	2	3	4	6
0.90	1	1	2	2

20.7.1.2   *Number of Patients Required for a Phase IIB Trial*

Table 20.10. Part I. Power $(1 - \beta) = 0.90$.

Additional number of patients (n_2) required for the phase IIB trial following completion of the phase IIA trial for which n_1 patients were accrued[1.6]. The value of n_2 depends on the therapeutic effectiveness (π), on n_1 and on the power $(1 - \beta)$ and on the specified precision (ϵ). Precision is defined as follows. An estimate of the true effectiveness is the proportion of patients in the sample who are treated successfully $(\hat{p})$ and the precision of this estimate is measured by its standard error SE$(\hat{p})$ where

$$SE\hat{p}) = \sqrt{\{[p(1 - p)]/n\}}$$

and n is the number of patients and p is the true proportion (which is unknown) of successes following treatment, then p can be substituted by $\hat{p}$ to obtain an estimate of the SE. The largest possible SE(p) is obtained when $p = 0.5$ and thus SE$(\hat{p})$ is less than or equal to $0.5/\sqrt{n}$.

Therapeutic effective-ness (π)	No. of patients in phase IIA trial (n_1)	No. of treatment successes (r_1) in phase IIA trial Power $(1 - \beta) = 0.90$					
		$r_1 = 1$	$r_1 = 2$	$r_1 = 3$	$r_1 = 4$	$r_1 = 5$	$r_1 = 6$
		for $\epsilon = 0.05$ and [$\epsilon = $ **0.10**]					
0.10	22	20 [0]	35 [0]	47 [0]	57 [0]	65 [0]	71 [0]
0.15	15	42 [0]	59 [4]	72 [7]	80 [9]	85 [10]	85 [10]
0.20	11	60 [7]	77 [11]	87 [14]	89 [14]	89 [14]	89 [14]
0.30	7	83 [16]	93 [18]	93 [18]	93 [18]	93 [18]	93 [18]

Note that the 2nd column in this table is the same as part of the 3rd column in Table 20.9

Table 20.10. Part II. Power $(1 - \beta) = 0.95$.

Therapeutic effective-ness (π)	No. of patients in phase IIA trial (n_1)	No. of treatment successes (r_1) in phase IIA trial Power $(1 - \beta) = 0.95$					
		$r_1 = 1$	$r_1 = 2$	$r_1 = 3$	$r_1 = 4$	$r_1 = 5$	$r_1 = 6$
		for $\epsilon = 0.05$ and [$\epsilon = $ **0.10**]					
0.10	29	4 [0]	17 [0]	28 [0]	38 [0]	46 [0]	53 [0]
0.15	19	29 [0]	45 [0]	58 [0]	67 [3]	75 [5]	79 [6]
0.20	14	46 [1]	64 [6]	76 [9]	84 [11]	86 [11]	86 [11]
0.30	9	71 [11]	87 [15]	91 [16]	91 [16]	91 [16]	91 [16]

Note that the 2nd column in this table is the same as part of the 4th column in Table 20.9

Figure 20.4 shows for a power of $(1 - \beta) = 0.95$ the variation in the total number $(n_1 + n_2$ as in Tables 20.9 and 20.10) of patients required for a Gehan design of phase II trial as a function of therapeutic effectiveness (π) for two different specified precisions ($\epsilon = 0.05$ and 0.10) and for one ($r_1 = 1$) and six ($r_1 = 6$) treatment successes in the phase IIA trial. It is seen that each of the four curves rises eventually to a plateau, $(n_1 + n_2) = 100$ for $\epsilon = 0.05$ and $(n_1 + n_2) = 25$ for $\epsilon = 0.10$. The values of $(n_1 + n_2)$ for $r_1 = 2, 3, 4, 5$ lie between those for $r_1 = 1$ and $r_1 = 6$. The minimum value of $(n_1 + n_2) = 15$ for any combination of parameters: see also Table 20.7.

Finally, a word of warning: the information in this chapter on patient numbers required for clinical trials are examples only to give the reader an idea of the order of magnitude of the numbers required, and also the complexity of the process of deciding on the numbers required. For instance for the phase II Gehan-design trials one has to consider $n_1, \pi, r_1, (1 - \beta)$ and the precision ϵ before reaching a conclusion on the patient number n_2 as seen from Table 20.10.

In practice, therefore, when planning a clinical trial, the reference with the complete set of tables/graphs should be consulted, e.g. those of Machin and Campbell[6], and interpolations/extrapolations should not be made from data such as in Tables 20.9 and 20.10.

20.8 PHASE III TRIALS

Phase III trials have already been referred to in tables 20.1–20.3 where their purpose and trial endpoints have been defined and typical parameters stated. These are recapitulated in Table 20.11.

Table 20.11. Summary of commentary on phase III trials in Tables 20.1–20.3.

Randomised studies of new treatment versus current standard therapy. Endpoint is survival or time to disease progression.

or

Comparison of effectiveness of new regime with standard regime or with another new method of management.

and

Number of patients:	50–5000+
Duration of phase III trial:	years
Randomised:	always
Multicentre:	often
Site-specific:	always

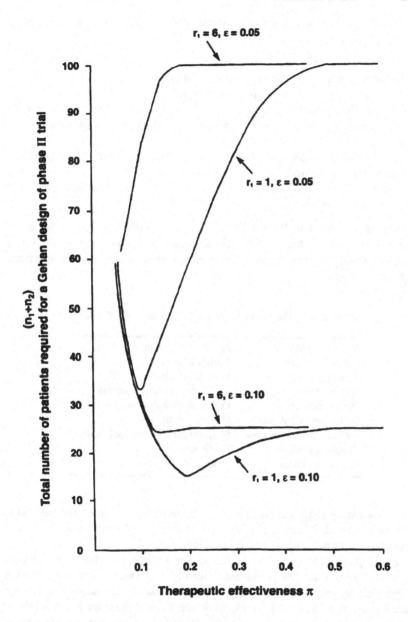

Figure 20.4. Total number of patients $(n_1 + n_2)$ required for a Gehan design of phase II trial, as a function of therapeutic effectiveness (π) which can also be described as the probability of an individual patient responding to a particular treatment. The power $(1 - \beta) = 0.95$ and the curves are drawn for two values of specified precision $(\epsilon = 0.05$ and $\epsilon = 0.10)$ and two values of the number of treatment successes in the phase IIA trial, $r_1 = 1$ and $r_1 = 6$. Curves constructed from data tables of Machin and Campbell[6].

Unequivocal answers must be available to all the following questions *before the start* of the phase III trial.

- What *population* is being studied?
- What *treatment* methods are being investigated?
- What *randomisation* methods are required?
- How is the *criterion of success* to be defined and measured, and for what *improvement in success* is it considered worthwhile organising a clinical trial?
- What level of *statistical signifcance* are we prepared to accept when analysing the results?
- Given the *number of patients* available each year what is likely to be the duration of the trial?

It is also helpful to have a list of headings, Table 20.12, which can be described as the main features of a clinical trial, and which can be used as a check list in trial planning discussions to ensure that nothing of major importance has been forgotten.

Table 20.12. Main features of a clinical trial, after Pocock[4].

A written protocol	Forms and data management
Controlled trials	Statistical analysis
Randomisation	Protocol violations
Size of trial	Monitoring of trial progress
Double blind trials	Ethical considerations
Definition of patients	Multicentre trials
Definition of treatments	Staff, responsibilities and funding
Endpoint evaluation	Publication
Prognostic factors	Truth and relevance

In discussing statistical significance, we have already considered (see Table 8.2) type I errors and α risks (e.g. $P < 0.05$) and type II errors and β risks and the concept of power of a test being $(1 - \beta)$.

α and β are also related to the size of the sample, N, which for a clinical trial would be the number of patients. Ideally, the values of α and β would be specified by the researcher before the trial, and these values would then determine the size, N, of the sample he would have to draw for computation of the test chosen for statistical significance. However, in practice it is usual for α and N to be specified *a priori* and this determines β. To reduce the possibility of both types of error, I and II, the value of N must be increased.

Different types of catastrophe will follow type-I and type-II errors. As an example, consider a clinical trial of an existing drug A and a new drug B where the null hypothesis, H_0, is that there is no difference and the alternative hypothesis, H_1, is the drug B is better than drug A (ignore H_2: drug A better than drug B for this example). If a type-I error occurs then H_0 is wrongly rejected and following the inference that H_1 should thus be accepted, drug A will be abandoned and drug B now given to patients. What are the consequences?

Since H_0 is really true and there is no difference between A and B then the patients will not have suffered since they have only been placed on another drug of similar efficacy.

Now consider a type-II error when H_0 is wrongly accepted. In this instance, the truth is that H_1 is correct and the consequences to the patient are that a drug with a better efficacy is denied to them, since the acceptance of H_0 would mean that the current drug A should remain in use and research be abandoned on the new drug B (or perhaps continued more stringently, if the value of N is relatively small).

Of the two types of error in this situation, a type-II error would seem to be worse than a type-I error since it could mean rejection of a new and useful drug—if the trial is not carefully planned.

Careful planning of phase III trials is, however, not limited only to decisions taken on the patient numbers and the problems most frequently encountered are listed in Table 20.13 after Buyse[13] and although the definition of clinical trial *illnesses* which follows this table is a joke, it nevertheless contains a warning to planners of clinical trials.

In the EORTC guidelines[9] for writing radiotherapy phase III trial protocols, the endpoints are given as overall survival, disease-free survival, acute and late side effects, local control and quality of life. In addition, EORTC state that in the near future cost-benefit aspects in public health may also be part of the study assessment. The 29 major headings given by EORTC for consideration of any radiotherapy protocol are listed in Table 20.14; for subheadings the reader is referred to the original publication[9].

Multicentre trials are not mentioned explicitly in Table 20.14 but this section closes with a word of warning, because the more centres involved in a trial the greater the possibilities for errors and miscommunication. Indeed, problems of multicentre trials most often arise from inadequate and unclear communications between participating investigators, all of whom must agree to follow a common protocol. Figure 20.5 gives examples of how not to set up clinical trial organisations!

20.8.1 Number of Patients Required for a Phase III Trial

The number of patients required for a trial depends on the α risk, the β risk (which is incorporated into the decision making as the power $1 - \beta$), and the difference in treatment responses between the two arms (denoted by subscripts 1 and 2) of the trial. This for example, may be a difference in T-year survival rate of 10%, 20%, ... etc, and Table 20.15 gives an example[20] where the T-year rate is a long-term survival rate which can be equated to a *cured proportion*

Table 20.13. Problems most frequently encountered with phase III trials[13].

Frequent source of bias	Frequent source of problems
• Design Post-randomisation exclusions	Optimistic treatment benefit Optimistic patient accrual Stringent selection criteria
• Conduct Losses to follow-up *Significant* interim results	Administrative burden Diminishing interest
• Analysis Exclusion of patients Analysis by actual treatment	Inadequate statistical tests Multiple comparisons
• Interpretation Data derived results Overemphasis of subgroups	$P > 0.05$ does not imply no benefit $P < 0.05$ does not imply large benefit

---

CLINICAL TRIAL ILLNESSES

Senility

Trials taking so long that when completed the information provided is of little use.

Sclerosis

Trials with extensive and complicated documentation which eventually will defy analysis.

Death

Trials never completed

Table 20.14. EORTC recommended major topics to be considered for a phase III clinical trial of radiotherapy[9].

1	Background and rationale
2	Objectives of the trial
3	Patient selection
4	Trial design
5	Pretreatment evaluation
6	Surgery (where applicable)
7	Radiation treatment
8	Drug therapy (including sensitisers)
9	Other therapies (if applicable)
10	Required clinical evaluation, laboratory tests and follow-up
11	Endpoints
12	Criteria for (disease) progression
13	Guidelines for second line treatment at relapse, if any
14	Patient registration and randomisation procedure
15	Forms and procedures for collection of data
16	Reporting adverse effects
17	Statistical considerations
18	Quality of life assessment
19	Cost evaluation assessment
20	Ethical considerations
21	Investigator commitment statement
22	Administrative responsibilities
23	Trial sponsorship/Financing
24	Trial insurance
25	Publication policy
26	Administrative signatures
27	List of participants with expected yearly accrual
28	References
29	Appendices

(C_2 and C_1), the α risk is specified by $P = 0.05$, there are two different values for the power $(1 - \beta)$ of 0.50 and 0.75, and four different computations of $(C_2 - C_1)$.

The notion of *difference in treatment responses between two arms* of a trial is the most important determinant in making a decision on sample size. In many trials, unrealistic differences are presented, in order to minimise the number of patients in a trial. The result, of course, is that the trial results are negative because of lack of power; see section 20.11 on meta-analysis.

Figure 20.6 includes the data in Table 20.15 and for $P = 0.05$ presents data for three different values of the power $(1 - \beta)$ of 0.50 (left), 0.75 (centre) and 0.90 (right). It is quite clear that for small differences in $(C_2 - C_1)$ of only

262   *Clinical Trials*

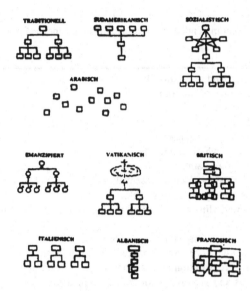

Figure 20.5. Organisation diagrams. From top left to bottom right: Traditional; South American; Socialistic; Arabian; Emancipated; Vatican; Albanian; French. (This organogram was originally given to me in German, which no doubt accounts for no German organogram being drawn!)

5% extremely large numbers of patients are required for a trial.

Tables of patient number requirements are given in various publications and three examples are listed below.

- Sample sizes for logrank test[12]. Number of patients (number of deaths) to detect an improvement ($P_2 - P_1$) in survival rate over a baseline rate (P_1 when

  {I} $\alpha = 0.05$ and $(1 - \beta) = 0.80$
  {II} $\alpha = 0.05$ and $(1 - \beta) = 0.90$
  {III} $\alpha = 0.01$ and $(1 - \beta) = 0.95$

  when the logrank is a two-tailed test. These data are reproduced from the publication of Freedman[22] and is for five values of ($P_2 - P_1$) of 0.05, 0.10, 0.15, 0.20 and 0.35 with values of P_1 in the range 0.05–0.90 in intervals of 0.05.

- Number of patients in each of two treatment groups, one-tailed and two-tailed tests[12] when

  {I} $\alpha = 0.05$ and $(1 - \beta) = 0.80$
  {II} $\alpha = 0.05$ and $(1 - \beta) = 0.90$

  with ($P_2 - P_1$) in the range 0.05–0.50 in intervals of 0.5 and the smaller success rate (P_1) in the range 0.05–0.50 in intervals of 0.05.

Table 20.15. Number of patients required for a clinical trial, as a function of the α-risk ($P = 0.05$), the β-risk ($1 - \beta = 0.5$ or 0.75), and the observed difference in C-values. C is a long-term survival rate[20].

Observed difference, $C_2 - C_1$, which should be statistically significant at the $P = 0.05$ (α-risk) level of significance	Total number of cases required for the clinical trial = $2N$ (N in group A and N in group B)	
	$1 - \beta = 0.5$	$1 - \beta = 0.75$
	A *1 in 2 chance* of getting a conclusive result in a single trial	A *3 in 4 chance* of getting a conclusive result in a single trial
$C_2 - C_1 = 5\%$ with $C_1 = 20\%$ and $C_2 = 25\%$	$2N = 1000$	$2N = 1000$
$C_2 - C_1 = 10\%$ with $C_1 = 40\%$ and $C_2 = 50\%$	$2N = 400$	$2N = 700$
$C_2 - C_1 = 15\%$ with $C_1 = 10\%$ and $C_2 = 25\%$	$2N = 100$	$2N = 170$
$C_2 - C_1 = 20\%$ with $C_1 = 20\%$ and $C_2 = 40\%$	$2N = 75$	$2N = 130$

- Total number of patients required to compare the response rate (comparison of proportions) in a treatment group with that of a control group[13] and total number of patients required to compare the survival rates of a treatment group with that in a control group[13]. Both tables are for two-tailed tests and for

{I} $\alpha = 0.05$ and $(1 - \beta) = 0.80$

for P_1 in the range 0.10–0.80 in intervals of 0.10 and P_2 in the range 0.15–0.95 in intervals of 0.05.

20.8.2 Power Curves

The power of a test is defined as the probability of rejecting the null hypothesis H_0 when it is in fact false. That is,

Power = $(1 -$ Probability of a type II error$) = (1 - \beta)$

The curves[23] in Figure 20.7 show that the probability of committing a type II error (β) decreases as the sample size (N) increases and hence that power

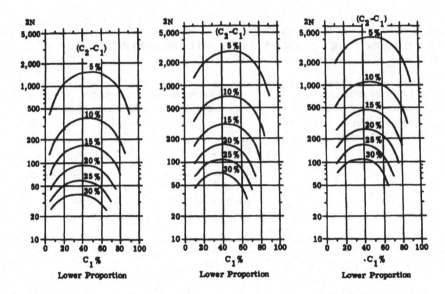

Figure 20.6. Charts[20] to determine the number of patients required in a clinical trial for different combinations of variables: α risk (i.e. $P = 0.05$), power $(1 - \beta)$ and the difference $(C_2 - C_1)$. (Left) Number of cases required in a clinical comparison of two treatments in order that the observed difference, $C_2 - C_1$, in the proportions cured should be statistically significant at the $P = 0.05$ level (N = number of cases, in each treatment group.) C_1, C_2 = the proportion cured in the first and second groups respectively. $(1 - \beta) = 0.50$. (Centre) Number of cases required in a clinical comparison of two treatments in order to stand a three in four chance of detecting (at the $P = 0.05$ level) a difference of $C_2 - C_1$ in the proportions curable in the whole population. $(1 - \beta) = 0.75$. (Right) Number of cases required in a clinical comparison of two treatments in order to stand a nine in ten chance of detecting (at the $P = 0.05$ level) a difference of $C_2 - C_1$ in the proportions curable in the whole population. $(1 - \beta) = 0.90$.

increases with N. Figure 20.7 also shows that when H_0 is true, i.e. when the true mean is μ_0 the probability of rejecting H_0 is 0.05. Which is as it should be since $\alpha = 0.05$ and α gives the probability of rejecting H_0 when it is in fact true.

20.8.3   Interim Analyses

The analysis of the results of a trial is usually undertaken at the end of the trial. Nevertheless, interim analyses can take place, but it is often forgotten that every time one looks at the results the statistical significance level being taken as meaningful (e.g. originally $P = 0.05$) needs to be made stricter.

This is because by definition at the $P = 0.05$ level of significance there

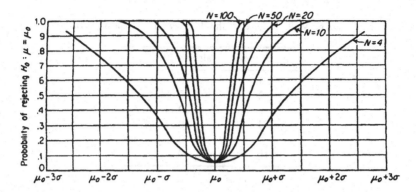

Figure 20.7. Power curves[23] of a two-tailed test with $\alpha = 0.05$ and sample sizes $N = 4, 10, 20, 50, 100$. The samples are taken from normal populations with variance σ^2 and a mean under H_0 of μ_0. The vertical axis of the diagram is power $(1 - \beta)$ and the horizontal axis is the size of the difference between the two treatments under study. Power is a function of N, α and the size of the difference between the success rates of the two treatments.

are five false positive results in every 100. Thus there are five chances in 100 of finding a *so-called difference* when no such difference really exists and the more often one looks at the results the greater is the chance of picking out one of these false positives by accident.

Zelen[11] has published the data in Table 20.16 which shows, for example, that if a clinical trial is planned to have a false positive rate of 5% then this would be changed to 14% if there were five interim analyses, because of the reasoning stated above.

Table 20.16 can also be used to determine an overall false positive rate of a trial when the analysis looks at several subgroups separately. For example, if there are five subgroups in a study, where each is analysed separately with a 5% false positive rate ($P = 0.05$), then the overall false positive rate is 14%.

To overcome this problem, it has been suggested[12,24,25] that interim analyses be discounted unless the difference is statistically significant at the two-tailed $P < 0.0025$ level. If the interim differences are not significant at this level the trial continues to its originally intended size. The final analysis is performed without referring to the interim analyses and the type I error, α risk, is almost unaffected by having performed interim analyses.

There have been many proposals on *how to spend* α during the trial, and how to keep the final α close to 0.05 for the last and therefore definitive analysis. a full discussion is to be found in the textbook of Friedman *et al*[3] in which they discuss alpha-spending functions and group sequential methods.

Table 20.16. Multiple *looks* at the data *versus* false positive rates[11].

No. of *looks*	False positive rate
1	0.05
2	0.08
3	0.11
4	0.13
5	0.14
10	0.19

20.8.4 Multiple Comparisons Between Subgroups: Bonferroni Method

The problem of multiple comparisons has already been referred to in the previous section. In Gelman's review[26] of this problem it is stated that for k non-overlapping (i.e. independent) subgroups, with each test having a 5% false positive rate, there is a probability of $(1 - 0.95^k)$ of finding at least one test to be statistically significant at $P < 0.05$, just by chance.

One of the methods of dealing with this problem of multiple comparisons is the Bonferroni method[27] to undertake each of the k tests with a cutoff corresponding to

$$\alpha' = 1 - \sqrt[k]{(1 - \alpha)}$$

where α is the required overall false positive rate. As an example, with $\alpha = 0.05$ and $k = 8$ only tests with $P < 0.0064$ would be declared significant. Equivalently one could multiply all significance levels by 7.8 (since $0.05/0.0064 = 7.8$) and use a cutoff of 0.05. It is also noted that since $1 - \{k^{\text{th}}$ root of $[1 - \alpha]\}$ is approximately equal to α/k the Bonferroni method is sometimes approximated by multiplying all significance levels by k.

20.8.5 Inflation Factor Due to Patient Refusals

The situation can arise in comparative trials of two treatments that some patients are not given the randomised allocated treatment. This can be the situation in pre-randomisation when patients are randomised after being found eligible for the trial but before the patient's consent to participate in the trial is sought[28]. Such patients must be followed-up and analysed by *intention to treat*, rather than by actual treatment.

If such deviations occur frequently they can cause a reduction in power and the number of patients must then be multiplied by an *inflation factor*, Table 20.17, to preserve the nominal power of the test[12,13].

It is never particularly easy to accrue large numbers of patients and if such a pre-randomised trial[28] is designed and there are a large number of patient refusals after the process of informed consent has been completed, such a trial is unlikely to accrue sufficient numbers for the required power.

Table 20.17. Factor by which the number of patients must be increased to maintain the statistical power of a two-arm clinical trial if some proportion of the patients refuse the treatment assigned by randomisation and choose the other treatment instead. This enables the avoidance of bias in analysis of results: patients must be compared *as randomised* rather than as treated. That is, a patient randomised to treatment A who refused and then received treatment B must be considered in group A for analysis. (This is from a proposal from Zelen[28]). However, if 50% of the patients were to cross over, any treatment effect is entirely diluted and can no longer be detected, regardless of the number of patients[12,13].

Refusal rate	Inflation factor
0.02	1.09
0.05	1.23
0.10	1.56
0.15	2.04
0.20	2.78
0.25	4 00
0.30	6.25
0.35	11.11
0.40	25
0.45	100

20.9 SEQUENTIAL TRIALS

Sequential trials have the advantage of a built-in stopping rule and the analysis continues throughout the duration of the trial, using a specially designed chart of which an example is given in Figure 20.8. To describe the use of the chart, assume first that (a) the patient intake has been *paired*; (b) each member of the pair has been *randomised* into either a treatment A or treatment B group such that sometimes the first member of a pair will have A and sometimes the second member of a pair will have A; (c) a *criterion of success* has been defined and the treatment result of a pair is given one of the three ratings: (1) Treatment A better, (2) Treatment B better, (3) No difference.

Now consider the *first* patient pair. If A is better, a cross is made in the square immediately *above* the black square in the charts; but if B is better, a cross is made in the square immediately *to the right* of the black square; and if there is no difference, no entry is made in the charts. The results for the *second* and subsequent pairs are entered on the chart in a similar way using the square above or to the right of that marked for the preceding pair.

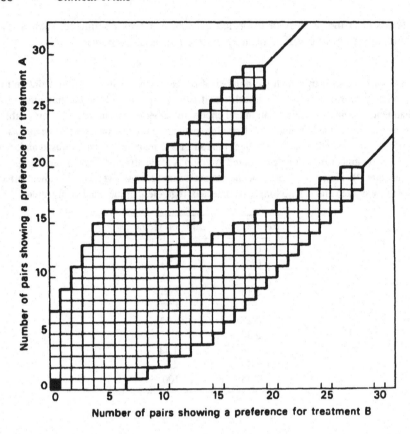

Figure 20.8. Sequential analysis chart with an α-risk of $2\alpha = 0.20$. Courtesy of Ciba-Geigy Ltd., Basle, Switzerland[29]. See also Bross[30].

Once a level of significance has been chosen, such as $2\alpha = 0.20$ in Figure 20.8, this means an α-risk of 0.10 for A *better than* B and an α-risk of 0.10 for B *better than* A. The total *patient intake* for the trial does not require prior specification and a built-in *stopping rule* is a feature of any sequential analysis trial.

The stopping rule is related to the *boundaries* of the square pattern of the chart, such that when the *upper boundary* is crossed the conclusion is that A is better than B, when the *lower boundary* is crossed, the conclusion is that B is better than A and when the *middle region* is entered, the conclusion is that *there is no difference between* A and B.

In section 8.2 when null, positive and negative results of studies were discussed it was noted that negative results hardly ever appear in the literature; only two could be quoted. One of these was a sequential analysis trial for head and neck cancer[31] and Figure 20.9 shows the results, which were rather surprising to the clinicians who set up the trial. Statistically the plan is interpreted[30] as follows.

- If placebo plus radiation is the superior treatment, the chance of this trial result occurring is 9 in 10.

- If there is no difference between treatments, the chance of this trial result occurring is 1 in 10 (i.e. $P = 0.10$).

- If Razoxane plus radiation is the superior treatment, the chance of this trial result occurring is 1 in 1000.

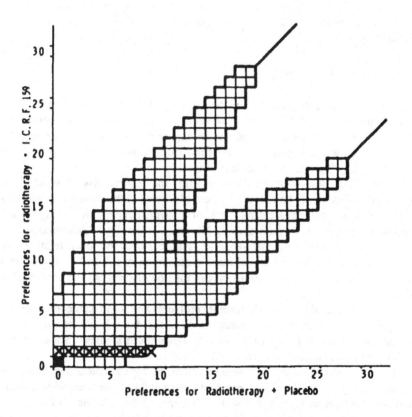

Figure 20.9. Sequential analysis of preferences for treatment type. A preferance for radiation plus Razoxane (I.C.R.F. 159) is marked in a vertical direction and for radiation plus placebo in a horizontal direction. The α risk in this plan was given by $2\alpha = 0.2$.

20.10   CROSSOVER TRIALS

A crossover trial is one in which the same group of patients are given both treatments. The two groups of patients may have different physical characteristics causing them to react differently even though the drugs are having similar effects. This may be avoided, after a certain period of time, by changing over the drugs under test to the alternative patient group. Each patient is thus their own control. (The alternative design, which has been the one discussed in the previous sections is a *parallel group design*). By using the crossover design, one advantage is that fewer patients are required.

However, such trials are of limited value because the condition of the patient changes with time, patients may drop out after the first treatment, there may be a carry over treatment effect from one period to another and crossover trials cannot be used for conditions which can be cured. Their use is controversial[12] and has been discouraged by the US Food and Drug Administration.

20.11   META-ANALYSIS

The technique of meta-analysis is the pooling of results from a number of small trials, none in itself (or at least very few) large enough to demonstrate statistically significant differences, but in aggregate capable of doing so. An example is shown in Figure 20.10 for trials of beta-blockers in the prevention of deaths following a myocardial infarction.

It is known that there is a tendency for the relative risk of death from myocardial infarction to be reduced by beta-blockers but the confidence intervals in most studies include a relative risk of 1, which means that a significant reduction in mortality was not shown. However, when the 11 randomised trial results are combined using meta-analysis, Figure 20.10, the mean relative risk is 0.65 and the confidence interval reach a maximum of 0.8 and do not contain a relative risk of 1. A clear conclusion can then be made about the preventative effect of beta-blockers[32].

The technique does, though, have its problems and for instance there are the two questions to be answered, bearing in mind that bias should be minimised. Should meta-analysis be restricted only to randomised studies? Should meta-analysis be restricted only to published studies?

These questions must be answered individually for each meta-analysis study by considering the *qualitative component* (application of predetermined criteria of quality) which takes into account such items as completeness of data and absence of bias; and the the *quantitative component* which takes into account the integration of numerical data.

The possibility of *competing risk bias* should also be considered. Thus for example, the effects of breast cancer treatment may differ in populations at different risk of non-cancer deaths, whether effects are evaluated for all causes

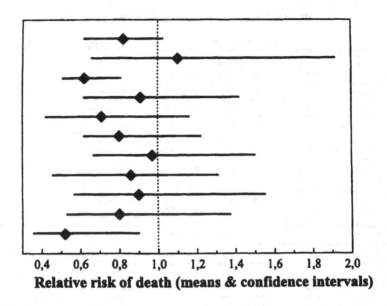

Relative risk of death (means & confidence intervals)

Figure 20.10. Randomised trials of beta-blockers[32] to study the prevention of deaths following a myocardial infarction. By meta-analysis it is found that the mean relative risk is 0.65 and the confidence interval maximum is 0.80.

(i.e. overall survival rate) or only breast cancer related deaths (i.e. cause-specific survival rate).

Nevertheless there are justifications for meta-analysis, including the problem of accruing adequate numbers of patients for a single randomised trial with for example a power $(1 - \beta) = 0.80$ and an α risk of $P = 0.05$.

It is also impossible to undertake *combined group analysis* where all the data on treatment A cases and all on treatment B cases are combined and an analysis is made on the two large groups, because there is not complete homogeneity for all studies.

A further justification is that in a common disease such as breast cancer, treatments of only small to moderate benefit could prolong many lives, since for example, breast cancer mortality in the USA is some 50,000 women per year. Very large numbers are required to detect small to moderate differences and most trials cannot accrue sufficient numbers.

A breast cancer meta-analysis study has been undertaken by the Early Breast Cancer Trialists' Collaborative Group[33]. The numbers which have been used for meta-analysis (from 133 randomised trials involving 31,000 recurrences and 24,000 deaths among 75,000 women) could obviously never have been accrued for a single randomised trial. Figure 20.11 illustrates the format of the meta-analysis results.

MORTALITY among women aged UNDER 50 at entry

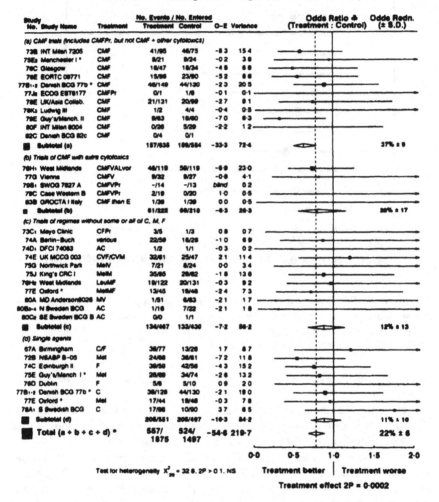

Figure 20.11. Example of the format of results of a meta-analysis study for breast cancer[33], see text for an explanation of the columns $O - E$ and Odds Ratio.

The columns are self-explanatory except for the last two[11]. The differences have been calculated between the observed number of deaths (O) and the expected number of deaths (E) for the treatment, assuming that there is no difference between treatment and control. This difference is written as $O - E$ and the results given for each trial. A negative value reflects that the treatment group had fewer deaths than expected.

The horizontal bar chart part of Figure 20.11 plots the ratio of treatment to control mortality rates (an *odds ratio*) with a 99% confidence interval (symbolised by ♣) for each trial. A value for the odds ratio of less than 1 indicates that mortality is less for the treatment than control. The odds reduction is the percentage improvement. The hollow diamond symbol is centred on the average ratio of mortality rates and its length represents a 95% confidence interval.

In conclusion, two summarised viewpoints of the benefits of meta-analysis are as follows. For such a disease as breast cancer, the large number of subsets of interest make meta-analysis mandatory because no trial has a large enough acocrual to evaluate treatment differences in each of these subsets. Meta-analysis is useful for hypothesis generation (but not as a hypothesis test) in that evaluation of the meta-analysis study differences can lead to the formation of important hypotheses deserving of further study.

20.12 REPORTING CLINICAL TRIAL RESULTS

The good reporting of clinical trial (sometimes now termed *clinical outcome studies*) results has been considered to be an implicit requirement throughout this Chapter, although it has seldom been explicitly mentioned, except in Table 20.14, EORTC[9] recommendations for *reporting adverse effects* and *publication policy*. It is a much neglected topic, as noted by Overgaard and Bentzen[34] and Bentzen[35] in 1988.

A major improvement in this field of reporting results is the publication of the Consolidationof Standards of Reporting Trials (CONSORT) guidelines[36]. Table 20.18 is reproduced from these guidelines, as published by Bentzen[35] with added radiotherapy items.

Also reproduced (Table 20.19) from Bentzen[35] is a useful summary of the most commonly used endpoints in radiotherapy trials, the topic listed in section 20.2 on the first page of this Chapter, in Table 20.12 after Pocock[4], in Table 20.14 after EORTC[14] and also discussed in Chapter 21. It was noted by Bentzen that a review[37] of 132 journal papers published between October and December 1992 found that in 62% of the 132 at least one endpoint was not clearly defined and in 39/64 papers it was unclear as to how death was included in the analysis of time to progression. This clearly shows up a problem in endpoint definition and reporting.

Table 20.18. Guidelines for reporting clinical outcome studies in radiotherapy and oncolgy, after Bentzen[35] and CONSORT[36] with the radiotherapy items added by Bentzen printed *italicised*.

Heading	Subheading	Item	Was it reported Yes/No/NA§	If Yes, on what page No.?
Title		1. Identify the study as a randomized trial	—	—
Introduction		2. State the prospectively defined hypothesis, clinical objectives and planned subgroup or covariate analyses	—	—
Methods	Study design	3. Define the patient population, inclusion and exclusion criteria	—	—
		4. Planned treatments and their timing	—	—
	Radiotherapy	5. *Radiotherapy dose prescription method, dose-planning procedure*	—	—
		6. *Target volume definition, critical organs considered, simulation and verification procedures*	—	—
		7. *Dose fractionation details*	—	—
		8. *Planned RT quality assurance procedures*	—	—
	Endpoints and analysis	9. Primary and secondary endpoints, specific follow-up procedures, the minimum clinically relevant difference, the target sample size and how it was decided	—	—
		10. Statistical analyses, their purpose and methods used, and whether the intention-to-treat principle was used	—	—
		11. Trial monitoring, early stopping rules	—	—
	Randomization	12. Method used for randomization	—	—
		13. Method of concealment and time of randomization	—	—
		14. Method to separate the generator of random treatment assignment from the treating physician	—	—
	Masking	15. Describe any blinding procedures (if relevant)	—	—

Table 20.18. continued

Heading	Subheading	Item	Was it reported Yes/No/NA[§]	If Yes, on what page No.?
Results	Patient flow and follow-up	16. Provide an overview of number of patients randomized, compliance with treatment, RT quality assurance results	——	——
	Analysis	17. State the effect of treatment on primary and secondary tumor outcome measures, including effect estimates with confidence intervals	——	——
		18. *Describe the incidence and grade of treatment-induced early and late toxicity by treatment group*	——	——
		19. State frequencies as absolute numbers when feasible (e.g. 10/20 and not just 50%)	——	——
		20. Present summary data with appropriate statistics to permit alternative analyses or interpretations and comparisons with other trials on the same problem	——	——
		21. Describe prognostic variables by treatment group, check if they were balanced, and if not describe attempts to adjust for them	——	——
		22. Describe protocol deviations from the study as planned, together with reasons	——	——
Discussion		23. State the interpretation of the study findings, including sources of bias and imprecision, discuss how this trial compares with other similar studies	——	——
Conclusion		24. State the general interpretation of the trial in view of all available evidence in the literature	——	——

[§]NA, not applicable, certain items apply to randomized studies only.

Table 20.19. Most commonly used clinical endpoints in radiotherapy trials, after Bentzen[35].

Endpoint	Definition of event	Comments
Local§ control	No evidence of disease at the primary site (T-position)	Statistically this is the absence of an event, namely local recurrence. Estimated at a given time as the local-recurrence-free rate
Local§ failure rate	Recurrence in T-position	Estimated as 1 minus the local-recurrence-free rate
Loco-regional control	No evidence of disease in T- and N-position	See above, but for T- or N-recurrence, whichever comes first. Sometimes defined as *in-field control*, i.e. control within the treated volume.
Survival or overall survival	Death irrespective of cause	
Cancer-specific survival or cause-specific survival	Death of cancer	Often defined as death of cancer or with active disease
Disease-free survival	Any recurrence or death from any cause, whichever comes first	A composite endpoint combining survival and disease control
Local§ recurrence-free survival	Local recurrence or death from any cause, whichever comes first	As above but with local recurrence as the relevant event
Disease-free rate	Any recurrence	Death without recurrence is a cause of censoring
Local§ relapse-free rate	Local recurrence	As above
Early reactions	Signs or symptoms of a specific early reaction Defined by measuring scale	Typically defined inside a time window Actuarial methods normally not used
Late reactions	Signs or symptoms of a specific late reaction Defined by measuring scale	Actuarial methods should be used
Quality of life	Typically defined by the measuring scale Typically defined by the measuring scale	May require special statistical considerations May require special statistical considerations

§All definitions involving local recurrence are defined for nodal recurrence (N-position) or distant recurrence (M-position) by analogy. Similarly for loco-regional response which means recurrence in T- or N-position, whichever comes first.

A BIBLICAL CLINICAL TRIAL

In the Bible in the book of Daniel, Chapter 1, it is related that King Nebuchadnezzar II, having invested Jerusalem and defeated Israel in 600 BC, took several youths back to his own country for indoctrination and training. They were carefully selected. All were of royal or princely blood and were physically fit and of high intelligence. They were put on a rigid diet of meat and wine for three years with one of the eunuchs acting as the trial monitor. Daniel persuaded the monitor to give him and three others a diet of pulse and water for 10 days, when it is recorded they were fairer in countenance and fatter in body than the other subjects who were given meat and wine. Daniel had ruined the trial, the eunuch had defied the king, and the trial had become uncontrolled. It is not recorded what Nebuchadnezzar did to the eunuch but he may well have been thrown to the lions!

Chapter 21

Cancer Treatment Success, Cure and Quality of Life

21.1 INTRODUCTION

For many years, until well into the 1960s, the term cure was used to signify the results of successful cancer treatment with the five-year survival rate often being equated to cure. However, for the earliest cancer treatment results following the discoveries of X-rays and radium, the textbooks and journals typically wrote of one-month to six-month cures. Figures 21.1–21.2 show early treatment results with X-rays and radium. In addition, there were also many spurious claims for curing cancer using quack remedies; that below is taken from the April 1896 issue of the *Windsor Magazine* which also carried the world's first illustrated journal paper on Röntgen's discovery of X-rays in Würzburg in the previous December.

Now Ready.   Price One Shilling.
CANCER:
ITS NATURE AND SUCCESSFUL TREATMENT.
Post Free One Shilling, from the Author,
H. KELWAY BAMBER, F.I.C., Westminster Chambers, 9, Victoria St., London, S.W.
E. W. ALLEN, 4, AVE MARIA LANE, E.C.

The disadvantage of equating cure with five-year survival is that although it might be acceptable for poor prognosis cancers such as lung cancer, for those with a much longer term prognosis, such as breast cancer or cancer of the cervix, a five-year survival is not appropriate.

The term *cured from cancer* also implies that no malignancy can be detected, yet it is known that some some breast cancer patients, for example, can for many years live a satisfactory life with virtually quiescent metastatic nodes present and eventually die from a cause such as heart attack or stroke. This can occur some 20 years after the initial treatment but *technically* they are not cured, and therefore by implication are a failed treatment case. This is philosophically unsound and the term *treatment success* should replace cure in many instances.

Figure 21.1. (*a*) One of the first two (the other was reported at the same meeting) histologically proven cancer cures using X-ray therapy[1]. This was in Stockholm for a basal cell carcinoma of the face in 1899. The photographs show the patient before treatment and 30 months after treatment.

Table 21.1. Karnofsky status: 10-point scale[4].

10	Normal, no complaints, no evidence of disease
9	Able to carry on normal activity, minor signs or symptoms of disease
8	Normal activity with effort, some signs or symptoms of disease
7	Cares for self, unable to carry on normal activity, or do active work
6	Requires occasional assistance, but is able to care for most needs
5	Requires considerable assistance and frequent medical care
4	Disabled, requires special care and assistance
3	Severely disabled, hospitalisation necessary, although death is not imminent
2	Very sick, hospitalisation necessary, active supportive treatment necessary
1	Moribund, fatal processes progressing rapidly

The introduction in 1949 by Karnofsky[4] of his concept of *quality of life* devised with chemotherapy in mind, went some way towards providing a method for specifying treatment success, Table 21.1 (i.e. Karnofsky status before and after treatment), and the quality of life concept has since been extended by many authors both specifically for individual cancers as well as for cancer in general. Quality of life as a measure of the wellbeing of a population is also relevant in other fields, using social indicators such as housing, employment and income, but this chapter is limited mainly to a discussion of those scales which can be used for cancer patients.

Treatment success, and whether it is worthwhile, can be viewed differently depending upon who is making the assessment, Table 21.2. The doctor's and patient's viewpoint may not always coincide and for example a patient might refuse a mutilating facial operation, even if plastic surgery can later ameliorate the cosmetic appearance and even if the primary tumour and nodes can viralally be guaranteed to be removed. In such a case a non-optimal treatment procedure might have to be accepted by the physician/surgeon.

Table 21.2. From whose point of view is treatment success defined?

Patient
Doctor
Statistician
Hospital Manager/Accountant

The statistician' s viewpoint might depend on a mathematical model which after the input of various parameters gives an estimate, with standard errors, of a *cured fraction of patients* in the population under study or a predicted long-term survival rate. Such a model might be appropriate for series of large numbers of patients, but could not be applied for small numbers.

With the world's population having a longer life expectancy; cancer treatment costs increasing significantly for more and more complex equipment (e.g. computer controlled (to a certain extent) radiotherapy linear accelerators with multileaf collimators for conformal treatments, 3D optimisation and virtual reality treatment planning computer software) and for new drugs; with cancer predominantly a disease of the elderly; and the worldwide trend in healthcare being towards managed care (managers and accountants in financial control) both in Europe and in North America, with a limited budget available; decisions must now be made as to whether certain sophisticated and expensive cancer treatments can be afforded. This leads to the use of Bayesian statistics and decision theory.

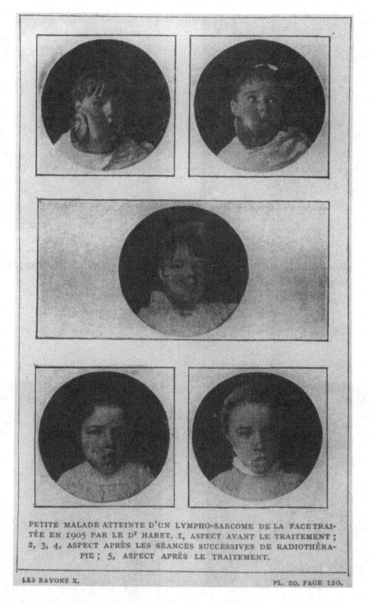

PETITE MALADE ATTEINTE D'UN LYMPHO-SARCOME DE LA FACE TRAI-
TÉE EN 1905 PAR LE Dr HARET. I, ASPECT AVANT LE TRAITEMENT ;
2, 3, 4, ASPECT APRÈS LES SÉANCES SUCCESSIVES DE RADIOTHÉRA-
PIE ; 5, ASPECT APRÈS LE TRAITEMENT.

LES RAYONS X. PL. 20, PAGE 120.

Figure 21.1. (*b*) Complete regression of a sarcoma of the face. Patient of Dr Haret, Paris, 1905: reported by Niewenglowski[2] in 1924. (Courtesy of Professor Jean-Marc Cosset, Institut Curie, Paris)

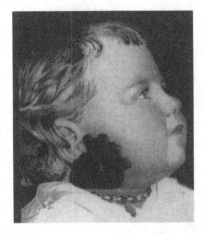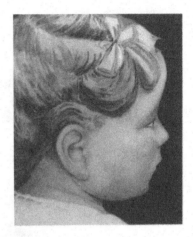

Figure 21.2. (*a*) Successful treatment of a haemangioma by Drs Wickham and Degrais[3] in 1906, Hospital St. Louis, Paris.

Figure 21.2. (*b*) Successful treatment of an epithelioma of the parotid by Drs Wickham and Degrais[3] in 1908, Hospital St.Louis, Paris.

21.2 QUALITATIVE ASSESSMENT

The ideal assessment is quantitative and it is seldom that qualitative or semi-quantitative assessments are satisfactory. Table 21.3 gives two examples of qualitative scales and Figure 21.3 a rare examples of where such scales are effective. In this particular example of Figure 21.3, the same radiation oncologist made all the assessments for the entire series of 76 patients.

An obvious conclusion can be drawn in favour of brachytherapy versus external beam radiotherapy for inoperable lung cancer and for *the particular treatment protocols*. The last four words in the previous sentence have been italicised since this emphasises that such results cannot be extrapolated to other treatment regimes.

Table 21.3. Two examples of qualitative scales.

Excellent	Better
Good	Same
Poor	Worse

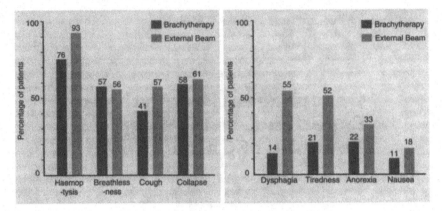

Figure 21.3. Histograms of the results from a trial of 76 patients with unilateral inoperable lung tumours who were treated using either high dose rate (HDR) endobronchial brachytherapy or by external beam radiotherapy. (Left) shows the percentage of patients who at two months follow-up had their symptoms assessed as *Better* on a scale of *Better, Same, Worse*. (Right) shows the percentage of patients who at two months follow-up had their symptoms assessed as *Worse* on a scale of *Better, Same, Worse*. These results indicate the superiority of HDR brachytherapy.

21.3   SEMI-QUANTITATIVE ASSESSMENT: COMPLETE, PARTIAL AND NO REGRESSION

The semi-quantitative scale of *Complete Regression (CR), Partial Regression (PR) and No Regression (NR)* is widely used and can be semi-quantitative in that the assessments can be made in some studies by a measurement of tumour size, (as distinct from a purely clinical assessment without any measurements taken). However, although an improvement on the purely qualitative method it still suffers from the disadvantage that there will be overlaps between CR & PR and PR & NR when the assessments are made by different physicians for patient subgroups. This can to a certain extent be overcome by having a panel of physicians to make the assessment rather than a single physician.

There is also the problem that CR may not last indefinitely and is therefore in effect an incomplete remisssion and cannot be used as a true measure of treatment success unless this assessment is made time dependent, e.g. at three months for all patients, see last line in Table 21.4.

21.4   QUALITY OF LIFE: 10-POINT OR FOUR-POINT SCALE?

The Karnofsky status scale[4], Table 21.1, is a 10-point scale and as such it is difficult, if not impossible, to guarantee similar assessments by different physicians/surgeons for an identical patient. Simpler scales have been devised

Table 21.4. Factors relevant to quality of life and treatment success assessment.

Which parameters should be measured
Self reporting *versus* external evaluation
Response alternatives (Yes/No or a scale)
Number of items to be assessed
Time frame of the questions

such as the ECOG performance scale, Table 21.5, which is also known as the Host, Zubrod or WHO scale[6] and is related to the Karnofsky scale as indicated in Table 21.6.

Table 21.5. ECOG performance status four-point scale.

0	Normal activity
1	Symptomatic and ambulatory, cares for self
2	Ambulatory more than 50% of time, occasionally needs assistance
3	Ambulatory 50% or less of time, nursing care needed
4	Bedridden, may need hospitalisation

Table 21.6. Equivalence between ECOG (Eastern Cooperative Oncology Group) and Karnofsky scales. The first two columns give the equivalence between the two scales. The third and fourth columns present clinical data[7] on the influence of performance status on patients with inoperable lung cancer. These median survival statistics are based on data for 5022 males, lung cancer of all histologies, entered onto Veterans Administration Lung Cancer Study Group (VALG) protocols from 1968–1978. The ECOG scale was first reported in 1960 by Zubrod[6].

ECOG (Zubrod) (Host) (WHO)	Karnofsky	Median survival (weeks)	Patients in group (%)
0	100	34	2
1	80–90	24–27	32
2	60–70	14–21	40
3	40–50	7–9	22
4	20–30	3–5	5

Table 21.7 is an example[8] of the use of the ECOG scale before and after treatment using HDR endobronchial brachytherapy. Treatment delivery is similar to the method for Figure 21.3 results but is using a different treatment protocol for Table 21.7 results. Figure 21.4 illustrates the technique of HDR endobronchial brachytherapy. These results in Table 21.7 clearly demonstrate improved quality of life when comparing the status immediately post-treatment and at six week follow-up for these lung cancer patients.

Table 21.7. ECOG performance status: 100 patients treated using HDR endobronchial brachytherapy (HDREB)[8].

ECOG status	No. of patients with a given ECOG status	
	After 1st. HDREB	At 6 weeks follow-up
0	3	40
1	39	44
2	15	11
3	18	0
4	25	5

21.5   FURTHER EXAMPLES OF QUALITY OF LIFE STATUS SCORING

The Karnofsky and ECOG scales allocate a single number to define patient status. Other quality of life status proposals have extended this concept so that several functions can be individually scored giving multidimensional aspects to quality of life assessment, and in some cases the summation of the separate scores for the major aspects (e.g. symptoms and side effects, psychological functioning, social functioning, physical functioning, as in the proposal by Tchekmedyian *et al*[9] for treatment of cancer anorexia) can be made to provide a global score.

However, it must be realised that problems can arise in using only a global score computed from for example four components, because the same global score can arise if in one case components 1 and 2 have a low score and 3 and 4 a high score, and if components 1 and 2 have a high score and 3 and 4 a low score.

A recent example mentioned in the literature is for the LENT SOMA scoring system published in 1995 both in the *International Journal of Radiation Oncology Biology & Physics*[10] and in *Radiotherapy & Oncology*[11] (LENT = Late Effects Normal Tissues; SOMA = Subjective, Objective, Management & Analytic).

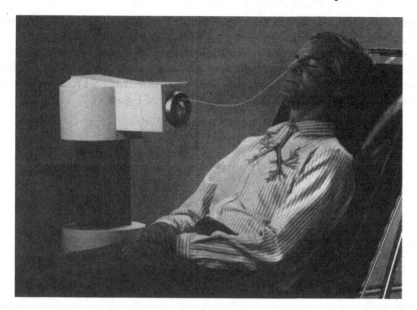

Figure 21.4. The technique of HDR endobronchial brachytherapy involves the use of a remote controlled afterloading machine in which the radioactive source is a miniaturised ^{192}Ir pellet on the end of a flexible cable which can be inserted within a catheter into a bronchus. This figure shows a microSelectron-HDR afterloading machine with the catheter inserted *via* the nose into a bronchus. A drawing of the bronchial tree is superimposed on the patient's shirt.

A corrrection in 1996[12] stated 'It was originally recommended that the scores should be summed and divided by the number of elements scored. This is not advocated'. This correction was made because it was realised that a situation could arise where the patient scores could be 0 or 1 for most elements in the SOMA scale but there could be a grade 4 for just one or two components of injury. These high scores would be diluted out by the low scores and give a misleading average score.

A statistical method of producing a valid global score for the LENT SOMA scales has as the time of writing (1998) not yet been recommended by RTOG and EORTC. Nevertheless, these tables are extremely useful with the following 10 major sections (each subdivided into several site groups): central nervous system, head and neck, breast, heart, lung, gastrointestinal, major digestive glands, genitourinary, gynecologic and, finally, bone, muscle and skin.

Three examples are now given for different cancer treatment sites: lung[8] (Table 21.8), limbs[13] (Table 21.9) and head and neck[14,15] (Table 21.10).

In addition the ASA (American Society of Anesthesiologists) classification of physical status[16] is given in Table 21.11. One example of when this has

288   *Cancer Treatment Success*

been used is for a preoperative study of two patient groups, one of which were to undergo a single knee replacement and the second group to undergo both knee replacements at the same operating session. To ensure no bias in patient status between the two groups both were analysed preoperatively (using the chi-squared test and a 2 × 4 contingency table) for ASA status I–IV. Some of the results from this study are shown in the box plot of Figure 21.5.

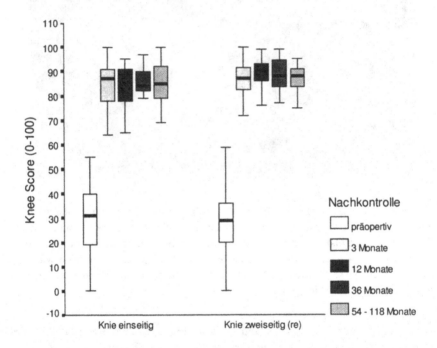

Figure 21.5. Knee status score (0–100) using the scoring scale of Insall *et al*[17] which awards positive points for pain (max= +50), motion (max= +25), anteroposterior stability (max= +10) and mediolateral stability (max= +15) and negative points for flexure contraction (max= −15), extension contraction (max= −15) and deviation (max= −20). This is an example of a scoring system which includes both positive and negative scores and for which it is theoretically possible for a global score to be negative. Although by convention when this ASA scale is used, all negative values are equated to zero. The results are shown for the right knee (Left): when only the right kneee was replaced; (Right): for the right knee when both right and left knees were replaced). The box plots show the 95% percentiles, the 25% and 75% quartiles and the median: preoperatively and at various follow-up times: 3, 12, 36 and during 54–118 months. (Courtesy: Dr.med. Christoph H. Kindler of the Department of Anasthesie, Kantonsspital, Basel.)

21.5.1 Lung

Table 21.8. Symptom scoring index, after Speiser and Spratling[8]. Four symptoms are scored and although an overall score can then be computed the results can be more effectively expressed for each symptom as in Figure 21.6.

	Dyspnea
0	None
1	Dyspnea on moderate exertion
2	Dyspnea with normal activity, walking on level ground
3	Dyspnea at rest
4	Requires supplementary oxygen
	Cough
0	None
1	Intermittent, no medication necessary
2	Intermittent, non-narcotic medication
3	Constant or requiring narcotic medication
4	Constant, requiring narcotic medication but without relief
	Haemoptysis
0	None
1	Less than two episodes per week
2	Less than daily but greater than two per week
3	Daily, bright red blood or clots
4	Decrease of Hb/Hct more than 10%, greater than 150 cc, requiring hospitalisation leading to respiratory distress, or requiring more than 2 units transfusion
	Pneumonia/Elevated temperature
0	Normal temperature, no infiltrates, WBC less than 10,000
1	Temperature greater than 38.5 and infiltrates, WBC less than 10,000
2	Temperature greater than 38.5 and infiltrates and/or WBC greater than 10,000
3	Lobar consolidation on radiograph
4	Pneumonia or elevated temperature requiring hospitalisation

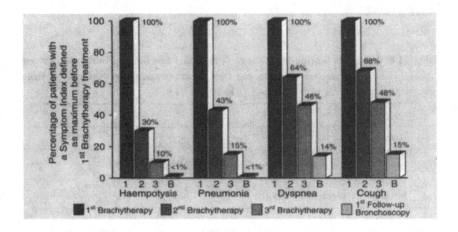

Figure 21.6. Symptom index response expressed as a percentage of a weighted index at each brachytherapy treatment (of which there were three) and at first follow-up bronchoscopy. The scores are weighted and normalised to 100% for the first score. These results from Speiser and Spratling in 1994[8] are more informative than their earlier ones from 1990 which used only ECOG status, see Table 21.7.

21.5.2 Limbs

The scale in Table 21.9 was developed by the Musculoskeletal Tumor Society and is taken from the paper by Bolek *et al* [13] for radiotherapy of Ewing's sarcoma of the extremeties. These authors quoted mean scores of 26.0 for radiotherapy given once daily (number of patients= 6) and 29.4 for radiotherapy given twice daily (number of patients= 9) and using a *t*-test obtained a *P*-value of 0.15. This is not surprising since the scores are close together and there were only a few patients in the study. Table 21.9 is also an example of a quality of life scale which (unlike Karnofsky or ECOG for example) can be used by the patient, i.e. for self assessment, see Table 21.4, or by the physician/surgeon.

Table 21.9. Functional limb evaluation scale with a total maximum score of 30. This is an example of a scale where the highest score represents the best outcome, whereas for some scales such as ECOG, Table 21.5, the highest score represents the worst outcome.

Lower limb	
Pain	0 =Severe, 1 =Moderate, 2 =Intermediate, 3 =Modest, 4 =Mild, 5 =None
Function	0 =Total disability, 1 =Partial disability, 2 =Intermediate, 3 =Recreational restriction, 4 =Mild restriction, 5 =No restriction
Emotional acceptance	0 =Dislikes, 1 =Accepts, 2 =Intermediate, 3 =Satisfied, 4 =Likes, 5 =Enthused
Supports	0 =Two canes/crutches, 1 =One cane/crutch, 2 =Intermediate, 3 =Brace, 4 =Occasional minor support, 5 =None
Walking	0 =Unable unaided, 1 =Inside only, 2 =Intermediate, 3 =Moderate limitation, 4 =Mild limitation, 5 =Unlimited
Gait	0 =Major HCAP, 1 =Major cosmetic/Minor HCAP, 2 =Intermediate, 3 =Minor cosmetic, 4 =Barely detectable, 5 =Normal
Upper limb	
Pain	0 =Severe, 1 =Moderate, 2 =Intermediate, 3 =Modest, 4 =Mild, 5 =None
Function	0 =Total disability, 1 =Partial disability, 2 =Intermediate, 3 =Recreational restriction, 4 =Mild restriction, 5 =No restriction
Emotional acceptance	0 =Dislikes, 1 =Accepts, 2 =Intermediate, 3 =Satisfied, 4 =Likes, 5 =Enthused
Hand positioning	0 =Flail, 1 =Not above waist, 2 =Intermediate, 3 =Not above shoulder or no Pro/Sup, 4 =Mild deficit only, 5 =Unlimited
Dexterity	0 =Cannot grasp, 1 =Cannot pinch, 2 =Intermediate, 3 =Loss of fine movement, 4 =Mild deficit only, 5 =Normal
Lifting ability	0 =Cannot, 1 =Helping only, 2 =Intermediate, 3 =Moderate limitation, 4 =Mild limitation, 5 =Normal

21.5.3 Head and Neck

Table 21.10 gives the performance status scale for head and neck cancer patients modified from List *et al*[14]. This scale was used by Moore *et al*[15] to assess outcome after primary radiotherapy for squamous cell carcinoma of base of tongue and the mean scores were reported by T-stage and were in the following ranges: eating in public (75–91), understandability of speech (83–100) and normalcy of diet (60–94).

Table 21.10. Performance status scale to assess post-treatment functional results for head and neck cancer patients.

	Eating in public
100	No restriction of place, food or companion
75	No restriction of place, but restricts diet when in public
50	Eats only in presence of selected persons in public
25	Eats only at home in presence of selected persons
0	Always eats alone
	Understandability of speech
100	Always understandable
75	Understandable most of the time, occasional repitition necessary
50	Usually understandable, face-to-face contact necessary
25	Difficult to understand
0	Never understandable, may use written communication
	Normalcy of diet
100	Full diet: no restrictions
90	Peanuts
80	All meats
70	Carrots, celery
60	Dry bread and crackers
50	Soft, chewable foods (macaroni, small pieces of meat)
40	Soft foods requiring no chewing (mashed potatoes, apple sauce)
30	Pureed foods
20	Warm liquids
10	Cold liquids

21.5.4   American Society of Anesthiologists Scale

Table 21.11. ASA classification of physical status[16].

Class	Physical status
I	A healthy patient with no systemic disease processes
II	A patient with a mild to moderate systemic disease process caused either by the condition to be treated surgically or other pathological process and which does not limit the patient's activities in any way, e.g. mild diabetic, treated hypertensive, or heavy smoker
III	A patient with a severe systemic disturbance from any cause, and which imposes on him or her e.g. ischaemic heart disease with a limited exercise tolerance, severe chronic obstructive airways disease with dyspnoea on exertion
IV	A patient with severe systemic disease which is a constant threat to life, e.g. the chronic bronchitic who is dyspnoeic at rest, advanced chronic liver failure
V	A moribund patient who is unlikely to survive 24 hours with or without surgery
E	Emergency operation. Any patient in any of the above classes who is operated on as an emergency is regarded as being in poorer physical condition, and the letter E is prefixed

21.6   SELF ASSESSMENT *VERSUS* EXTERNAL EVALUATION

Whether or not self assessment by the patient is appropriate (see Table 21.4) will depend on the particular study and the choice and complexity of the quality of life scale. One example where self assessment is compared to assessment by the healthcare provider is given in Table 21.12 which in this instance is seen to be well correlated for 300 evaluable patients, (a Spearman rank correlation coefficient of 0.55 and $P < 0.0001$). Although 36% of the time patients judged their dermatitis to be more severe than did their health care provider, whereas only 7% of the time patients judged their dermatitis as less severe than did the physician (McNemar test $P < 0.0001$).

Table 21.12. Comparison of self assessment by the patient and external evaluation (by the healthcare provider) of the maximum dermatitis severity on a four-point scale (0 =None, 1 =Mild, 2 =Moderate, 3 =Severe) for studies of an aloe vera gel *versus* placebo for radiation induced skin toxicity, data from Williams *et al*[18].

No of cases in which there was an external evaluation report of a given severity score	No. of cases in which the patient reported a given severity score			
	0	1	2	3
0	4	8	9	2
1	3	57	51	10
2	3	6	77	28
3	0	1	10	31

21.7   DEMONSTRATION OF THE EXISTENCE OF A CURED GROUP OF PATIENTS

In these two examples, subsections 21.7.1 and 21.7.2, the term *cure* is uniquely defined in each case and if the concept is acceptable then cure can be demonstrated. However, these methods do have certain problems in obtaining sufficient data in order to be able to use the methods. For this reason, they have not been widely used.

21.7.1   Relative Survival Rate and Cohort Interpolated Life Tables

The *relative survival rate* (sometimes termed the *age corrected survival rate*) is defined below where the normal population is that of the county/region/country from which the cancer patient series has been drawn:

> 100× ((Crude T-year survival rate)/(Expected T-year survival rate in the normal population for a group of people with the same age and sex distribution as the treated group)) = Relative T-year survival rate

The problem with this equation is that in many instances the data to determine the denominator is not available. In England and Wales it is fortunate that such data do exist, in the form of tables by sex for birth year cohorts every five years (e.g. 1896, 1901, 1906 etc) which give the yearly expected survival probabilities for one, two, three, four and five years, conditional on having attained a certain age. Such tables are called *cohort interpolated life tables* and Table 21.13 is an example of a part of such a table.

Table 21.14 illustrates how values of $_np_x$ can be calculated from the probabilities given in Table 21.13 using as an example the computation of the

Table 21.13. Example of part of a cohort interpolated life table quoting survival probabilities for 1, 2, 3, 4 or 5 years for each year from age 50 to 59, conditional on having attained that age. These tables were published by the Chester Beatty Research Institute, London, in 1974 for a series of birth cohorts every 5 years, e.g. 1881, 1886, 1891, 1896. These years were chosen since the second year in every decade since the 1840s (i.e. 1841, 1851 ...) is the year chosen for the census returns of England and Wales which take place every 10 years. The survival probabilities are given the notation $_n p_x$ where x is the attained age (50, 51, 52, 59) and n is the number of years survived ($n = 1, 2, 3, 4, 5$) corresponding to the particular probability.

From age x	Survival probability for n years ($_n p_x$)				
	$n = 1$	$n = 2$	$n = 3$	$n = 4$	$n = 5$
50	0.98962	0.97868	0.96714	0.95483	0.94171
51	0.98894	0.97728	0.96484	0.95159	0.93748
52	0.98821	0.97563	0.96223	0.94796	0.93279
53	0.98727	0.97371	0.95927	0.94392	0.92762
54	0.98626	0.97164	0.95609	0.93957	0.92169
55	0.98517	0.96941	0.95266	0.93453	0.91493
56	0.98400	0.96700	0.94859	0.92870	0.90724
57	0.98273	0.96402	0.94381	0.92200	0.89853
58	0.98097	0.96039	0.93820	0.91432	0.88869
59	0.97903	0.95641	0.93206	0.90593	0.87831

probability of surviving from age 50 to 53, i.e. $_3 p_{50}$, having been born in 1891 (or indeed during the five year period of which 1891 is the central value, i.e. 1889–1893: the 1896 birth cohort is for the period 1894–1898).

Table 21.14. Computation of $_3 p_{50}$ using the $_n p_x$ values in Table 21.13.

Probability of survival from age 50 to 51 $= _1 p_{50} = 0.98962$
Probability of survival from age 51 to 52 $= _1 p_{51} = 0.98894$
Probability of survival from age 52 to 53 $= _1 p_{52} = 0.98821$

Thus probability of survival from age 50 to age 53 is the product
$0.98962 \times 0.98894 \times 0.98821 = 0.96714$
Now note that in Table 21.13, $_3 p_{50} = 0.96714$
which agrees with the above computation.

21.7.2   Method of Easson and Russell

Easson and Russell[19] of the Christie Hospital, Manchester reviewed in 1968 cases treated during the period 1932–49 and defined the term *cure* in the following manner: 'We may speak of cure of a disease when there remains a group of disease-free survivors, probably a decade or two after treatment, whose annual death rate from all causes is similar to that of a normal population group of the same sex and age distribution'.

Their method of presentation, see Figure 21.7, was to use semi-logarithmic graph paper to plot survival curves of the treated group of patients (the *observed*) and of the normal population group (the *expected*) with the same age and sex structure. If a time T exists where the two curves become parallel then it has been demonstrated at time T that there exists a group of patients who are cured.

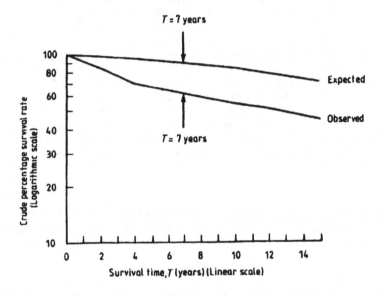

Figure 21.7. Graphical demonstration that cure has been demonstrated at $T = 7$ years. Data for 262 stage I cancer of the cervix patients[19].

Easson and Russell also commented that if parallelism was clearly not observed then one of two conclusions may be drawn. 'Either the treatment is not successful, or the expected survival of the normal population is not the appropriate yardstick for comparison with the observed survival rate. This would be true if those patients who have cancer in a specific site also have a higher risk of dying from some other cause. For example with laryngeal cancer there is some evidence to suggest that the incidence of deaths from cancer of the alimentary canal may be higher in these patients'.

21.7.3 Parametric Statistical Modelling: Statistical Cure

The use of parametric statistical models (where values of parameters for a given frequency distribution are required before the model can be used) for predicting long-term cancer survival rates and the proportion of patients cured were proposed in the late 1940s but have only been used occasionally. This might originally have been due to the lack of computing facilities in the 1950s and 1960s as the method requires intensive calculations, but a more serious problem is that such a method has to be validated first for a given cancer site before it can be put into general use.

Such a validation requires knowledge of cancer patient case histories over a long period of time, e.g. 1944–1962 (some 5000 were available for cancer of the cervix by stage in the late 1960s from hospitals in London, Manchester, Houston and Oslo)[20]. The validation procedure is then to take a cohort of case histories for an early period, e.g. 1944–1949, and use a series of parameter values to predict the medium to long-term survival rates, e.g. at 5, 10, 15 and 20 years. Then, because the follow-up on the patients is known (in many cases to death with the cause determined as cancer or an intercurrent disease) the predictions can be checked against the known actual experience at medium to long-term. The optimum statistical model with optimum parameter values can then be found.

Figure 21.8 shows schematically one type of model[20]. The basis of this particular model is a known formula for the distribution curve of the unsuccessful fraction of patients $(1 - C)$. For cancer of the cervix this was found to be a lognormal distribution, which has two parameters, mean and standard deviation, and for which the formula is given in section 3.6. The statistically cured proportion of patients, C, was determined for the various hospital series and was found to be in the range 0.51–0.71 for stage I, 0.36–0.43 for stage II and 0.14–0.27 for stage III.

21.8 ENDPOINTS: DISEASE-FREE SURVIVAL *VERSUS* OVERALL SURVIVAL

Probabably the first improvement on overall survival (OS) at five yeaxs as an indicator of cure/treatment success was the specification of a disease-free survival (DFS). One of the obvious advantages of DFS is that it can be assessed sooner than duration of response or OS but as stated by Buyse *et al*[22], 'Whatever endpoints are finally chosen to assess treatment efficacy, one should not forget to also include a measure of the patient's quality of life as one of the parameters studied'.

The relevant dates required to calculate DFS are generally available in a patient's case history notes but one must be careful when using such information if the time lapse between the last recorded no sign of recurrence (NSR) and the first recorded recurrence (REC) is large. For cancer treatment

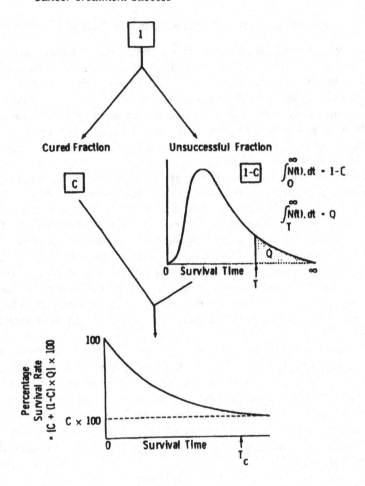

Figure 21.8. Schematic diagram of a statistical model which can be used to estimate a cured fraction (C) of patients treated for cancer. The model has been validated for cancer of the cervix using a lognormal distribution for the $(1 - C)$ fraction. Also, several other cancers, e.g. many in the head and neck[21], have been shown to have a lognormal distribution for the fraction of cases which were unsuccessfully treated in that they died with cancer present, but a complete validation procedure as for the cervix has never been made. The long-term T-year survival rate, using the notation in the diagram, is $100 \times (C + (1 - C) \times Q)$ where the total area under the lognormal survival distribution curve is 1.

Data collection

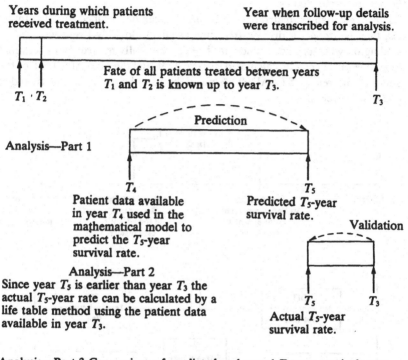

Years during which patients received treatment.

Year when follow-up details were transcribed for analysis.

Fate of all patients treated between years T_1 and T_2 is known up to year T_3.

T_1 · T_2

T_3

Prediction

Analysis—Part 1

T_4
Patient data available in year T_4 used in the mathematical model to predict the T_5-year survival rate.

T_5
Predicted T_5-year survival rate.

Validation

Analysis—Part 2
Since year T_5 is earlier than year T_3 the actual T_5-year rate can be calculated by a life table method using the patient data available in year T_3.

T_5 T_3

Actual T_5-year survival rate.

Analysis—Part 3 Comparison of predicted and actual T_5-year survival rates.

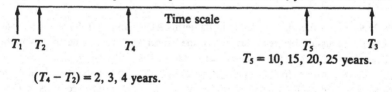

Time scale

T_1 T_2 T_4 T_5 T_3

$T_5 = 10, 15, 20, 25$ years.

$(T_4 - T_2) = 2, 3, 4$ years.

Figure 21.8. (Continued) Schematic diagram of the validation schedule for a long-term survival rate prediction model, assuming a specified analytical form (e.g. lognormal) for the distribution of survival times for those patients who die with cancer present.

sites/stages/histologies which have a good long-term prognosis, patient follow-up after 1–2 years post-treatment is often as long as 12 months and therefore any estimated date of recurrence as the mid-time between the NSR and REC dates may be accurate only to within six months or greater. This may not be sufficient for some studies.

However, if an estimated date of recurrence can be obtained for a treatment series, then an *apparent* disease free time (ADFT), possibly even two ADFTs depending on whether a recurrence has been successfully treated, and a terminal disease time (TDT) can be calculated, Figure 21.9, and the ratio ADFT/ST or the percentage difference (ST−ADFT)% might be useful for computing an index of treatment success.

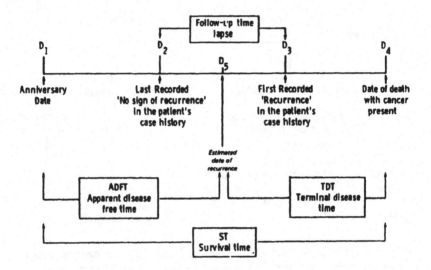

Figure 21.9. Schematic diagram defining the terms ADFT and TDT for a patient who experiences a remission before recurrence of the cancer and eventual death. D_5 is the mid-date between D_2 and D_3. For the patient who dies with cancer present without experiencing any period free from the disease, ADFT = 0. It is also possible to experience a second ADFT after successful treatment of a first recurrence of the cancer. In this case, the second ADFT is calculated using the mid-date of the second follow-up time lapse, the last recorded *no sign of second recurrence* and the first recorded *second recurrence*.

It is, however, important to calculate survival time and ADFT from the *start* of initial planned treatment and not from the *date of diagnosis* or from the *end* of the initial planned treatment. In the case of the date of diagnosis this could be defined as the date when the patient was initially diagnosed by a general practitioner, or the date a clinical diagnosis was made in a hospital, or the date when the diagnosis of cancer was confirmed histologically.

Although the time lapse between date of diagnosis and date of initial planned treatment is a measure of the efficiency of a particular hospital organisation and a measure of the length of patient waiting lists

If survival time is calculated from the end of the initial planned treatment this can vary widely with the treatment regime. Possibilities include a single day if the treatment is only surgical, maybe 6–7 weeks if it is radiotherapy or even six months if initially planned courses of chemotherapy are included. Thus if it is not the start date of the treatment which is used then for a given cancer site/stage/histology there can be a bias introduced when trying to compare like-with-like for results from different oncology centres.

Returning to DFS, two advantages of its use are as follows. If recurrence of the disease invariably leads to the patient's death after some period of time then DFS can be appropriate. When DFS is a good surrogate for OS then differences in DFS will eventually translate into OS: although there are not many instances clinically when this is positively known *a priori*. An example of good correlation between DFS and OS is lung cancer, and an example of a weak correlation is early stage colon cancer such as Duke's A or B1.

A disadvantage of the use of DFS as an appropriate endpoint is if a highly toxic treatment produces a lot of treatment related mortality. Another is when a good salvage treatment is available, since a relapse may be completely reversed (e.g. malignant melanomas and low grade lymphomas) thus making DFS an inappropriate endpoint. However, it must be emphasised that OS requires the least personal judgement and is therefore least vulnerable to bias.

21.9 SURVIVAL OF UNTREATED CANCER

Very little data have been published on this subject because of course most patients receive some form of treatment unless they adamanetly refuse or the disease has progressed to such an extent that treatment would be inappropriate. Theoretically, a comparison between treated and untreated cancer series could provide an indication of treatment success, but because of the lack of data this is only possible for one cancer site: lung, Figure 21.10.

21.10 SCORING OF COMPLICATIONS

There are many cancer site-specific proposals for complication scoring systems some of which are defined relatively simply: 1 = Complete recovery, 2 = Incomplete recovery, 3 = Surgery required, 4 = Intensive treatment required; and as a second example, 1 = Requires symptomatic treatment or advice only, 2 = Symptoms interfere with the quality of life but require conservative treatment only. 3 = Requires operative correction or otherwise causes death.

However, there are also far more detailed systems for reporting complications of which one of the most detailed is the French–Italian glossary[24]

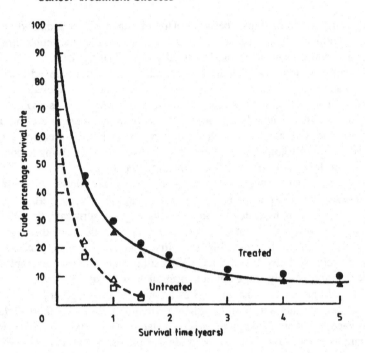

Figure 21.10. Comparison between untreated (data compiled from the literature from the 1920s–1950s) and treated (1975–79 in the N.W. Thames region of England: all histologically proven cases) lung cancer: 667 *versus* 3039 cases[23]. One-year survival rates for 2094 treated males are 30.8% (solid circles), for 463 untreated males are 9.9% (hollow triangles), for 945 treated females are 25.1% (solid triangles) and for 204 untreated females are 6.1% (hollow squares). Thus for both sexes the improvement in one-year survival when treated, compared to untreated, is some 20%.

for gynaecological cancers. This allows an accurate identification of the grade of early and late normal tissue damage using five grades of increasing severity, G0–G4, and describing each grade of complications per tissue or organ at risk. For example, rectal complications G1 include five definitions of signs and symptoms, G1a–G1e, corresponding to different types of morbidity of about the same inconvenience for the patient.

This glossary was also designed to register and analyse signs and symptoms induced by different types of treatment: radiotherapy, surgery, chemotherapy. This, as stated by Chassagne *et al*[24], then enables a comparison of the incidence and severity of the sequelae and complications, and moreover identify unambiguously each type of complication that otherwise would remain hidden behind the scoring of complications. This should at least lead to objective comparisons between series treated with different strategies when authors comply with the rules and definitions of the glossary. An example from the glossary is given in Table 21.15 for rectal complications by grade and type.

Table 21.15. Rectal complications by grade and type: French–Italian glossary[24].

G1a	Acute proctitis during radiotherapy with 10% increases of treatment time or lasting more than two months
G1b	Minor rectorragia
G1c	Rectocele
G1d	Preoperative injury with immediate repair
G2a	Rectorragia. Transfusion
G2b	Rectal ulcerations. Pain
G3a	Recto-vaginal fistula
G3b	Severe rectorragia. Surgery needed
G3c	Rectal stenosis. Surgery needed
G4	Death from complications

21.11   QUALITY OF LIFE OF CANCER PATIENTS IN DEVELOPING COUNTRIES

Pain as a complication of treatment is given as grade G2b in Table 21.15, but it should be remembered that cancer pain is a problem not only after treatment, but also before treatment and is sometimes stated to be the most frequently occurring cancer symptom.

This is particularly true in developing countries where the cancer problem is somewhat different from that in developed countries, Table 21.16. In the publication[25] on *Perspectives on quality of life* by Stjernswärd and Teoh the WHO global estimates for cancer were given as: 7×10^6 new cancer patients diagnosed annually. Slightly less than 50% are in the developing countries. About 5×10^6 will die of their disease. In developed countries 67% of male and 60% of female cancer patients will die of their disease but in developing countries the figure is much higher.

These WHO authors[25] also state that for cancer patients in developing countries no curative cancer treatment exists and that the quality of life of these patients would be better if they had access to palliative care and pain relief: (termed the *waste paper basket alternative* to anti-cancer treatment) from the start.

Table 21.16. Availability of effective strategies in developing countries for the control of the eight most common cancers worldwide. Curative therapy is classified as *curative for the majority of cases with a realistic opportunity of finding these cases in the early stage of the disease.* (3.88 × 10⁶ cases). EFFV=Effective. PE=Partially effective. NE=Not effective.

Tumour site	No. of cases per year	Primary prevention	Early diagnosis	Curative therapy	Palliative care
Stomach	670,000	PE	NE	NE	EFFV
Lung	660,000	EFFV	NE	NE	EFFV
Breast	572,000	NE	EFFV	EFFV	EFFV
Colorectal	572,000	PE	PE	PE	EFFV
Cervix	466,000	PE	EFFV	EFFV	EFFV
Mouth/pharynx	379,000	EFFV	EFFV	EFFV	EFFV
Oesophagus	310,000	NE	NE	NE	EFFV
Liver	251,000	EFFV	NE	NE	EFFV

21.12 CONCEPT OF COMPLICATED CURE & FAILURE AND UNCOMPLICATED CURE & FAILURE

21.12.1 Introduction

This concept was discussed by Bush[26] of the Princess Margaret Hospital, Toronto, as long ago as 1979 when he applied decision tree analysis to determine optimum treatment for cancer of the ovary, Table 21.17.

Table 21.17. Cure and failure possibilities, after Bush[26].

Cured without complications:	A_1
Cured with complications:	A_2
Died/no cure with complications:	A_3
Died/no cure without complications:	A_4

More recently Brahme, Agren and colleagues[27] in Sweden have applied this concept when discussing the optimisation of radiotherapy treatment plans for maximising tumour control in head and neck tumours. These authors illustrated this concept in the schematic diagram in Figure 21.11.

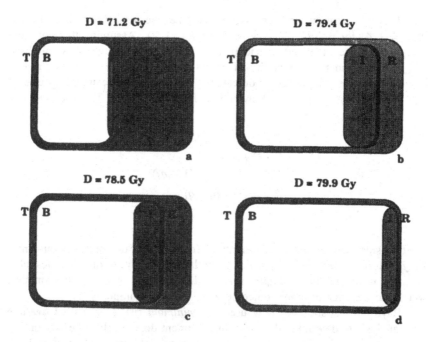

Figure 21.11. Schematic diagrams after Agren *et al*[27] are for four different treatment dose levels, where (*B*) is the probability of treatment control, (*R*) is the probability of treatment recurrence and (I) is the probability of radiation injury for the entire clinical material (T). The small cross-hatched areas correspond to patients suffering both injuries and a recurrent tumour, i.e. complicated failure.

21.12.2 Bayesian Statistics Application: Treatment Optimisation

Bayesian probabilities are used in statistical decision theory and differ from classical probabilities. A classical probability can be described by a statement such as 'Probabilities are relative frequencies, e.g. number of heads or tails when a coin is tossed. This takes into account the state of the coin'.

On the other hand, a Bayesian probability is a quantitative measure of the strength of one's knowledge or one's beliefs. It is a statement of a *personal* probability and as such, focusses as much attention on the decision maker as on the process or phenomenon under study. A recommended review is that by Raeside[28] entitled *Bayesian statistics: a guided tour*.

Bayesian statistics incorporates prior probability laws into decision strategies and the procedures take into account the consequences of incorrect decisions; classical probability procedures do not. As an example consider optimisation in treatment planning and the four possible outcomes given in Table 21.17 which have been symbolised A_i ($i = 1, 4$).

The optimisation problem is to compute the probability of outcome A_i ($i = 1, 4$) for two treatment levels D_j ($j = 1, 2$). Assuming that cure and complication are independent and that the probability of cure is symbolised p_j ($j = 1, 2$) and the probability of complications q_j ($j = 1, 2$) for the two dose levels. The probability of outcome A_i ($i = 1, 4$) at treatment dose level D_j ($j = 1, 2$) is given by the following four P_{ij} ($i = 1, 4$; $j = 1, 2$) probabilities:

$$P_{1j} = p_j \cdot (1 - q_j)$$
$$P_{2j} = p_j \cdot q_j$$
$$P_{3j} = (1 - p_j) \cdot (1 - q_j)$$
$$P_{4j} = (1 - p_j) \cdot q_j$$

Next consider an assignment of weights w_i ($i = 1, 4$) for the treatment outcomes A_i ($i = 1, 4$). One assigns w_1 which is the largest weight, to the most desirable outcome which in this example is A_1 and assigns w_4 which is the smallest weight, to the least desirable outcome which in this example is A_4.

Bayesian statistics are now used for *minimum* risk (also called *average loss*) analysis to determine the optimum treatment dose level: either D_1 or D_2. Assuming a risk R_i ($i = 1, 2$) is associated with treatment level D_i ($i = 1, 2$), the risk R_i ($i = 1, 2$) is the summation of the product ($w_j \cdot P_{ji}$) for $j = 1, 2, 3, 4$, for each of the two values of R_i which are defined in Table 21.18.

Table 21.18. Calculation of R_1 and R_2.

$R_1 = [w_1 \cdot p_1 \cdot (1 - q_1)] + [w_2 \cdot p_1 \cdot q_1] + [w_3 \cdot (1 - p_1) \cdot (1 - q_1)] + [w_4 \cdot (1 - p_1) \cdot q_1]$
$R_2 = [w_1 \cdot p_2 \cdot (1 - q_2)] + [w_2 \cdot p_2 \cdot q_2] + [w_3 \cdot (1 - p_2) \cdot (1 - q_2)] + [w_4 \cdot (1 - p_2) \cdot q_2]$

Treatment dose schedule D_2 is chosen over schedule D_1 if $R_1 > R_2$ since D_2 will thus have the least risk, or D_1 is chosen over D_2 if $R_2 > R_1$ since D_1 will then have the least risk.

21.13   ECONOMIC THEORY APPLICATION AND QUALITY OF LIFE

In the 1990s and into the next century in the field of oncology it will rapidly become more and more necessary to numerically quantify applications for funding for example for new equipment or an expensive chemotherapy regime by some form of cost assessment linked to patient benefit.

21.13.1   Quality Adjusted Life Years

One measure which has already been used is QUALYs: *quality adjusted life years*. These are a measure of the reduction or increase in health-centred utility, due respectively to disease and treatment. The value of future years benefit has to be discounted when QUALYs are used.

This has been explained by the Office of Health Economics[29] in the following manner. 'The reason for discounting future benefits is that even if money keeps its constant value (without inflation), a gain in future is worth less in real terms than a gain today. For example a gift of £100 now is worth much more than the same gift of £100 some 10 years hence. This is because if it is received now, it can be invested (or used in other ways) so that its value is increased as the years go by. If the £100 were invested, for instance to obtain a compound interest rate of 10% per annum, it would be worth about £260 in 10 years time. Conversely a gift of £100 some 10 years hence would be *worth* only £38 today. Hence discounting is getting at the present value of a future gain'.

21.13.2   Cost Benefit, Cost Effective and Cost Utility Analyses

There are three classes of cost analysis which are now used in healthcare and these are defined as follows. *Cost benefit analysis*: does the treatment pay off? *Cost effective analysis*: what is the most effective treatment using given resources? *Cost utility analysis*: how does treatment affect the length and quality of life?

21.13.3   Net Benefit and Net Detriment

Figure 21.12, after Bush[26], shows schematically the net benefit for a treatment A compared with a treatment B when survival outcome is the same for both treatments. The area between the two curves when the health index is better for B than A (in a short period of time after the initial treatment) is subtracted from the area between the two curves when the health index is better for A than for B (a much longer period of time, ending at death) to represent the net benefit for treatment A over treatment B.

The current problem, as discussed earlier in this chapter, is that a single *global* value for a health index/quality of life score has several disadvantages and can obscure important aspects of quality of life assessment. However, it would certainly be practical to construct graphs such as Figure 21.12 for the most important symptoms/complications and for example, present six graphs, some of which might give A better than B and some give B better than A.

The survival times do not necessarily have to be the same and in Figure 21.13 there are two treatments C and D for which the survival times are different. This is also an example where the health index can be negative, e.g. patient on a life support machine and with a deteriorating and terminal prognosis.

However, one must remember that statistics is only a tool (and statistical significance is not neccessarily the same as clinical significance) and the viewpoints of patient and doctor may differ (Table 21.2). Also, it can easily be imagined (using diagrams such as Figures 21.12 and 21.13) that a situation can exist of a treatment with a better quality of life but a shorter survival prognosis *versus* a treatment with a worse quality of life but a longer survival prognosis, and the patient preferring the former. Nevertheless, such analyses as in Figures 21.12 and 21.13 can be helpful and they are more easily understood by hospital managers than qualitative assesments.

The units of area beneath the curves in Figures 21.12–21.13 are 'Health Index–Time' which represents a measure of quality of life and treatment success. However, similar curves can also be drawn for complications and Maciejewski[30] has pointed out that the vertical axis of such a curve can be a scale of severity of complications, which can vary with time, and the areas beneath the curves are then 'Severity–Time' units for complications. This is a useful approach since it can quantify the importance of late complications, but of course the follow-up must be available.

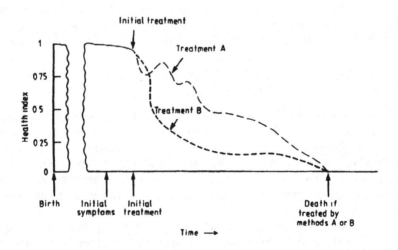

Figure 21.12. Schematic representation, after Bush[26], of the benefit/detriment of treatment A *versus* treatment B when both have the same survival prognosis.

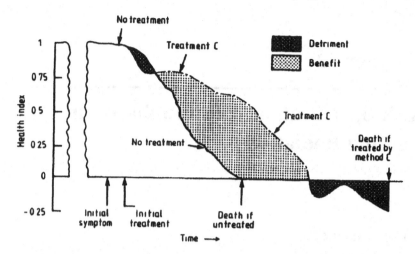

Figure 21.13. Schematic representation illustrating that a health index/quality of life score can be negative and that this type of net benefit/detriment analysis can be used for treatments with different survival prognoses.

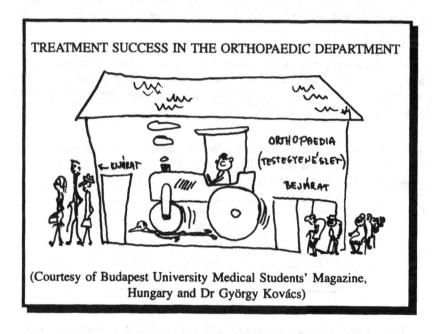

Chapter 22

Risk Specification with Emphasis on Ionising Radiation

22.1 INTRODUCTION

Risk can be specified in many different ways, but not all are as informative as each other. For example, the simple qualitative statement that the risk of a fatal acccident when crossing the road in busy traffic in a city is greater than when crossing the road in the countryside, is stating the obvious, but it gives no comparison betwen the two situations in terms of how much greater is the actual risk in the city.

Some possible methods of specifying mortality risks are by stating the risk per population sample per year, Table 22.1, or a fatal accident frequency rate, Table 22.2, which is useful for risk comparisons in different industries. For industrial risk comparisons, what is termed as an average reduction in lifespan is also sometimes used, Table 22.3.

Table 22.1. Mortality risks per population per year: from the 1970 *Statistical Abstract of the United States*[1]. The probability of dying from a particular cause of death is calculated as the number of cause-specific deaths divided by the population number.

Cause of death	Risk/year /100,000 population	Probability
Heart disease	364/100,000	0.00364
Cancer	157/100,000	0.00157

Comparative risks are of interest, not just for industrial workers, but also for everyday activity hazards such as smoking and air travel. Tables 22.4 and 22.5 give two examples of methods of quantifying such risks.

310

Table 22.2. Fatal accident frequency rate (FAFR): defined as the number of fatal accidents in a group of 1000 men in a working lifetime, assumed to be 100 million hours, for various United Kingdom industries[1]

FAFR	Industry
67	Construction
36	Fishing
14	Coal mining
10	Agriculture
5	Chemical
1.3	Vehicles
0.15	Clothing and footwear

Table 22.3. Average reduction in lifespan in days, of one year working life of a person age 40 years[1].

Lifespan reduction (days)	Industry
31.9	Deep-sea fishing
3.6	Coal mining
2.6	Oil refinery
2.2	Railways
2.1	Construction

Table 22.4. Activities which incur one in a million risk of death[1].

600 km air travel
100 km car travel
Smoking 75% of a cigarette
1.5 minutes of mountaineering
20 minutes of life at age 60
Drinking half a bottle of wine

Table 22.5. Risk of death per person per year for a given activity[1]. These numerical values of risk can be compared with the probabilities given in Table 22.1 where that for heart disease can be written as 364×10^{-5} and for cancer as 157×10^{-5}.

Risk/person /year ($\times 10^{-5}$)	Activity
500	Smoking 20 cigarettes/day
200	Motor cycling
120	Car racing
75	Drinking 1 bottle of wine/day
17	Car driving
4	Rock climbing
2	Taking contraceptive pills

For the atomic bomb survivors of Hiroshima and Nagasaki, and since 1986 for the survivors of the Chernobyl accident, much work has been undertaken in mathematical modelling of risk estimates in irradiated populations (not only high dose-high risk but also low dose-lower risk) for leukaemia and for solid tumours including thyroid cancer. These results are generally expressed as relative risk, excess relative risk and absolute risk.

22.2 DEFINITIONS OF ABSOLUTE RISK AND RELATIVE RISK

Absolute risk can be defined[2] as the excess risk attributed to irradiation and is usually expressed as the numerical difference between irradiated and non-irradiated populations: e.g. 1 excess case of cancer per 1 million people irradiated annually for each Gy (or rad). Absolute risk may be given on an annual basis or a lifetime (70 year) basis.

It is the magnitude of risk in a group of people with a certain exposure, but it does not take into account the risk of disease in unexposed individuals. It cannot therefore help to discover if the exposure is associated with an increased risk of the disease.

Relative risk (RR) is[2] the ratio between the number of cancer cases in the irradiated population to the number of cases expected in the unexposed population. A relative risk of 1.1 indicates a 10% increase in cancer due to radiation, compared with the *normal* incidence of the baseline/reference group.

Excess relative risk (ERR) is thus relative risk minus 1.0. Relative risk is more appropriate to use when considering selected population groups.

The NCRP report[3] on *Induction of thyroid cancer by ionizing radiation* defines an *absolute risk coefficient R* as the number of cases attributable to radiation exposure per million person-rad-years at risk, in the formula

$$R = (C/n) \cdot (10^6/[D \cdot y])$$

where C is the number of cases attributable to radation exposure, n is the number of subjects at risk in the irradiated population, D is the average radiation dose (rad) to the thyroid gland, and y is the average number of observed years at risk per subject.

Relative risk, unlike absolute risk, therefore gives an estimate of how strong an association exists between exposure to a factor and the development of a disease. There are of course many different factors that may be associated with increased risk of disease and for example, Table 22.6 lists the factors considered as being of importance in the UICC publication[4] *Cancer risks by site*, although not all factors are of significance for every cancer site. For each cancer site listed, this publication summarises the high risk areas, factors and conditions, providing a very useful overview, Table 22.7.

Table 22.6. Possible risk factors for various cancers[4].

Host factors
Sex
Age
Genetic predisposition
Precancerous lesions
Predisposing morbid conditions
Multiple primary lesions
Environmental factors
Socioeconomic status
Tobacco
Drugs
Alcohol
Diet
Radiation
Occupation
Air pollution
Sex life and pregnancy
Biological agents

Table 22.7. High risk factors [F] and conditions [C] for selected cancers[4].

Oesophagus
  Alcohol & tobacco [F]. Dysplasia of the epithelium [C].
Stomach
  Familial clustering, diet of salted pickles & fish, tobacco [F].
  Intestinal metaplasia, pernicious anaemia [C].
Colon
  High meat & fat diet, familial polyposis [F].
  Adenomatous polyps, villous adenoma, ulcerative colitis [C].
Rectum
  High meat diet, multiple polyposis [F].
  Villous adenoma, ulcerative colitis [C].
Pancreas
  Tobacco, high fat diet [F]. Diabetes [C].
Lung
  Tobacco, familial history, arsenic, asbestos, steel, nickel, chrome,
  radiation [F]. Metaplasia of bronchial epithelium [C].
Bone
  Paget's disease of bone, radiation [F].
Breast
  Familial history, early menarche, first parity late, high fat diet,
  obesity [F]. Hyperplastic lesions with cellular atypia [C].
Cervix uteri
  Low socio-economic class, sexual promiscuity [F]. Severe dysplasia [C].
Corpus uteri
  Higher socio-economic class, obesity, nulliparity, late onset of fertility [F].
Ovary
  Higher socio-economic class, radiation [F].
Prostate
  Familial history, cadmium oxide, prostatic enlargement [F].
Bladder
  Dye stuffs, tobacco, rubber, abnormality of tryptophane metabolism [F].
  Papillomas [C].
Brain & spinal cord
  Genetic predisposition [F]
Thyroid
  Radiation [F]. Papillary adenoma, autoimmune thyroiditis,
  Hashimoto's disease [C].
Hodgkin's disease
  Infectious mononucleosis [C].
Leukaemia
  Radiation, Down's syndrome, Fanconi's anaemia, Bloom's syndrome,
  ataxia telangiectasia [F].

22.3 ODDS RATIO

Odds is the ratio of the probability of occccurrence of an event to that of the non-occurrence of the event, i.e. the ratio of the probability of something that is true to the probability that it is not true.

This has been explained[5] in the following manner. If 60 smokers develop a chronic cough and 40 do not, the odds among these 100 smokers in favour of developing a cough are 60:40, or 1.5. This may be contrasted with the probability that these smokers will develop a cough, which is 60/100 or 0.6.

An *odds ratio* which is also sometimes termed a *relative odds* or a *cross-product ratio*, is the ratio of two odds where the odds are defined depending on the situation being studied. Hence there can be an *exposure-odds ratio* (odds in favour of exposure among cases and among non-cases), and a *disease-odds ratio* (odds in favour of disease among the exposed and among the non-exposed).

One example of the use of odds is in the review[6] of 19 randomised trials of post-mastectomy radiotherapy for the treatment of early breast cancer, in which was calculated the typical reduction in the odds of treatment failure. In this instance, the odds ratio is the ratio of the odds of an unfavourable outcome among the treatment allocated patients, to the corresponding odds among the controls.

Thus an odds ratio of 0.8 would correspond to a reduction of about 20% (*about* because the odds ratio will have an associated standard deviation) in the odds of an unfavourable outcome. In practice[6] the odds ratio was found to be 0.99 with 95% confidence limits of 0.92 to 1.06. This indicates that the result is not conventionally significant since the odds ratio of 1.00 is within the confidence interval.

22.4 HIROSHIMA AND NAGASAKI

Absolute and relative risks for atomic bomb (ATB) survivors are published at regular intervals by the Radiation Effects Research Foundation (RERF)[7] to show, for example, for leukaemia and other cancers, the effects of age, sex and time since ATB exposure. Figure 22.1 is from RERF data[8,9] and is a summary of the excess relative risk (ERR) for various cancers, and also a summary of relative risks.

Figure 22.1 shows statistically significant risks not only for leukaemia and multiple myeloma, but also for the following solid tumours: cancers of the stomach, colon, lung, breast, ovary, urinary bladder and thyroid. Table 22.8 is also for ATB data and lists for leukaemia both relative and absolute risks.

The incidence of leukaemia peaked in 1951, Figure 22.2, and the excess risk at all dose levels was always higher in Hiroshima than in Nagasaki[9]. There has been a recent reclassification of leukaemia among the ATB survivors and the radiation effects differ depending on the type of leukaemia. The four types are acute myeloid leukaemia (AML), acute lymphoid leukaemia (ALL), chronic

316   *Risk Specification*

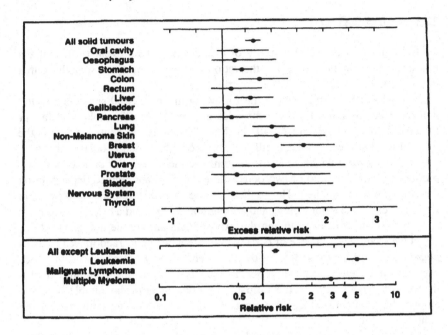

Figure 22.1. (Top) Excess relative risk at 1 Sv (RBE 10) and 95% confidence interval, 1958–1987[8]. (Bottom) Relative risk at 1 Gy (shielded kerma) and 90% confidence interval, 1950–1985[9], for selected cancers which are not given in the earlier (top) bar chart.

Table 22.8. Estimated relative risks at 1 Gy, and absolute risks stated as an excess risk per 10^4 person-years·Gy for leukaemia[9] for the period 1950–1985. The figures given in brackets are 90% confidence intervals.

Estimated RR at 1 Gy	Excess risk per 10^4 PY·Gy
4.92 [3.89, 6.40]	2.29 [1.89, 2.73]

myeloid leukaemia (CML) and chronic lymphocytic leukaemia (CLL).

The A-bomb exposure effects are significantly greater for ALL and CML, than for AML. ALL and CML were also observed earlier than AML and it also appears[2] that the leukaemia type most characteristic of the A-bomb survivors in CML. However, there does not appear to be a difference in the shape of the dose-response curve.

Figure 22.3 is a schematic representation[2] of the induction period and risk of leukaemia as a function of age at exposure.

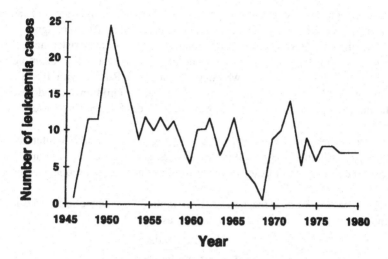

Figure 22.2. Leukaemia among A-bomb survivors proximally exposed within 2000 metres from ground zero who received more than 1 rad radiation dose, Hiroshima and Nagasaki 1945[9].

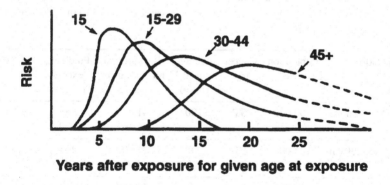

Figure 22.3. Schematic respresentation of the induction period and risk of leukaemia as a function of age at exposure.

22.5 CHERNOBYL

There are many problems in estimating risks of cancer induction following the Chernobyl accident, not least of which is that very few direct measurements of radiation exposure at the time of the accident are available, and that the Hiroshima–Nagasaki data cannot be automatically used for Chernobyl because the types and patterns of radiation exposure are very different. Reliance is therefore placed on estimates from mathematical models which are still largely unproven for the Chernobyl accident which occured only 10 years ago.

Nevertheless, best estimates for relative and absolute risks have been modelled, particularly for thyroid cancer and leukaemia. For example for the Bryansk oblast in Russia, which is the area in Russia with the highest contamination, the relative risk of thyroid cancer in children and adolescents for the dose range 0.6–1.4 Gy is 7.3, but the 95% confidence limits are wide at (1.7, 38.9)[10].

Some of the more recent estimates[11] were given at the IAEA International Conference in Vienna in April 1996 on *One decade after Chernobyl* and Table 22.9 is taken from the presentations summing up the consequences of the accident.

The term *attributable fraction* (AF) in Table 22.9 is sometimes used in the presentation of risk estimates and is defined as:

$$AF = \frac{\text{Excess deaths}}{\text{Total deaths from the same cause}}$$

Table 22.9. Predicted background and excess deaths[11] for a lifetime period, from solid tumours (ST) and leukaemia (L) in selected populations exposed as a result of the Chernobyl accident. AF = attributable fraction. EAWs = emergency accident workers, also termed *liquidators*. Evacuees: 30km = evacuees from the 30 km zone. SCZs = specially controlled zones.

Population	Population size	Cancer type	Average dose (mSv)	Background no. of cancer deaths	Predicted excess number	AF (%)
EAWs 1986–87	200,000	ST	100	41,500	2000	5
		L		800	200	20
Evacuees:30km	135,000	ST	10	21,500	150	0.1
		L		500	10	2
Residents:SCZs	270,000	ST	50	43,500	1,500	3
		L		1000	100	9

The lifetime risk estimates in Table 22.9 have followed the methods of the UNSCEAR 1994 Report[12]. However, at the November 1995 World Health Organisation International Conference in Geneva[13] on *Health consequences of the Chernobyl and other radiological accidents*, the following conclusions were drawn for thyroid cancer, Table 22.10, although for leukaemia it was stated that there was 'No significant increase in leukaemia or other blood disorders so far'.

At the time of writing, the most recent data[14] on excess relative risk of thyroid cancer relates to the cohort of emergency accident workers (liquidators)

Table 22.10. Summary commentary[13] on thyroid cancer incidence in Belarus, Ukraine and Russia post-Chernobyl accident: WHO, Geneva, November 1995.

- More solid papillary type of cancer more frequently observed than in European countries
- Causative relationship with Chernobyl based on geographical distribution: see Table 22.11
- Cancer type particularly aggressive
- Extra-thyroidal invasion with lymph node invasion in more than 60%
- Some 7% with lung metastases
- Thyroid cancer cases conceived before but not after the accident
- Increase in thyroid cancer in children
  - Belarus: 400-600 depending on reports
  - Ukraine: more than 200
  - Russia: less than 60

Table 22.11. Incidence of thyroid cancer in children[11] (age < 15 years) in Belarus. There is a geographical correlation with the incidence in Gomel since the radioactive cloud passed directly over this town when it was raining.

Region	Year										
	86	87	88	89	90	91	92	93	94	95	86–95
Gomel	1	2	1	3	14	43	34	36	44	48	226
Remainder of Belarus	1	2	4	4	15	16	32	43	38	43	198

who reside in Russia, as distinct from Belarus and Ukraine. This analysis showed that there is an excess relative risk (ERR) of thyroid cancer per Gy of 5.31 (95% confidence interval 0.04–10.58) and an excess absolute risk of thyroid cancer per 10^4 person-years per Gy of 1.15 (95% confidence interval 0.08–2.22). Where is good agreement between these results and those of BEIR V[15].

The liquidators' mean radiation dose was 140 mGy and in 67.3% (33/47) of cases they were in the age range 35–49 years when the thyroid cancer was diagnosed.

22.6 BREAST CANCER

The BEIR V committee[15] has estimated the risk of radiation induced cancers using a linear quadratic dose-response function with a relative risk function that depends on sex, age at exposure and time elapsed since exposure. Most of the data which have been used to fit the model parameters are from single dose/high dose rate exposures such as for the A-bomb survivors. Hence the BEIR V parameters might not be very accurate for highly fractionated and/or low dose rate exposure.

For further reading on limitations of the model see the work of Hendee and Edwards[16] on health effects of exposure to low level ionising radiation and for risk estimates for specific sites the work of Mettler and Upton[2]. Figure 22.4 shows the relative risk of breast cancer per 0.1 Gy and Figure 22.5 the excess deaths from breast cancer per 10,000 person-years·Gy

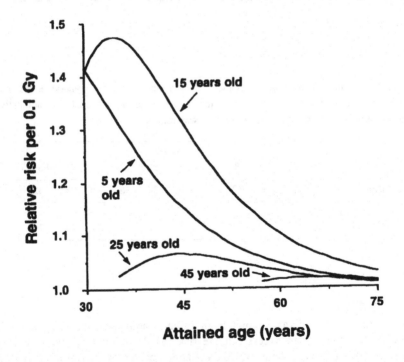

Figure 22.4. Replotted BEAR V data[15,16] of the relative risk of breast cancer per 0.1 Gy, as a function of attained age, for four different ages (5, 15, 25 and 45 years) at time of exposure. The time since exposure is the attained age minus the age at the time of exposure.

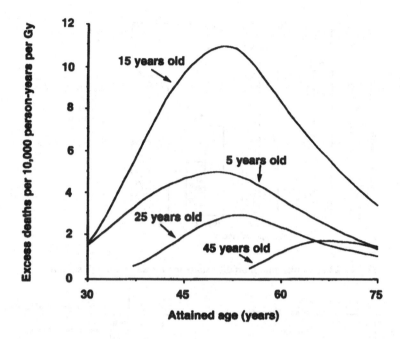

Figure 22.5. Replotted BEIR V data[15,16] of excess deaths from breast cancer per 10,000 person-years·Gy, as a function of attained age for four different ages at the time of exposure. This is an absolute risk.

22.7 MORTALITY RATIO

Relative and absolute risks are used in predictive modelling beyond the period of observation but it is only from the 1980s that much work has been undertaken with BEIR and other models. Prior to such modelling for A-bomb survivors, virtually all that was stated was essentially mortality ratios (MR) for various population subgroups where the MR was defined as follows.

$$MR = \frac{\text{Observed deaths}}{\text{Expected deaths}}$$

In one of the earlier RERF reports[17] in 1972, the observed and expected deaths from leukaemia were tabulated by time period and the values of MR are seen in Figure 22.6 as a vertical bar chart[17]. MR peaks for the period 1950–54 and as seen in Figure 22.2 this agrees with the peak incidence year of 1951.

Mortality ratios are also shown in Figure 22.7 for radiologists[18] and include the period 1896–1920 when radiation protection procedures were only slowly being adopted worldwide. This is reflected in the reduction of the mortality ratio with time.

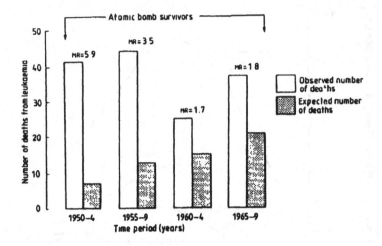

Figure 22.6. Leukaemia mortality statistics for atomic bomb survivors.

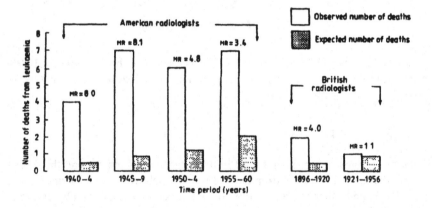

Figure 22.7. Leukaemia mortality statistics for radiologists. The mortality ratio (MR) is the ratio of the observed number of deaths to the expected number of deaths.

22.8 LUNG CANCER AND SMOKING

Another example of the use of mortality ratios is shown in Table 22.12 for lung cancer and different types of smoking. These data are for American males in a 1975 study[18,19] but the mortality ratio is defined slightly differently to those for Figures 22.6 and 22.7. In this instance it is:

$$MR = \frac{\text{Age-standardised death rate of smokers with a given type of smoking}}{\text{Age-standardised death rate of those who never smoked regularly}}$$

Table 22.12. Lung cancer mortality ratios for different types of smoking for American males aged 35–84 years[18,19].

MR	Type of smoking
1.00	Never smoked regularly
1.23	Pipe and cigar
2.15	Cigar only
2.23	Pipe only
8.23	Cigarette and other
10.08	Cigarette

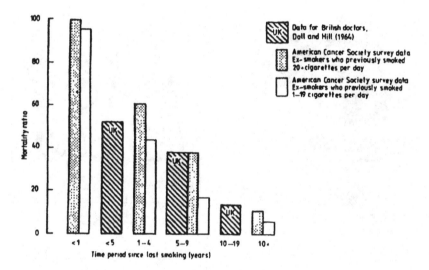

Figure 22.8. Lung cancer mortality statistics for ex-smokers. Mortality ratio is the ratio of the age-standardised mortality rate in current cigarette smokers and the age-standardised mortality rate for ex-smokers.

Mortality ratios also provide useful information when tabulated for the number of cigarettes smoked daily. For men who smoke 30 cigarettes per day, compared with non-smokers, it has long been demonstrated[20] that the increase in mortality risk is in the range 12-fold to 25-fold, depending on the population considered. It has also been demonstrated both in America and the United Kingdom that lung cancer mortality starts to decline in smokers who cease smoking and this can conveniently be expressed[18] in terms of mortality ratios, Figure 22.8.

22.9   EDUCATIONAL CARTOONS ON CANCER RISK

Prevention is better than cure and this is the aim of educational cartoons and probably one of the most effective series is that in the late 1980s in Hungary which together formed a large poster detailing the seven main warning signs and symptoms for cancer. The introduction to this poster is 'Our body sends messages to us with symptoms and signs which are warning us about the development of dangerous illnesses. Naturally, the symptoms don't mean cancer in every case. If you find any suspicious malformation on your body, see your doctor without delay. Don't forget that cancer which has been diagnosed in time, can certainly be cured. You have to learn the **Seven Warning Signs**. Early symptoms rarely give pain. If you have any of these symptoms for more than two weeks go to see a doctor straight away'.

1. Szemmel látható változások szemölcsön, anyajegyen

2. Széklettel és vizelettel kapcsolatos változások

3. Makacs köhögés vagy rekedtség

The seven warning signs are described as follows.

1. Visual changes to any wart or birthmark.
2. Changes in connection with stool and urine.
3. Persistent cough or hoarseness.
4. Constant swallowing problems.
5. Not recovering from ulcers or skin damage.
6. Unusual bleeding or secretion dripping.
7. Palpable mass in the breast or anywhere else in your body.

(I am grateful to Erika Bender and Lodi Fox
for translation from the Hungarian)

4. Állandóan fennálló nyelési nehézségek

5. Nem gyógyuló fekély vagy sérülés a bőrön

6. Szokatlan vérzés, váladékfolyás

7. Tapintható duzzanatok a mellben
vagy a test más részén

Many cartoons have been designed for anti-smoking campaigns and that shown is a badge from the UK Health Education Bureau in the 1970s. Excessive exposure to ultra violet radiation is a known cause of skin cancer and the two car stickers shown are from the Anti-Cancer Council of Victoria, Melbourne, Australia in the early 1980s.

Chapter 23

Types of Epidemiological Study:
Case-Control, Cohort and Cross-Sectional

23.1 INTRODUCTION

Epidemiology has already been defined in section 19.1 as the *study* of the *distribution* and *determinants* of health related states or events in specified populations and the application of this study to control health problems. Expanding the terms study, distribution and determinants we have the following[1].

- *Study* includes surveillance, observation, hypothesis testing, analytical research and experiments.
- *Distributions* refer to analysis by time, place and classes of persons affected.
- *Determinants* are all the physical, biological, social, cultural and behavioural factors that influence health.

The term *analytical study* is one which attempts to explain the observed pattern of occurrence of a disease and a *descriptive study* describes the occurrence of the disease or disease-related phenomena in populations.

Epidemiological studies are not generally amenable to being investigated by randomised trials and observational studies are therefore the more practical to study factors or exposures which cannot be controlled by the investigators[2]. The two main types of observational study, are the case-control study and the cohort study but the terminology can be confusing, and has been described by Gordis[3] as a *terminology jungle*, Table 23.1.

There are several possible sources of bias in these studies and one such source, which is also relevant to clinical trials, see section 20.6 on historical controls, is the inaccuracies of retrospective data. Another is a detection bias in which the cases receive more frequent screening for a disease, eg. cervical cancer, than do the controls. There is also potential bias in the assessment of the outcome (when this is other than death) if the assessor knows if the

Table 23.1. *Terminology jungle*: after Gordis[3].

Case-control study		=	Retrospective study
Cohort study	= Longitudinal study	=	Prospective study
Concurrent cohort study	= Prospective cohort study	=	Concurrent prospective study
Retrospective cohort study	= Historical cohort study	=	Non-concurrent prospective study
Randomised trial		=	Experimental study
Cross-sectional study		=	Prevalence study

person was exposed and the assessment is not *blind*. The quality and extent of the information for the exposed and non-exposed may be different and this could produce an information bias. Losses to follow-up can also cause serious problems.

23.2 CASE-CONTROL STUDIES

Figure 23.1 is a WHO schematic diagram[4] of the design of a case-control study. They include people with a disease (or other outcome variable) of interest and a suitable control group (comparison or reference group) of people unaffected by the disease or outcome variable. The occurrence of the possible cause is compared between cases and controls.

Data concerning more than one point in time are collected and thus the study is termed *longitudinal* and it is also *retrospective* since the study looks backwards from the disease to a possible cause. However, terminology as seen from Table 23.1 can be confusing, and a case-control study may be either *retrospective* when all data deal with the past, or *prospective* in which data collection continues with the passage of time[4].

Furthur information to that given in this chapter can be obtained from several textbooks devoted to the topic of epidemiology[3-6] and for cancer research in particular, from the IARC publication by Breslow and Day[7].

23.2.1 Selection of Cases

Cases can be selected from various sources including hospital case notes, patient records in a general practitioner's practice, and any specialist registries such as a cancer registry. However, several problems in case selection must be avoided to prevent bias and ensure that the final results of the study can be generalised to all patients with the disease.

Thus for example if hospital cases are used it must be ensured that there is no patient referral pattern which results in these patients having a risk factor

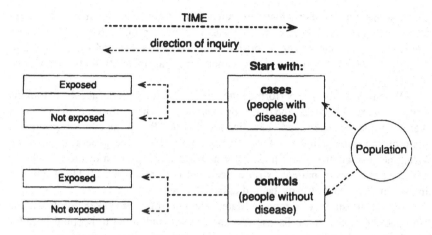

Figure 23.1. Design of a case-control study (Courtesy: World Health Organisation[4]).

unique to that hospital caseload. It is good practice to select from several hospitals and not only a single hospital.

One method of ensuring that cases and controls are as comparable as possible is to *individually match* them by for example, age, sex and occupation. However, this is only useful for variables which are known to be strongly related to both exposure and outcome. They should certainly not be matched for possible risk factors since this matching would mean that no association could be found, if one exists, between the disease and this possible risk factor.

A decision must also be made as whether to use *incident* cases, i.e. those who are new cases of the disease or *prevalent* cases, i.e. those who have already had the disease for a period of time. One advantage of prevalent cases is that the case numbers already exist, whereas with incident cases one has to wait for the cases to accrue.

However, it is preferable to use incident cases because it will not be certain if prevalent cases are used in a case-control study of disease aetiology if the risk factors which will be identified are related more to *survival* with the disease than to its development.

23.2.2 Selection of Controls

The controls should represent people who would have been designated study cases if they had developed the disease. They may be selected from non-hospitalised patients living in the community or from hospitalised patients without the disease under study but care must be taken to avoid bias.

One good example of an incorrect method of choosing controls is quoted by Gordis[3] and refers to a 1929 study[8] at Johns Hopkins University to test the hypothesis that tuberculosis protected against cancer. The study data were taken from autopsy reports with 816 cases with cancer and 816 controls without cancer. The percentage of autopsies with TB was 6.6% (54/816) cases and 16.3% (133/816) controls and the conclusion was that TB had a protective effect against cancer.

When the study was performed the majority of the patients at Johns Hopkins Hospital were TB cases and what had happened when choosing the controls were that many of them had been diagnosed with TB and the control group was therefore not representative of the distribution of TB in the general population. What the case-control study had performed was a comparison of the prevalence of TB at autopsy in patients with cancer and the prevalence of TB at autopsy in patients who had already been diagnosed with TB.

When the study was repeated later[9] the controls were taken from patients who had died from heart disease and had not been admitted to hospital because of TB. This study showed no difference in the prevalence of TB in the two groups.

Hospital records are more accurate than patients' memories as demonstrated in Table 23.2. When choosing the controls this shows that total reliance on memory would not be an acceptable option. Sensitivity and specificity (see Chapter 19) can be calculated from the data and are quoted below the data in Table 23.2.

Table 23.2. Comparison of hospital records and patient's statements as to the presence or absence of a prenatal abdominal X-ray examination. Data from Harvard School of Public Health[5].

Hospital record	Patient's statement X-rayed	Not X-rayed	Don't know	Total
X-rayed	**24**	10	3	37
Not X-rayed	2	**31**	5	38
Total	26	41	8	75

Sensitivity is the probability of the patient stating they had been X-rayed when in fact this was true: 24/37 = 65%.

Specificity is the probability of the patient stating they had not been X-rayed when in fact this was true: 31/38 = 82%

The *positive predictive value* is the probabilitiy that persons who say they manifest a characteristic truly do: 24/26, i.e. 92%. The *negative predictive value* is the probability that persons who say they do not have the characteristic, truly do not: 31/41, i.e. 76%.

23.2.3 Example: Thalidomide and Limb Defects

A classic example[10] of a case-control study was undertaken in 1961; the discovery of the relationship between thalidomide and unusual limb defects in babies born in the Federal Republic of Germany in 1959 and 1960. This case-control study compared affected children with normal children and of 46 mothers whose babies had typical malformations, 41/46 had taken thalidomide between the fourth and ninth weeks of pregnancy whereas none of the 300 control mothers, whose children were normal, had taken the drug during this period.

23.2.4 Example: Meat Consumption and Enteritis Necrotans

The association between recent meat consumption and enteritis necrotans in Papua New Guinea[4,11] is shown in Table 23.3. The cases were people with the disease and the controls were those who did not have the disease and the results of this study are given below by describing the association of an exposure and an outcome by calculating the odds ratio (see also section 22.3).

Table 23.3. Association between recent meat consumption (exposure) and the disease enteritis necrotans (outcome) in Papua New Guinea[11].

Outcome	Exposure Yes	No	Total
Yes	50	11	61
No	16	41	57
Total	66	52	118

The odds ratio is the ratio of the odds of exposure among the cases to the odds in favour of exposure among the control. Thus for the data in Table 23.3

$$\text{Odds Ratio} = (50/11)/(16/41) = 11.6$$

This shows that the cases were 11.6 times more likely than the controls to have recently ingested meat.

23.3 COHORT STUDIES

The term *cohort* has been encountered in section 21.7.1 when cohort interpolated life tables were discussed. The term originates from the Latin *cohors* meaning warriors and referred to one-tenth of a Roman legion.

In medical statistics it refers to a component of the population born during a particular period so that its characteristics (e.g. numbers still alive, see $_nP_x$

survival probabilities in Table 21.13) can be described as it enters successive time and age periods. It is also now broadened to describe any designated group of persons who are followed-up over a period of time, as in cohort (prospective) study.

Figure 23.2 is a WHO schematic diagram[4] of the design of a cohort study. This begins with a group of persons, the cohort, free of disease, who are classified into subgroups according to exposure to a potential disease or outcome.

Variables of interest are specified and measured and the entire cohort is followed-up to assess how the development of the disease, or other outcome, differs between the groups with and without exposure. However, although they provide the best information about the causation of disease and the most direct measurement of risk of developing a disease, they suffer the disadvantage that they usually require long periods of follow-up.

A *concurrent cohort study*, Table 23.1, is one which first identifies the cohort to be studied, hypothetically for example school children, follows them up for say 10 years until smokers and non-smokers can be identified, then both sub-cohorts are followed up to determine who develops lung cancer and who does not. The term *concurrent* is used because the investigator identifies the original population at the start of the study and in effect accompanies the members of the cohort *concurrently* through time until the disease (lung cancer) has or has not developed.

A hypothetical example of a *retrospective cohort study*, Table 23.1, is one in which retrospective data is already available on smoking habits and the study commences from this point in time until lung cancer has developed or not.

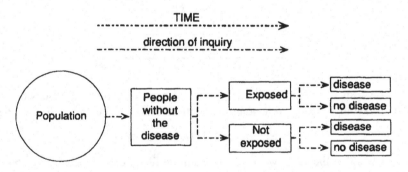

Figure 23.2. Design of a cohort study (Courtesy: World Health Organisation[4]).

23.3.1   Example: The Framingham Study

One of the most well known cohort studies is the Framingham study[12,13] of cardiovascular disease which commenced in 1948. The town of Framingham is some 20 miles from Boston and had a population in 1948 of some 30,000 of which the cohort for study consisted of 5127 men and women in the age range 30–62 years who when entered into the study were free of cardiovascular disease. Table 23.4 gives some of the details of this study.

Table 23.4. Details of the Framingham coronary heart disease (CHD) study[3,12,13].

Exposures

These included smoking, obesity, elevated blood pressure, elevated cholesterol levels and low levels of physical activity

Cohort surveillance for new coronary events

Cohort examined every two years and by daily surveillance of hospitalisations at the only hospital in Framingham

Hypothesis testing aims and objectives

- Incidence of CHD increases with age and occurs earlier and more frequently in males
- Persons with hypertension develop CHD at a greater rate than those who are not hypertensive
- Elevated blood cholesterol level is associated with an increased risk of CHD
- Tobacco smoking and an habitual use of alcohol are associated with an increased incidence of CHD
- Increased physical activity is associated with a decrease in the development of CHD
- An increase in body weight predisposes a person to CHD
- An increased rate of development of CHD occurs in patients with diabetes mellitus

Derivation of the study cohort of 5127 persons

	No. of men	No. of women	Total
Random sample	3074	3433	6507
Responders from sample	2024	2445	4469
Volunteers	312	428	740
Responders free of CHD	1975	2418	4393
Volunteers free of CHD	307	427	734
Total free of CHD	2282	2845	5127

In conclusion, it is noted that much of our knowledge about CHD today, which is currently taken for granted, was obtained as a result of this Framingham study which commenced some 50 years ago.

23.4   NESTED CASE-CONTROL STUDY

One of the disadvantages of a cohort study is the lengthy time period required and the high cost to implement the study. This can be partially overcome by using a nested case-control study design, Figure 23.3, in which the cases and controls are chosen from a defined cohort (the population in Figure 23.3) and for which some information on exposures and risk factors are already available. The case-control study is *nested* within the cohort.

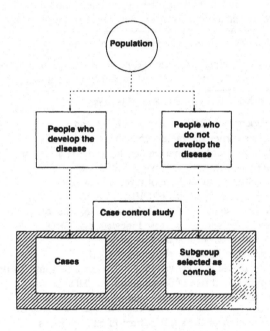

Figure 23.3. Design of a nested case-control study.

23.5   CROSS-SECTIONAL STUDIES

Cross-sectional studies measure the prevalence of a disease, see Table 23.1, and the measurements of exposure and outcome (i.e. effect) are made *simultaneously*, Figure 23.4. It is therefore not always easy to interpret the reasons for associations found in cross-sectional studies. The major question to be asked is whether the exposure precedes or follows the effect.

Figure 23.4. Schematic diagram of a cross-sectional study. The study can be analysed using 2 × 2 tables in the two ways shown below.

• The prevalence of the disease is compared in exposed and non-exposed This will be, using the notation below: $\{a/[a+b]\}$ *versus* $\{c/[c+d]\}$

• The prevalence of exposure is compared in diseased and non-diseased. This will be $\{a/[a+c]\}$ *versus* $\{b/[b+d]\}$

	Disease	No disease
Exposed	a	b
Not Exposed	c	d

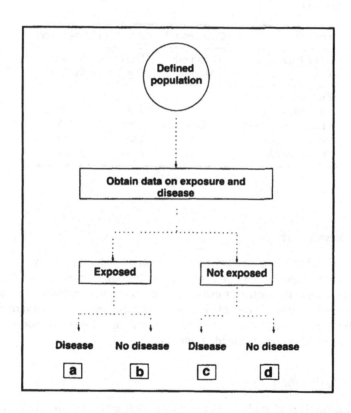

23.6   SUMMARY OF ADVANTAGES AND DISADVANTAGES OF THE DIFFERENT STUDIES

Table 23.5 summarises, after WHO[4], the advantages and disadvantages of the three types of study so far considered in this chapter. Possibilities of selection bias have already been discussed in sections 23.2–23.5 and sources of bias in general at the end of section 23.1. Recall bias and counfounding, Table 23.5, are two special types of bias which can be encountered in epidemiological studies and are defined in the next two sections.

Table 23.5. Advantages and disadvantages of different types of observational study (NA: not applicable).

Feature	Case-control	Cohort	Cross-sectional
Time required	Medium	High	Medium
Cost	Medium	High	Medium
Probability of:			
Selection bias	High	Low	Medium
Loss to follow-up	Low	High	NA
Recall bias	High	Low	High
Confounding	Medium	Low	Medium

23.6.1   Recall Bias

The word recall refers to memory recall of a person entering a study. In practice this has been shown to be not always good in many persons: one generally accepted example is a patient's estimate of duration of symptoms. An example where numerical data is available is from an assessment by the Harvard School of Public Health[5] and refers to memory recall of whether or not an X-ray examination had been performed, Table 23.2.

23.6.2   Confounding

In an epidemiological study of an association between exposure and occurrence of a disease, confounding can occur when another exposure exists in the study population and is associated both with the disease and the exposure being studied.

If this confounding factor is unequally distributed between the exposure groups under analysis then incorrect conclusions can be drawn from the study. Age and social class are often confounding factors[4].

Figure 23.5 (see also section 17.1, page 198) illustrates the fact that confounding may be the explanation for the relationship demonstrated between coffee consumption and the risk of coronary heart disease, since it is known that coffee consumption is associated with cigarette smoking. People who drink coffee are more likely to smoke than people who do not drink coffee. It is also well known that cigarette smoking is a cause of coronary heart disease. It is thus possible that the relationship between coffee consumption and heart disease merely reflects the known causal association of smoking with the disease. In this situation, smoking confounds the apparent relationship between coffee consumption and coronary heart disease.

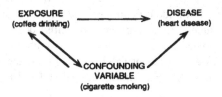

Figure 23.5. Confounding: coffee drinking, cigarette smoking and coronary heart disease (Courtesy: World Health Organisation[4]).

23.7   OTHER TYPES OF EPIDEMIOLOGICAL STUDY: CLUSTER ANALYSES OF HISTORICAL INTEREST

Epidemiological studies are not limited to the three types of *observational* study listed in the title of this chapter. They also include *experimental* studies (also termed *interventional* studies), such as *randomised controlled trials* (i.e. clinical trials) in which the population under study consists of patients, *field trials* using healthy persons, and *community trials* (also termed community intervention studies) on communities.

Observational studies allow nature to take its course and the investigator measures but does not intervene: they may be *descriptive* or *analytical*. In the former the study only describes the occurrence of a disease in a population whereas the latter extends to analysing relationships between health status and other variables.

For further details of epidemiological studies reference should be made to textbooks devoted solely to epidemiology, such as those already referred to[3-6] and a recently published workbook of epidemiology[14]. Three epidemiological studies, each with a historical aspect, and each *detective stories*, are now described as examples of *cluster analysis*, of which John Snow's in section 23.7.2 is the classic example.

23.7.1 The Unique Hostel

Standardised mortality rates (SMRs) are useful when mortality data for several subregional populations within a region, such as a city or county, have to be compared. There is perhaps less likelihood of the subregional populations having widely different age and sex structures from the standard regional population, as in the paper by Freedman and Rubin[15] when SMRs were used to compare lung cancer mortality in 40 voting constituencies in the city of Liverpool.

The paper[15] concerns the limitations of lung cancer SMRs as a guide to lung cancer incidence, but it was also noted in these data that only two of the 40 SMRs exceeded 165, one for Central Liverpool (SMR = 222) and the highest (SMR = 326) for Everton with a population of some 5000. Further study, using a cluster analysis technique involving identification of the addresses at which pulmonary deaths (lung cancer, tuberculosis, chronic bronchitis) occurred in Everton and surrounding wards, showed that the reason for this high lung cancer SMR was due to the presence of a 200-bed privately owned registered common lodging house for men with the somewhat appropriate name *The Unique Hostel*, Figure 23.6.

The Liverpool Echo, Friday, March 17, 1978   15

Hostel plea after shock city report

A PLEA that hostels for the homeless be improved is to be made after a shock report highlighting the death rate from lung cancer in Liverpool's Unique hostel.

Although the plea will be made to the Department of the Environment, Health Secretary David Ennals is to be asked to look into links between health services hospitals

By Lynn May

In Liverpool as a whole, one in five men die from lung cancer or chronic bronchitis.

Figure 23.6. The Unique Hostel, Shaw Street, Everton, Liverpool. The lettering on the door states *All Enquiries for Beds to be Made in the Cafe Basement.* It was later found that the cook, an ex-sailor, suffered from pulmonary tuberculosis.

The buildings were in the centre of a decaying 150-year old terrace of houses which had an architectural preservation order placed on it because the houses had cast iron balconies with a distinctive Prince of Wales feathers design[17]. Table 23.6 shows the pattern of pulmonary disease mortality in Everton, 1969–76 including the data for the lodging house which artificially inflated the SMR to 326.

Table 23.6. Chest disease in Everton, 1969–76.

Disease	Year	Area X			Everton ward excluding area X
		200-bed lodging house	Six high-rise blocks of flats	Other housing	
Lung	1969–72	2	5	7	5
cancer	1973–76	8	1	7	12
Chronic	1969–72	2	4	3	6
bronchitis	1973–76	9	2	4	5
Pulmonary	1969–72	1	0	1	1
tuberculosis	1973–76	6	0	0	0
Totals	1969–76	28	12	22	29

The fact that the Unique Hostel was identified was due to a knowledge of the folklore of Everton which said that in the 19th century if one lived at the bottom of the hill (Everton valley) which ran from St. George's Church (Figure 23.7: interior view of the cast iron structure) then you would die young, but if you lived at the top of the hill on which stood the church, you would remain healthy.

Figure 23.7. Areas mentioned in the sketch map in Table 23.6. [Top] View of Everton in 1819. [Centre left] Interior of St. George's church, Everton: the world's first cast iron church. It was built in 1813 by subscription to consist of 110 shares at £100 each. It was also agreed (at the founding meeting in a Coffee House in 1812 and later legalised by an 1813 Act of Parliament) that each seat in the church be put up for sale to the highest bidder among the subscribers and the monies arising from the sale to be divided among the proprietors according to their shares. In the early days it was an Everton joke that 'a gentleman's best dividends came from St. George's'. [Centre right]. Architectural detail of the roof. [Bottom left]. In the 19th century Everton was famous for Everton Toffee which in 1862 was recommended by Charles Dickens when he visited Liverpool. [Bottom right]. The Stone Jug which was Everton's 18th century gaol and was known by Liverpool urchins at the turn of the 20th century as the Stewbum's Palace. It forms the central motif (often mistaken for a beehive) of the badge of Everon Football Club. [Photographs courtesy Mr Rick Houghton.]

This pattern repeated itself in the 1970s with the data in Table 23.6. It was of course realised that the high death rate in the 1800s in Everton valley was due to over-crowding (Figure 23.7: The church is shown faintly at top left in this 1930s photograph), unhygienic living conditions and epidemics (Table 23.7) whereas near St. George's were the villas of the Liverpool business magnates of the era. Nevertheless, it was an intriguing pattern for a local historian[17] and this is what prompted the study!

For football fanatics, the sketch map in Table 23.6 shows the grounds of both Liverpool FC and Everton FC. The latter are known as *The Toffees* because of the famous Everton toffee (Figure 23.7: advertisement of 1832) and the team badge is usually considered to include a beehive. Actually it is the local 19th century gaol (Figure 23.7: photograph in 1977), *The Stone Jug* which is only some two minutes walk from the Unique Hostel.

Table 23.7. Liverpool epidemic statistics 1832–1866. 20,000 deaths representing one in 15 of the entire 1847 population of Liverpool, occurred during four epidemics of cholera, three of typhoid (also known as *gaol fever*, *Irish fever* and *famine fever*) and one of smallpox.

Years	Epidemic	Statistics
1832	Cholera	1523 deaths,
		4912 registered cases (600/month)
1837	Typhoid	524 deaths
1837–39	Smallpox	880 deaths
1838–39	Typhoid	1000 deaths
1840	Smalllpox	400 deaths
1847	Typhoid	7000 deaths
1849	Cholera	5308 deaths
		(600 in one August week), 20,000 registered cases
1854	Cholera	1290 deaths
1866	Cholera	2122 deaths

23.7.2 The Broad Street Pump

The technique of cluster analysis in the identification of mortality patterns, not only from cancer, is well known and the most famous example is illustrated in Figure 23.8. This map was published by John Snow (1813–58) in 1855 and shows the concentration of cholera deaths during the period 19 August 1854 to 30 September 1854 in the vicinity of the Broad Street water pump in London. Snow confirmed his theory that the cholera epidemic was due to this water supply by tying the handle of the pump, after which the epidemic ceased.

Figure 23.8. Map showing the distribution of cholera deaths in the Broad Street epidemic of 1854. The map formed the frontispiece of John Snow's book *On the Mode of Communication of Cholera*, published in London by Churchill in 1855. (Courtesy of the Wellcome Trustees.)

23.7.3 Typhoid Mary

Epidemiologieal studies also led to the discovery of the *carrier* concept. A carrier of an infectious disease is an individual who does not suffer from any ill effects of the disease, but nevertheless is able to transmit the disease to others. This concept was first proposed in 1884 for diptheria and then in 1893 for cholera[18].

The most famous carrier was an Irish born cook, Mary Mallon (ca. 1868–1938), working in households in New York. She is now known as *Typhoid Mary*, Figure 23.9, and her detection had a major impact in changing the 19th century theory of transmission of infectious diseases[18,19], which was considered to be due to filthy living conditions, including sewage disposal, when in fact it was due to micro-organisms.

Figure 23.9. Typhoid Mary[19].

19th century public health measures had concentrated on providing programmes for urban sanitation projects to bring clean water into cities and to institute sewage disposal, garbage collection and disposal, vaccination measures and in keeping the environment clean. By the end of the 19th century, following the work of Louis Pasteur and Robert Koch, the science of bacteriology had begun and the *dirt theory* of the cause of epidemics (such as typhoid, smallpox, cholera and yellow fever) would eventually be abandoned.

The leading proponent in the USA at the turn of the century, for applying bacteriological theory in the field of public health was Charles V. Chapin (1856–1941),‡ the public health officer in Providence, Rhode Island for the years 1884–1931. His interest in public health statistics led him to devise a *points scale* with the total of city activities scored at 100 and among the items he scored a maximum of 36 points for communicable disease work but only 9 points for sanitation. The control of water supplies and garbage disposal eventually passed from a city public health department to the department of public works.

Mary Mallon was the first typhoid fever carier identified in North America. In August 1906 she was employed in a rented summer home in Oyster Bay, Long Island, when typhoid struck 6 out of 11 persons. The owner thinking he would be unable to rent out his property again unless the mystery of the cases could be solved, hired a civil engineer, George Soper, known for his epidemiological analyses of typhoid, to investigate.

Soper eventually found a clue in the fact that the cooks had changed during the weeks of the outbreak. He then found eight families who had previously employed Mary Mallon before August 1906 and in 7 families out of 8, typhoid

‡ Chapin is credited with the statistical quotation from his paper *Pleasures and hopes of the Health Officer* of 'We cannot expect that figures will ever cease to lie, but we may hope that vital statisticians will'.

had followed her stay.

Eventually she was found in March 1907 at a Park Avenue home in which she was employed. Soper appeared unannounced and tried to explain that Mallon was carrying the disease (carriers transmit typhoid through water or food contaminated by their faeces or urine. She promptly threw Soper out of the house!

Soper continued and convinced the New York City health department that the evidence was strong enough for them to pursue her and gather specimens of blood, faeces and urine to confirm her as the source of these outbreaks. The New York police department was called in to assist and had to restrain her before taking her to hospital[18] and she ended up with a public health official sitting on her chest[19].

She was duly identified as the typhoid carrier and kept in health department custody but in 1909 she successfully sued a court for her release and was freed in 1910. However, in 1915, typhoid outbreaks were traced to her kitchen in a New Jersey sanatorium and a New York maternity hospital[19]. She was then kept in custody until her death in 1938, a total of 26.5 years confinement.

Figure 23.10 shows the increase in identification of typhoid carriers in the city of New York 1908–1950 including the detection of Typhoid Mary in 1908 who had been a proven carrier of the disease at least from the year 1900 and who is considered to have caused at least 10 outbreaks which included 51 cases and three deaths.

The effect of epidemics such as typhoid and smallpox was noticeable not only in the populations of crowded urban cities but also in armies. For example, the French and German armies adopted sanitation procedures as a result of typhoid epidemics and the annual French typhoid morbidity of some 380 per 100,000 was reduced by one-third in a single year when water filtration was introduced in 1888 and in 1889 this mortality was reduced by a further 50% because of additional hygiene and sanitation procedures. The German army had a typhoid death rate in 1870 of 170 per 100,000 whereas by 1914 it was only 8 per 100,000.

The British army were slow to learn this lesson and in the Boer War of 1899–1901 where were more than 57,000 cases and some 8000 deaths, with typhoid being responsible for more deaths than those in combat!

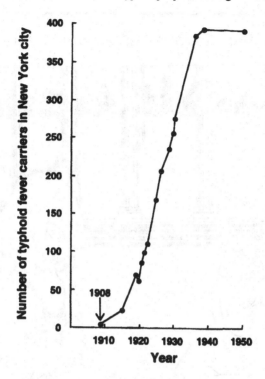

Figure 23.10. Statistics[18] for health carriers of typhoid fever in New York city 1908–1950. The outbreaks identified by George Soper in 1908 as having been caused by Typhoid Mary were as follows. **1900:** 1 case in Maranaroneck, NY. **1901:** 1 case in New York city. **1902:** 9 cases in Dark Harbor, Maine. **1904:** 4 cases in Sands Point, NY. **1906:** 6 cases in Oyster Bay, NY and 1 case in Tuxedo, NY. **1907:** 2 cases in New York city. The case of Typhoid Mary and those in the previous two sections: the Broad Street pump and the Unique Hostel, are examples of investigating the pattern of a cluster of sites where a disease has been observed and identifying the reason for the cluster. It should, though, also be noted that the term *cluster analysis* is used to describe a type of complex multivariate analysis where the relationship between several variables are being studied.

ROMAN ADVICE ON DRUG TESTING, 1000 AD

Avicenna recommended that drugs should be tried on opposed cases and he stressed the importance of human pharmacology when he warned that 'testing a drug on a lion or a horse might not prove anything about its effect on man'.

Glossary of Rates and Ratios: Terminology in Vital Statistics

There are many different rates and ratios in medical statistics and those which fall into the category of *vital statistics* may be grouped as follows: **Demographic statistics** (population, marriages and fertility), **Mortality statistics** (number and causes of death), **Morbidity statistics** (illnesses and injuries, incapacity, hospitalisations etc). This short glossary contains some of the rates and ratios encountered in vital statistics.

Index of Terms

Age-adjusted incidence rate
Age-corrected mortality rate
Age-corrected survival rate
Age-specific rates
Age-specific incidence rate
Age-specific mortality rate
Age-standardised incidence rate
Age-standardised mortality rate
Case fatality rate
Crude rate
Crude incidence rate
Crude mortality rate
Hazard rate
Infant mortality
Life expectation

Likelihood ratio
Mortality
Neonatal mortality
Odds ratio
Perinatal mortality
Period prevalence rate
Point prevalence rate
Prevalence
Relative survival rate
Standardised rate
Standardised mortality ratio
Standardised registration ratio (SRR)
Still birth rate
Survival fraction
Survival rate

347

Age-adjusted incidence rate This is an alternative term for *age-standardised incidence rate*. An example is given for the state of Kentucky, USA, for lung cancer rates per 100,000 population, presented in a map format for the 15 area development districts of Kentucky which are formed from 120 counties[8]. The highest age-adjusted rates are seen in two areas. These are rural districts where most of the tobacco farming takes place and they are also high coal mining areas. In a separate study[9] of self-reporting of smoking habits the areas with the highest number of smokers correlated well with these two areas.

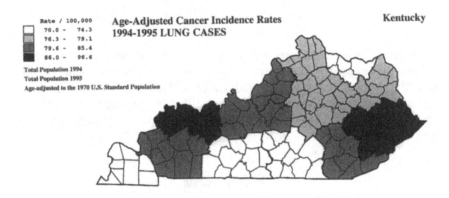

Figure G.1. Age-adjusted lung cancer rates for Kentucky, USA[8]. (Courtesy: Kentucky Cancer Registry.)

Age-corrected mortality rate *see* Age-standardised mortality rate

Age-corrected survival rate *see* Relative survival rate.

Age-specific rates Refer to the rates for specific age groups for each sex. The definition of the age groups will depend on the nature of the disease and its distribution in the population. However, it is wise to ensure that the age groups fall into one or more of the five year age intervals used by a country in official publications on population.

Age-specific incidence rate An age-specific incidence rate refers to a population in a specified age range, usually a 5-year or 10-year range. Thus for the age range 40–45 years, the annual age-specific incidence rate for a specified disease in males is per 100,000 population in the defined population group equal to

$$\left[\frac{\text{Number of new cases of the disease in males aged 40–45 years registered in year } Y}{\text{Average number of males aged 40–45 years at risk in year } Y} \right] \times 10^5$$

A similar rate can be defined for females aged 40–45 years. The advantage of the age-specific incidence rate over the crude incidence rate is that any peculiarities in the disease incidence pattern which are related to particular age groups can be shown. Variations with age would not he apparent using only a crude incidence rate. An example of age-specific incidence rates is seen in Figure 1.11.

Age-specific mortality rate An age-specific mortality rate is defined for a population in a specified age range, say 50–55 years. The annual age-specific mortality rate for a specified disease for year Y, for males, per million population in the defined population group is

$$\left[\frac{\text{Number of deaths from the disease in males aged 50–55 years which occurred during year } Y}{\text{Average number of males aged 50–55 years at risk in year } Y} \right] \times 10^6$$

A similar rate can be defined for females aged 50–55 years.

Age-standardised incidence rate Age-standardised rates have been designed to enable comparisons to be made of rates in different places and for different registration periods. To calculate an age-standardised incidence rate for a disease, a standard population must first be defined. For a given age group i, for example, 40–44 years, the following data will be available:

For the observed population Population in 100,000s for age group $i = n_i$. Number of disease registrations for age group $= r_j$. Hence the age-specific incidence rate per 100,000 population $= r_i/n_i$.

For the standard population Population in 1000s for age group $i = N_i$ ($N_t = WS_i$ in Table G.2). Total population $= 100,000$. One method of calculating an age-standardised incidence rate for the observed population is by the following summation for all age groups i:

$$\frac{\sum N_i \times (r_i/n_i)}{\sum N_i}$$

This is known as *direct* standardisation and is the rate which would have occurred if the observed age-specific rates had operated in the standard population defined by the arbitrary proportion of people in each age group. The use of age-standardised incidence rates is rather limited and the figures of greatest value are the age-specific rates (see **Standard population**).

Age-standardised mortality rate (sometimes termed **Age-adjusted**) Defined in a similar manner to age-standardised incidence rate. An example of standardisation of mortality rates is given in Figure G.2 and is for mortality from cancer at selected sites 1970–1993 in the USA. The rates are termed age-adjusted

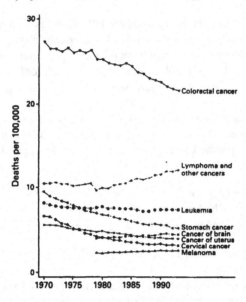

Figure G.2. Age-adjusted mortality rates[1] in the total United States population. The rates have been age-adjusted to the USA resident population of 1990.

rather than age-standardised in this *New England Journal of Medicine* 1997 paper[1].

WHO has defined four different Standard Populations which can be used for standardisation of incidence rates and mortality rates. These are the World, African, European and Truncated, Table G.1. Table G.2 shows this method of calculation with the World standard data from Table G.1 for the age-specific incidence rate per 100,000 for a given country C being $AS_i = r_i/n_i$.

Case fatality rate The death rate amongst those known to have a specific disease.

Crude rate refers to the average rate for the whole population. Unless populations have a similar age structure and similar sex structure, the crude rate can be misleading when making comparisons. See **Age-specific rates**.

Crude incidence rate An annual incidence rate is a measure of the new cases of a disease in a particular year. It is usually quoted as a proportion per 100,000 of a defined population at risk, but can also be stated per million or per thousand population at risk. The adjective *crude* refers to the fact that the rate is not modified to take into account such factors as age or reference year. The crude

Table G.1. WHO defined World and African standard populations, each of 100,000 persons.

Age	African	World	Age	African	World
0–	2	2.4	45–49	5	6
1–4	8	9.6	50–54	3	5
5–9	10	10	55–59	2	4
10–14	10	9	60–64	2	4
15–19	10	9	65–69	1	3
20–24	10	8	70–74	1	2
25–29	10	8	75–79	0.5	1
30–34	10	6	80–84	0.3	0.5
35–39	10	6	85+	0.2	0.5
40–44	5	6	Totals	100	100

(Populations for given age groups are in 1000s)

annual incidence rate per 100,000 for a specified disease in males is equal to

$$\left[\frac{\text{Number of new cases of the disease in males registered in year } Y}{\text{Average number of males at risk in year } Y} \right] \times 10^5$$

A similar rate can be defined for females. An example of crude incidence rates is seen in Figure 1.7(a).

Crude mortality rate The crude annual male mortality rate per million population, for year Y, is

$$\left[\frac{\text{Number of deaths from the disease among males which occurred in year } Y}{\text{Average number of males at risk in year } Y} \right] \times 10^6$$

A similar rate can be defined for females.

Hazard rate *see* section 18.2

Infant mortality The number of deaths of infants in the *first year* of life stated per 1000 live births.

Life expectation Number of years a person is expected to live after birth. Figure G.3 shows how the expectation of life in England and Wales has varied from 1841 (the year of the first census in England and Wales) to 1968–1970, by sex. Life expectation estimates as of April 1997 for selected countries[7] are given in Table G.3.

Table G.2. Calculation schedule for an age-standardised incidence rate for country C.

Age group (years) i	WS_i World standard population (thousands)	AS_i Age-specific incidence rate per 100,000 for country C	$WS_i \times AS_i$
0–	2.4	AS_1	$2.4 \times AS_1$
1–4	9.6	AS_2	$9.6 \times AS_2$
5–9	10	AS_3	$10 \times AS_3$
10–14	9	AS_4	$9 \times AS_4$
15–19	9	AS_5	$9 \times AS_5$
20–24	8	AS_6	$8 \times AS_6$
25–29	8	AS_7	$8 \times AS_7$
30–34	6	AS_8	$6 \times AS_8$
35–39	6	AS_9	$6 \times AS_9$
40–44	6	AS_{10}	$6 \times AS_{10}$
45–49	6	AS_{11}	$6 \times AS_{11}$
50–54	5	AS_{12}	$5 \times AS_{12}$
55–59	4	AS_{13}	$4 \times AS_{13}$
60–64	4	AS_{14}	$4 \times AS_{14}$
65–69	3	AS_{15}	$3 \times AS_{15}$
70–74	2	AS_{16}	$2 \times AS_{16}$
75–79	1	AS_{17}	$1 \times AS_{17}$
80–84	0.5	AS_{18}	$0.5 \times AS_{18}$
85+	0.5	AS_{19}	$0.5 \times AS_{19}$
	$\sum = 100,000$		$\sum(WS_i \times AS_i) =$ Age-standardised incidence rate per 100,000

Likelihood ratio *see* section 18.7.2

Mortality The death rate or mortality is the proportion of persons dying from a disease or set of diseases (i.e. a *cause* or *multiple causes* or *all causes*). The rate might be expressed per 1000; 10,000; 100,000 or per million population at risk.

Neonatal mortality Deaths of infants in the *first four weeks* of life stated per 1000 live births.

Odds ratio *see* section 22.3

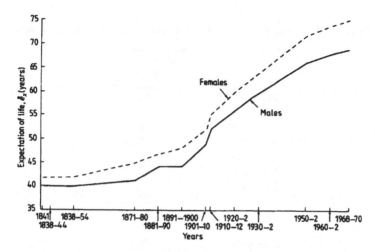

Figure G.3. Expectation of life, England and Wales, from English Life Table data[10].

Perinatal mortality Number of stillbirths *plus* the number of deaths in the first week of life stated per 1000 total live *and* stillbirths.

Period prevalence rate A period prevalance rate per 100,000 male population refers to the number of cases of disease existing at any time within the specified period. It is equal to

$$\left[\frac{\text{Number of cases of disease present in the male population at any time during a specified period}}{\text{Number of males in the population at the midperiod time}} \right] \times 10^5$$

Whereas the point prevalence rate will describe the cases of a given disease existing on a particular day, the period prevalence rate will give the cases during a specified interval, for example, a month. Prevalence refers to *all* cases and not, as for incidence, only *new* notifications. Thus for a period prevalence rate persons must be included whose illness began and ended during the period, or began during the period and still existed at the end, or began before the period started and ended either during the period or after the end of the period.

Point prevalence rate A point prevalence rate per 100,000 male population is equal to

$$\left[\frac{\text{Number of cases of disease present in the male population at a specified time}}{\text{Number of males in the population at that specified time}} \right] \times 10^5$$

A similar rate can be defined for females.

Table G.3. Life expectancy at birth for selected countries, April 1997 data[7].

Country	Life expectancy (1995–2000)
Algeria	69
Argentina	73
Australia	78
Austria	77
Bangladesh	58
Botswana	50
Canada	79
China	70
Denmark	76
Egypt	66
Ethiopia	50
France	79
Gambia	47
Germany	77
India	62
Japan	80
Kenya	54
Malawi	41
Nigeria	52
Russia	64
South Africa	65
United Kingdom	77
USA	77
Zaire	53
Zambia	43

Prevalence Number of cases of a disease in a defined population. See *Period prevalence rate* and *Point prevalence rate*. Figure G.4 (top right) is an example published[7] by the World Health Organisation for selected cancers of five-year (period) prevalence. These figures can be compared with 1996 statistics for worldwide mortality (top left) and incidence for both the developed world (bottom left) and the developing world (bottom right). Examples of annual (period) prevalence for 1996 for other diseases/conditions are given[7] in Table G.4.

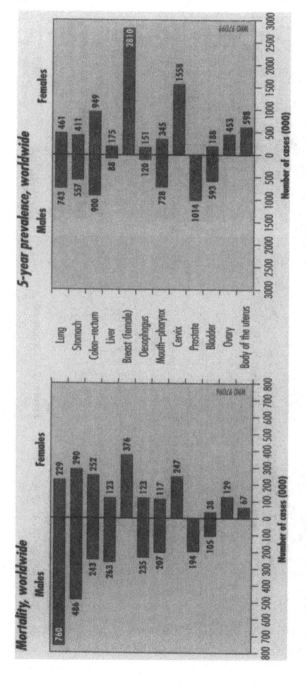

Figure G.4. Worldwide five-year prevalence, 1996 incidence and 1996 mortality, in thousands of cases, for selected cancers[7]. Cancer now accounts for about 20% of all deaths in developed regions and about 10% of all deaths in developing regions. In 1996 there were an estimated 17.9 million persons with cancer surviving up to five years after diagnosis. Of these, 10.5 million were women. 5.3 million of whom had cancer either of the breast, cervix or colon–rectum. Among men, prostate, colorectal and lung cancer were the most prevalent.

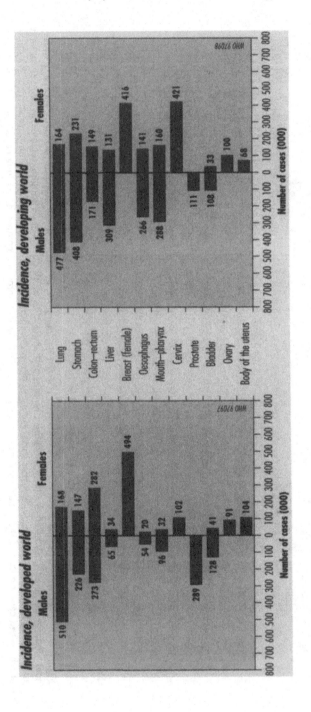

Figure G.4. (continued)

Table G.4. Estimated incidence (new cases) and prevalence (all cases) for selected diseases/conditions worldwide in 1996[7]. (NS: not stated[7].) The numbers are given in thousands and thus for example the prevalence of HIV/AIDS is 22.6 million worldwide in 1996.

Disease/condition	Incidence	Prevalence
Tuberculosis	7400	NS
HIV/AIDS	3100	22600
Leprosy	530	1260
Syphilis	12000	28000
Gonorrhoea	62000	23000
Haemophilia	10	420
Epilepsy	2000	40000
Asthma	NS	155000

Relative survival rate

$$\text{Relative } T\text{-year survival rate} = 100 \times \left[\frac{\text{Crude } T\text{-year survival rate}}{\substack{\text{Expected } T\text{-year survival rate in the normal} \\ \text{population with the same age and sex structure} \\ \text{as the group under observation}}} \right] \%$$

A relative survival rate is also sometimes termed an age-corrected survival.

Figure G.5 shows the crude and relative 5-year to 15-year survival rates for all the cancer of the cervix and cancer of the tongue patients who were registered in England and Wales in 1954–1955. If the patients were all relatively young at treatment, the expected T-year survival rate in the normal population would be high and the crude and age corrected T-year survival rates would be similar. It is seen, though, that for the two cancers shown, there is a large difference between crude and age corrected rates. This emphasises the need to be aware of exactly which type of rate is quoted when comparing survival results from different publications. See also section 21.7.1 for further comments on relative survival rates.

Standardised rates are rates which have been compounded to take into account differences in the age and sex structures of populations over several areas and are therefore considered to refer to *average* or *standard* populations.

Standardised mortality ratio is a comparison of actual deaths in a particular population compared with those which would be expected in the *standard* population. See also section 22.7.

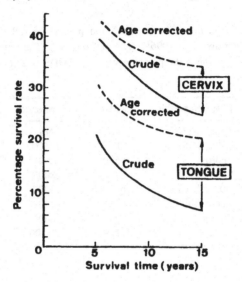

Figure G.5. Crude and relative (age corrected) survival rates for cancer registrations in England and Wales, 1954–1955, which have been followed-up for a period of 15 years. The number of cancer of the cervix patients is 3014 and the number of the cancer of the tongue patients, all males, is 378.

Standardised registration ratio (SRR) The standardised registration ratio is an index routinely used by the Office of Population Censuses and Surveys (OPCS) in their cancer statistics publications for England and Wales. OPCS compute SRR for a given cancer size relative to a standard set of age-specific incidence rates for a standard year. However, SRRs can also be calculated for different countries with respect to a standard country and year, or for different regions within a country or for diseases other than cancer. For cancer registration data in England and Wales the OPCS has chosen as standard years 1968[2–4] and 1979[5]. The SRR equals 100 for the standard year 1968 and for 1970 and cancer of the lung, for example, the ratio is calculated by the formula

$$SRR = \left[\frac{\text{Total registrations of lung cancer in 1970}}{\sum_{\text{All } i} \left(\begin{array}{c} \text{Population for} \\ \text{age group } i \\ \text{in 1970} \end{array} \times (r_i/n_i)_{\text{Standard year}} \right)} \right] \times 100$$

where population for age group i in 1968 (the standard year) $= n_i$, number of registration of lung cancers for age group i in 1968 $= r_i$, and age-specific registration rate in 1968 $= r_i/n_i$.

Thus

$$SRR = \left[\frac{\text{Observed number of registrations of lung cancer in 1970}}{\substack{\text{Expected number of registrations of lung cancer in} \\ \text{1970, assuming that the registration distribution through} \\ \text{the age groups } i \text{ in 1970 is in the same proportion as in} \\ \text{the standard year 1968}}} \right] \times 100$$

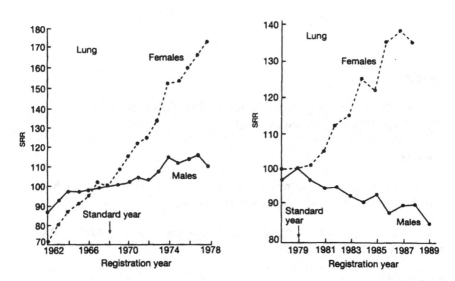

Figure G.6. (Left) SRRs for the period 1962–1977 for lung cancer with the SRR standardised as 100 for 1968. These data are from OPCS publications 1979–1983[2–4] but those from later OPCS publications such as for 1987[5] have standardised on the year 1979, (right). This use of SRR clearly demonstrates for England and Wales the continuing increase in lung cancer incidence in women. It is noted that April 1996 the Office of Population Censuses and Surveys (OPCS) merged with the Central Statistics Office to form the Office for National Statistics (ONS) and published their first data for cancer registrations in 1997[6] and at this point in time SRRs are only available to 1989.

Still birth rate Number of still births expressed per 1000 total live *and* still births.

Survival fraction A *T*-year survival fraction is the proportion of persons surviving to *T*-years, stated as a number between 0 and 1.

Survival rate Proportion of persons surviving to *T*-years, stated either as a percentage between 0 and 100 per cent or as a number between 0 and 1; see also **Survival fraction**.

References

Chapter 1

[1] Mould R F and Tungsubutra K (eds) 1985 *Carcinoma of the Cervix in Developing Areas* (Bristol: Adam Hilger)

Chapter 4

[1] Altman D G 1982 in *Statistics in Practice* (ed) S M Gore and D G Altman (London: British Medical Association) pp 1–2

[2] Swinscow T D V 1983 *Statistics at Square One* 8th edn (London:British Medical Association) pp 22–23

Chapter 7

[1] Lipschutz S 1974 *Theory and Problems of Probability* Schaum's Outline Series (New York: McGraw-Hill) p 119

[2] Kreyszig E 1977 *Statistische Methoden und ihre Anwendungen* (Göttingen: Vandenhoeck & Ruprecht) p 118

Chapter 8

[1] Bakowski M T, Macdonald E, Mould R F, Cawte P, Sloggem J, Barrett A, Dalley V, Newton K A, Westbury G, James S E and Hellmann K 1978 *Double blind controlled clinical trial of radiation plus Razoxane (ICRF 159) versus radiation plus placebo in the treatment of head and neck cancer* Int. J. Radiation Oncology Biology Physics **4** 115–119

[2] Medical Research Council (multiple authors) 1993 *A trial of Ro 03-8799 (pimonidazole) in carcinoma of the uterine cervx: an interim report from the*

Medical Research Council Working Party on advanced carcinoma of the cervix Radiotherapy & Oncology **26** 93–103

[3] Scott C and Wasserman T 1998 *When is a negative study not negative?* Int. J. Oncology Biology Physics (in press)

[4] Peto R, Pike M C, Armitage P, Breslow N E, Cox D R, Howard S V, Mantel N, McPherson K, Peto J and Smith P G 1976 *Design and analysis of randomized clinical trials requiring prolonged observation of each patient: I. Introduction and design* Brit. J. Cancer **34** 592–593

Chapter 9

[1] International Atomic Agency 1984 *Quality Control of Nuclear Medicine Instrumentation* TECDOC-317 (Vienna: IAEA)

[2] Siegel S 1956 *Non-Parametric Statistics for the Behavioural Sciences* (New York:McGraw-Hill) p 45

[3] Mould R F and Boag J W 1975 *A test of several parametric statistical models for estimating success rate in the treatment of carcinoma cervix uteri* Brit. J. Cancer **32** 529–550

Chapter 10

[1] Nicholson R A, Nicholson J P and Mould R F 1986 *Westminster Hospital post-Chernobyl results with a single probe sodium iodide counter*, in: Heywood J (ed) *Chernobyl: the response of Medical Physics Departments in the United Kingdom* (London: Institute of Physical Sciences in Medicine) pp 27–32

[2] Armitage P 1971 *Statistical Methods in Medical Research* (Oxford: Blackwell Scientific) pp 136–138

Chapter 11

[1] Mould R F and Williams R J 1980 *Age distribution of cancer of the cervix uteri* Brit. Medical Journal **280** 366

Chapter 12

[1] Schiffman S S, Buckley C E, Sampson H A, Massey E W, Baraniuk J N, Follett J V and Warwick Z S 1987 *Aspartame and susceptibility to headache* New England J. Medicine **317** 1181–1185

Chapter 13

[1] Robertson E M, Baily B, Hearnden T M and Mould R F 1981 *An assessment of intravenous urogram imaging using 100 mm and conventional films* Brit. J. Radiology **54** 944–947

[2] Bakowski M T, Macdonald E, Mould R F, Cawte P, Sloggem J, Barrett A, Dalley V, Newton K A, Westbury G, James S E and Hellmann K 1978 *Double blind controlled clinical trial of radiation plus Razoxane (ICRF 159) versus radiation plus placebo in the treatment of head and neck cancer* Int. J. Radiation Oncology Biology Physics **4** 115–119

[3] Raven R W, Hanham I W F and Mould R F 1986 *Cancer Care: an International Survey* (Bristol: Adam Hilger)

[4] Solomon R L and Coles M R 1954 *A case of failure of generalization of imitation across drives and across situations* J. Abnormal & Social Psychology **49** 7–13

[5] Mann H B and Whitney D R 1947 *On a test of whether one of two random variables is stochastically larger than the other* Ann. Mathematics Statistics **18** 52–54

[6] Auble D 1953 *Extended tables for the Mann–Whitney statistic* Bulletin of the Institute of Educational Research at Indiana University **1** No. 2

[7] Lentner C (ed) 1982 *Geigy Scientific Tables* 8th. edn **2** (Basle: Ciba-Geigy) pp 156–162

[8] Altman D G 1991 *Practical Statistics for Medical Research* (London: Chapman & Hall)

Chapter 14

[1] Kaplan E L and Meier P 1958 *Non-parametric estimation from incomplete observations* J. Amer. Statistical Association **53** 457–482

[2] Greenwood 1926 *A Report on the Natural Duration of Cancer* Reports on Public Health and Medical Subjects No.33 (London: HMSO)

[3] Ederer F 1961 *A parametric estimate of the standard error of the survival rate* J. Amer. Statistical Association **56** 111–118

[4] Merrell M and Shulman L E 1955 *Determination of prognosis in chronic disease, illustrated by systemic lupus erythematosus* J. Chronic Diseases **1** 12–32

[5] Berkson J and Gage R P 1950 *Calculation of survival rates for cancer* Proc. Staff Meetings of the Mayo Clinic **25** 270–286

[6] Wood C A P and Boag J W 1950 *Researches on the radiotherapy of oral cancer* Medical Research Council Special Report Series No. 267 (London: HMSO)

[7] Takahashi M, Abe M, Ri N, Dodo Y, Doko S and Nishidai T 1979 *Uneven fractionation in bronchogenic carcinoma* in: Abe M, Sakamoto K and Phillips T L *Treatment of Radioresistant Cancers* (Amsterdam: Elsevier/North-Holland Biomedical Press) p 200

[8] Caplan R J, Pajak T F and Cox J D 1994 *Analysis of the probability and risk of cause-specific failure* Int. J. Radiation Oncology Biology Physics **29** 1183–1186

[9] Bentzen S M, Vaeth M, Pedersen D E and Overgaard J 1995 *Why actuarial estimates should be used in reporting late normal tissue effects of cancer treatment ......now!* Int. J. Radiation Oncology Biology Physics **32** 1531–1534

[10] Chappell R 1996 *Re: Caplan et al IJROB 29: 1183–1186 1994 and Bentzen et al IJROB 32: 1531-1534 1995* Int. J. Radiation Oncology Biology Physics **36** 988–989

Chapter 15

[1] Schwartz D, Flamant R and Lellouch J 1980 *Clinical Trials* (London: Academic Press) p 223

[2] Buyse M E, Staquet M J and Sylvester R J 1984 *Cancer Clinical Trials Methods and Practice* (Oxford: Oxford University Press)

[3] Peto R, Pike M C, Armitage P, Breslow E, Cox D R, Howard S V, Mantel N, McPherson K, Peto J and Smith P G 1977 *Design and analysis of randomised clinical trials requiring prolonged observation of each patient: II. Analysis and examples* Brit. J. Cancer **35** 1–39

[4] Mantel N 1966 *Evaluation of survival data and two new rank order statistics arising in its consideration* Cancer Chemotherapy Reports **50** 163–170

[5] Mantel N and Haenszel W 1959 *Statistical aspects of the analysis of data from retrospective studies of disease* J. National Cancer Institute **22** 719–748

[6] Friedman L M, Furberg C D and DeMets D L 1982 *Fundamentals of Clinical Trials* (Bristol: John Wright) pp 185–191

Chapter 16

[1] Jones B and Tan L T 1996 *Long-term complications following brachytherapy for gynaecological cancers* in: Cadel C (ed) *Long-term results of GYN brachytherapy* Activity Nucletron Journal Special Report 8, p 48

[2] Box G E P, Hunter W G and Hunter J S 1978 *Statistics for Experimenters: an Introduction to Design, Data Analysis and Model Building* (New York: Wiley)

[3] Hunter W G 1981 *Six statistical tales* The Statistician **30** 107–117

[4] Trichopoulos D, Zavitsanos X, Koutis C, Drogari P, Proukakis C and Petridou E 1987 *The victims of Chernobyl in Greece: induced abortions after the accident* Brit. Medical Journal **295** 1100

[5] Kendall M G 1955 *Rank Correlation Methods* 2nd edn (London: Charles Griffin)

[6] Snedecor G W and Cochran W G 1967 *Statistical Methods* 6th edn (Ames: Iowa State University Press)

[7] Sachs L 1978 *Applied Statistics* 5th edn (New York: Springer)

Chapter 17

[1] Barron S L and Vessey M P 1966 *Birth weight of infants born to immigrant women* Brit. J. of Preventative and Social Medicine **20** 127-134

[2] Last J M 1995 *A Dictionary of Epidemiology* 3rd. edn (Oxford: Oxford University Press)

[3] Beaglehole R, Bonita R and Kjellström 1993 *Basic Epidemiology* (Geneva: World Health Organisation) pp 48–50

[4] Bourke G J and McGilvray J 1975 *Interpretation and Uses of Medical Statistics* 2nd. edn (Oxford: Blackwell Scientific) pp 140–142

[5] Kirkwood B R 1988 *Essentials of Medical Statistics* (Oxford: Blackwell Scientific) p 66

Chapter 18

[1] Cox D R 1972 *Regression models and life tables* J. Royal Statistical Society **34B** 187–220

[2] Breslow N E 1975 *Analysis of survival data under the proportional hazards model* Int. Statistical Review **43** 45–58

[3] Norwegian Multicenter Study Group 1981 *Timolol-induced reduction in mortality and reinfarction in patients surviving acute myocardial infarction* New England J. Medicine **304** 801–807

[4] Kalbfleisch J D and Prentice R L 1980 *The Statistical Analysis of Failure Time Data* (New York: Wiley)

[5] O'Quigley J 1982 *Regression models and survival prediction* The Statistician **31** 107–116

[6] Breslow N E, Day N E, Halvorsen K T, Prentice R L and Sabai C 1978 *Estimation of multiple relative risk functions in matched case-control studies* Amer. J. Epidemiology **108** 299–307

[7] International Agency for Researech on Cancer 1980 *Statistical Methods in Cancer Research* **1** *The Analysis of Case-Control Studies* ed N E Breslow and N E Day, IARC Scientific Publication No. 32 (Lyon: IARC)

[8] Cox D R 1962 *Renewal Theory* Methuen Monograph on Applied Probability and Statistics (London: Methuen)

[9] Hastings N A J and Peacock J B 1975 *Statistical Distributions* (London: Butterworths)

[10] Nelson W 1972 *Theory and applications of hazard plotting for censored failure data* Technometrics **14** 945–966

[11] Gross A J and Clark V A 1975 *Survival Distributions: Reliability Applications in the Biomedical Sciences* (New York: Wiley)

[12] Mould R F 1983 *Cancer Statistics* (Bristol: Adam Hilger) pp 170–175

[13] Rubens R D, Armitage P, Winter P J, Tong D and Hayward J L 1977 *Prognosis in inoperable stage III carcinoma of the breast* Eur. J. Cancer **13** 805–811

[14] Gore S M and Pocock S J 1981 Lecture at the Institute of Statisticians Conference *Statistics in Medicine* Cambridge *The statistical modelling of survival in breast cancer*

[15] Langlands A O, Pocock S J, Kerr G R and Gore S M 1979 *Long-term survival of patients with breast cancer: a study of the curability of the disease* Brit. Medical J. **2** 1247–1251

[16] Willner J, Kiricuta I C and Kölbl O 1997 *Locoregional recurrence of breast cancer following mastectomy: always a fatal event? Results of univariate and multivariate analysis* Int. J. Radiation Oncology Biology Physics **37** 853–863

[17] Halverson K J, Perez C A, Kuske R R, Garcia D M, Simpson J R and Fineberg B 1992 *Survival following locoregional recurrence of breast cancer: univariate analysis and multivariate analysis* Int. J. Radiation Oncology Biology Physics **23** 285–291

[18] Schwaibold F, Fowble B L, Solin L J, Schultz D J and Goodman R L 1991 *The results of radiation therapy for isolated local regional recurrence after mastectomy* Int. J. Radiation Oncology Biology Physics **21** 299–319

[19] Christensen E 1987 *Multivariate survival analysis using Cox's regression model* Hepatology **7** 1346–1358

[20] Auquier A, Rutqvist L E, Host H, Rotstein S and Arriagada 1992 *Postmastectomy megavoltage radiotherapy: the Oslo and Stockholm trials* Eur. J. Cancer **28** 433–437

[21] Saeter G, Hoie J, Stenwig A E, Johansson A K, Hannisdal E and Solheim O P 1995 *Systemic relapse of patients with osteogenic sarcoma* Cancer **75** 1084–1093

[22] Haybittle J L, Blamey R W, Elston C W, Johnson J, Doyle P J, Campbell F C, Nicholson R I and Griffiths K 1982 *A prognostic index in primary breast cancer* Brit. J. Cancer **45** 361–366

[23] Clark G M and Hilsenbeck S G 1994 *Integration of prognostic factors: can we predict breast cancer recurrences?* in: Schmitt M, Graeff H, Kindermann G, Jänicke F, Genz T and Lampe B (eds) *Prospects in Diagnosis and Treatment of Breast Cancer* International Congress Series 1050 Excerpta Medica (Amsterdam: Elsevier Science) pp 177–186

[24] Pinchon M F, Broet P, Magdelenat H, Delarue J C, Spyratos F, Basuyau J P, Saez S, Rallet A, Courriere P, Millon R and Asselain B 1996 *Prognostic value of steroid receptors after long-term follow-up of 2257 operable breast cancers* Brit. J. Cancer **73** 1545–1551

Chapter 19

[1] Kramer M S 1988 *Clinical Epidemiology and Biostatistics* (Heidelberg: Springer) pp 201–205

[2] Rang M 1972 *The Ulysses syndrome* Can. Med. Assoc. J. **106** 122–123

[3] Elveback L R, Guillier C L and Keating F R 1970 *Health, normality and the ghost of Gauss* JAMA **211** 69–75

[4] Brawer M K 1995 *How to use prostate-specific antigen in the early detection or screening for prostatic carcinoma* CA A Cancer Journal for Physicians **45** 148–164

[5] American Cancer Society 1996 *Cancer Facts and Figures-1996* (Atlanta: ACS)

[6] Murphy G P, Lawrence W and Lenhard R E (eds) 1995 *American Cancer Society Textbook of Clinical Oncology* 2nd edn (Atlanta: ACS)

[7] Souchkevitch G N, Asikainen M, Bäuml A, Bergmann H, Busemann-Sokole E, Carlsson S, Delaloye B, Dermentzoglou F, Herrera N, Jasinski W, Karanfilski B, Mester J, Oppelt A, Perry J, Skretting A, van Herk G, Volodin V, Wegst A and Mould RF 1988 *The World Health Organisation and International Atomic Energy Agency second interlaboratory comparison study in 16 countries on quality performance of nuclear medicine imaging devices* European J. Nuclear Medicine **13** 495–501

[8] Hermann G A, Herrera N and Sugiura H T 1982 *Comparison of interlaboratory data in terms of receiver operating chanracteristic (ROC) indices* J. Nuclear Medicine **23** 325–531

[9] Altman D G 1991 *Practical Statistics for Medical Research* (London: Chapman & Hall) pp 404–409

[10] Boyd N F, Wolfson C and Moskowitz M 1982 *Observer variation in the interpretation of xeromammograms* J. Nat. Cancer Inst. **68** 357–363

[11] Landis J R and Koch G G 1977 *The measurement of observer agreement for categorical data* Biometrics **33** 159–174

[12] Mussurakis S, Buckley D L, Coady A M, Turnbull L W and Horsman A 1996 *Observer variability in the interpretation of contrast enhanced MRI of the breast* Brit. J. Radiology **69** 1009–1018

Chapter 20

[1] Burdette W J and Gehan E A 1970 *Planning and Analysis of Clinical Studies* (Springfield: Charles C Thomas)

[2] Schwartz D, Flamant R and Lellouch J 1980 *Clinical Trials* (Healy M J R translator) (London: Academic Press)

[3] Friedman L M, Furberg C D and DeMets D L 1996 *Fundamentals of Clinical Trials* 3rd. edn (St. Louis: Mosby)

[4] Pocock S J 1983 *Clinical Trials: a Practical Approach* (Chichester: Wiley)

[5] Buyse M E, Staquet M J and Sylvester R J 1984 *Cancer Clinical Trials, Methods and Practice* (Oxford: Oxford University Press)

[6] Machin D and Campbell M J 1987 *Statistical Tables for the Design of Clinical Trials* (Oxford: Blackwell Scientific) *see also* Machin D, Campbell M, Fayers P and Pinol A 1997 *Sample size for clinical studies* 2nd edn (Oxford: Blackwell Scientific)

[7] Peto R, Pike M C, Armitage P, Breslow N E *et al* 1976 & 1977 *Design and analysis of randomized clinical trials requiring prolonged observation of each patient* I *Introduction and design* Brit. J. Cancer **34** 585–612 (1976) II *Analysis and examples* Brit. J. Cancer **35** 1–39 (1977)

[8] American Cancer Society 1994 *National conference on clinical trials* (proceedings) Atlanta, November 1993 Cancer Supplement **74** 2603–2744

[9] Bolla M, Bartelink H, Garavaglia G, Gonzalez D *et al* 1995 *EORTC guidelines for writing protocols for clinical trials of radiotherapy* Radiotherapy & Oncology **36** 1–8

[10] Rubin P, Keys H and Salazar O 1987 *Innovative designs for radiation oncology research in clinical trials* in: *Principles and Practice of Radiation Oncology* Perez C A and Brady L (eds) (Philadelphia: Lippincott) pp 291–297

[11] Zelen M 1993 *Theory and practice of clinical trials* in: *Cancer Medicine* Holland J F, Frei E, Bast R C, Kufe D W *et al* (eds) 3rd. edn (Philadelphia: Lea & Febiger) pp 340–360

[12] Simon R 1993 *Design and conduct of clinical trials* in: *Cancer Principles & Practice of Oncology* DeVita V T, Hellman S and Rosenberg S A (eds) 4th. edn 1 (Philadelphia: Lippincott) pp 418–440

[13] Buyse M E 1995 *Clinical trial methodology* in: *Oxford Textbook of Oncology* Peckham M, Pinedo H and Veronesi U (eds) (Oxford: Oxford University Press) pp 2377–2395

[14] Carter S K, Selawry O and Slavik M 1977 *Methods of Development of New Anticancer Drugs* (Bethesda: US Department of Health, Education & Welfare)

[15] Shibamoto Y, Takahashi M and Abe M 1996 *A phase I study of hypoxic cell sensitizer KU-2285 in combination with conventional radiotherapy* Radiotherapy & Oncology **40** 55–58

[16] Friedman M A 1995 *Clinical trials* in: *American Cancer Society Textbook of Clinical Oncology* Murphy G P, Lawrence W and Lenhard R E (eds) 2nd. edn (Atlanta: American Cancer Society) pp 194–197

[17] Byar D P, Simon R M, Friedewald W T, Schlesselman J J *et al* 1976 *Randomized clinical trials: perspectives on some new ideas* New England J. Medicine **295** 74–80

[18] Sylvester R 1989 *Phase I, II and III trials: role, description and statistical design* in: *Data Management and Clinical Trials, EORTC Study Group on Data Management* Rotmensz N, Vantongelen K and Renard J (eds) (Amsterdam: Elsevier) pp 9–22

[19] Gehan E A 1961 *The determination of the number of patients required in a preliminary and a follow-up trial of a new chemotherapeutic agent* J. Chronic Diseases **13** 346–353

[20] Boag J W, Haybittle J L, Fowler J F and Emery E W 1971 *The number of patients required in a clinical trial* Brit. J. Radiology **44** 122–125

[21] Gelber R D and Goldhirsch A 1989 *Statistics in clinical trials* in: *Data Management and Clinical Trials EORTC Study Group on Data Management* Rotmensz N, Vantongelen K and Renard J (eds) (Amsterdam: Elsevier) pp 175–204

[22] Freedman L S 1982 *Tables of the number of patients required in clinical trials using the logrank test* Statistics in Medicine **1** 121–129

23 Siegel S 1956 *Nonparametric Statistics for the Behavioural Sciences* (New York: McGraw-Hill) p 10

24 Haybittle J L 1971 *Repeated assessment of results in clinical trials of cancer treatment* Brit. J. Radiology **44** 793–797

25 Peto R 1978 *Clinical trial methodology* Biomedicine **28** 24–36

26 Gelman R 1996 *Techniques in the interpretation of clinical trials* in: Harris J R, Lippman M E, Morrow M and Hellman S (eds) *Diseases of the Breast* (Philadelphia: Lippincott-Ravel) pp 998–999

27 Ingelfinger J A, Mosteller F and Thibodeau L A 1983 *Biostatistics in Clinical Medicine* (New York: Wiley) p 169

28 Zelen M 1979 *A new design for randomised clinical trials* New England J. Medicine **310** 1404

29 Lentner C (ed) 1982 *Geigy Scientific Tables* **2** (Basle: Ciba-Geigy)

30 Bross I 1952 *Sequential medical plans* Biometrics **8** 188–205

31 Bakowski M T, Macdonald E, Mould R F, Cawte P, Sloggen J, Barrett A, Dalley V, Newton K A, Westburg G, James S E and Hellmann K 1978 *Double blind controlled trial of radiation plus Razoxane (ICRF 158) versus radiation plus placebo in the treatment of head and neck cancer* Int. J. Radiation Oncology Biology Physics **4** 115–119

32 Beaglehole R, Bonita R and Kjellström T 1993 *Basic Epidemiology* (Geneva: World Health Organisation) p 77

33 Early Breast Cancer Trialists' Collaborative Group 1990 *Treatment of Early Breast Cancer* **1** *Worldwide Evidence* 1985–1990 (Oxford: Oxford University Press)

34 Overgaard J and Bentzen S M 1998 *Evidence based radiation oncology* Editorial, Radiotherapy and Oncology **48** 1–3

35 Bentzen S M 1998 *Towards evidence based radiation oncology: improving the design, analysis and reporting of clinical outcome studies in radiotherapy* Radiotherapy and Oncology **48** 5–18

36 Begg C, Cho M, Eastwood *et al* 1996 *Improving the quality of reporting randomized controlled trials* JAMA **276** 637–639

[37] Altman D G, De Stavola B L, Love S B and Stepniewska K A 1995 *Review of survival analyses published in cancer journals* Brit. J. Cancer **72** 511–518

Chapter 21

[1] Mould R F 1993 *A Century of X-rays and Radioactivity in Medicine with Emphasis on Photographic Records of the Early Years* (Bristol: Institute of Physics Publishing) p 110

[2] Niewenglowski G H 1924 *Les Rayons X et le Radium* (Paris: Librairie Hachette) p 127

[3] Wickham L and Degraid P 1910 *Radiumtherapy* (London: Cassell)

[4] Karnofsky D A and Burchenal J H 1949 *The clinical evaluation of chemotherapeutic agents in cancer* in: MacLeod C M (ed) 1949 *Evaluation of Chemotherapeutic Agents* (New York: Columbia University Press) pp 191–205

[5] Stout R, Barber P and Burt P 1994 *Single dose brachytherapy for endobronchial cancer* in: Mould R F, Battermann J J, Martinez A A and Speiser B L (eds) *Brachytherapy from Radium to Optimization* (Veenendaal: Nucletron) 196–199

[6] Zubrod C G, Schneiderman M and Frei E 1960 *Appraisal methods for the study of chemotherapy of cancer in man: comparative therapeutic trial of nitrogen mustard and triethylene thiophosphate* J. Chronic Disease **11** 7–33

[7] Speiser B L 1995 *Oncological assessment using the four-tiered scoring system* Current Oncology **2** 54–59

[8] Speiser B L and Spratling L 1994 *Remote afterloading brachytherapy for the local control of endobronchial carcinoma* in: Mould R F, Battermann J J, Martinez A A and Speiser BL (eds) *Brachytherapy from Radium to Optimization* (Veenendaal: Nucletron) 180–195

[9] Tchekmedyian N S, Hickman M, Siau J, Greco A and Aisner J 1990 *Treatment of cancer anorexia with megestrol acetate: impact on quality of life* Oncology **4** 185–192

[10] American & European LENT Working Committees 1995 *LENT SOMA scales for anatomic sites* Int. J. Radiation Oncology Biology Physics **31** 1049–1092

[11] American & European LENT Working Committees 1995 *LENT SOMA tables: table of contents* Radiotherapy & Oncology **35** 17–60

[12] Denekamp J, Bartelink H and Rubin P 1996 *Correction for the use of the SOMA LENT tables* Int. J. Radiation Oncology Biology Physics **35** 417–420

[13] Bolek T W, Marcus R B, Mendenhall N P, Scarborough M T, Graham-Pole J 1996 *Local control and functional results after twice-daily radiotherapy for Ewing's sarcoma of the extremities* Int. J. Radiation Oncology Biology Physics **35** 687–692

[14] List M A, Ritter-Sterr C and Lansky S B 1990 *A performance stastus scale for head and neck cancer patients* Cancer **66** 564–569

[15] Moore G J, Parsons J T and Mendenhall W M 1996 *Quality of life outcomes after primary radiotherapy for squamous cell carcinoma of the base of tongue* Int. J. Radiation Oncology Biology Physics **36** 351–354

[16] Nimmo W S and Smith G (eds) 1989 *Anaesthesia 1* (Oxford: Blackwell Scientific) p 436

[17] Insall J N, Dorr L D, Scott R D and Scott W N 1989 *Rationale of the Knee Society clinical rating system* Clinical Orthopedics **248** 13–14

[18] Williams M S, Burk M, Loprinzi C L, Hill M, Schomberg P J, Nearhood K, O'Fallon J R, Laurie J A, Shanahan T G, Moore R L, Urias R E, Kuske R R, Engel R E and Eggleston W D 1996 *Phase III double-blind evaluation of an aloe vera gel as a prophylactic agent for radiation-induced skin toxicity* Int. J. Radiation Oncology Biology Physics **36** 345–349

[19] Easson E C and Russell M H 1968 *The Curability of Cancer in Various Sites* (London: Pitman Medical)

[20] Mould R F and Boag J W 1975 *A test of several parametric statistical models for estimating success rate in the treatment of carcinoma cervix uteri* Brit. J. Cancer **32** 529–550

[21] Mould R F, Hearnden T, Palmer M and White G C 1976 *Distribution of survival times of 12,000 head and neck cancer patients who died with their disease* Brit. J. Cancer **34** 180–190

[22] Buyse M E, Staquet M J and Silvester R J 1984 *Cancer Clinical Trials, Methods and Practice* (Oxford: Oxford University Press)

[23] Mould R F and Williams R J 1982 *Survival of histologically proven carcinoma of the lung registered in the North West Thames Region 1975–1979* Brit. J. Cancer **46** 999–1003

[24] Chassagne D, Sismondi P and Horiot J C 1993 *A glossary for reporting complications of treatment in gynaecological cancers* Radiotherapy & Oncology **26** 195–202

[25] Stjernswärd J and Teoh N 1991 *Perspectives on quality of life and the global cancer problem* in: Osoba D (ed) *Effect of Cancer on Quality of Life* (Boca Raton: CRC Press) pp 1–5

[26] Bush R S 1979 *Malignancies of the Ovary, Uterus and Cervix* (London: Edward Arnold)

[27] Agren A, Brahme A and Turesson I 1990 *Optimization of uncomplicated control for head and neck tumours* Int. J. Radiation Oncology Biology Physics **19** 1077–1085

[28] Raeside D E 1976 *Bayesian statistics: a guided tour* Medical Physics **3** 1–11

[29] Office of Health Economics 1985 *Measurement of Health* (London: OHE)

[30] Maciejewski B 1998, February, Personal communication, Warsaw conference, *100 years after the discovery of polonium and radium*

Chapter 22

[1] International Atomic Energy Agency 1980 *Representative risks of different energy sources* Adapted from presentations at a symposium in Paris, January 1980 IAEA Bulletin **22** No. 5/6

[2] Mettler F A and Upton A C 1995 *Medical effects of ionising radiation* 2nd edn (Philadelphia: Saunders)

[3] National Council on Radiation Protection & Measurements 1987 *Induction of thyroid cancer by ionizing radiation* NCRP Report 80 (Bethseda: NCRP)

[4] Hirayama T, Waterhouse J A H and Fraumeni J F 1980 *Cancer risks by site* UICC Technical Report Series **41** (Geneva: Union Internationale Contre le Cancer)

[5] Last J M 1995 *A dictionary of epidemiology* 3rd edn (Oxford: Oxford University Press)

[6] Early Breast Cancer Triallists' Collaborative Group 1990 *Treatment of early breast cancer, worldwide evidence 1985–1990* **1** (Oxford: Oxford University Press)

[7] Radiation Effects Research Foundation Hiroshima and Nagasaki 1994 *Cancer incidence in atomic bomb survivors* Radiation Research **137** Supplement S1–S112

[8] Thompson D E, Mabuchi K, Ron E and Soda M 1994 *Cancer incidence in atomic bomb survivors, Part II: Solid Tumours 1958–1987* Radiation Research **137** Supplement S17–S67

[9] Kato H and Shimizu Y 1990 *Cancer mortality risk among A-bomb survivors* in: *Health effects of atomic radiation, Hiroshima–Nagasaki, Lucky Dragon, Techa River and Chernobyl* Proc. Japan–USSR seminar on Radiation Effects Research, Tokyo, pp 225–236

[10] Ivanov V K, Remmenik L V, Tsyb A F, Starinsky V V, Chissov V I, Maksyutov M A, Gorsky A I, Korelo A M, Nilova E V, Efendiyev V A, Leshakov S Y, Shirayaev V I, Proshin A V, Pochtennaya G T, Kvitko B I and Mould R F 1996 *Possible role of radiation in the induction of cancers in Russia following the Chernobyl accident* Current Oncology **3** 112–117

[11] International Atomic Energy Agency 1996 International Conference *One decade after Chernobyl: summing up the consequences of the accident* Vienna, April 1996

[12] United Nations Scientific Committee on the Effects of Atomic Radiation 1994 *Sources and effects of ionising radiation* UNSCEAR Report (New York: United Nations)

[13] World Health Organization 1995 International Conference on: *Health consequences of the Chernobyl and other radiological accidents* Geneva, November 1995; see also World Health Organization 1995 *Health consequences of the Chernobyl accident: results of the IPHECA pilot projects and related national programmes; summary report* (Geneva: WHO); see also: Souchkevitch G N, Tsyb A F, Repacholi M N and Mould R F (eds) 1996 *Scientific Report* (Geneva: WHO)

[14] Ivanov V K, Tsyb A F, Gorsky A I, Maksyutov M A, Rastopchin E M, Konogorov A P, Biryukov A P, Matyash V A and Mould R F 1997 *Thyroid cancer among 'liquidators' of the Chernobyl accident* Brit. J. Radiology **70** 937–941

[15] National Academy of Sciences, National Research Council, Report of the Committee on the Biological Effects of Ionizing Radiations (BEIR) 1990 *Health effects of exposure to low levels of ionizing radiation* (BEIR V) (Washington, DC: National academy of Sciences)

[16] Hendee W R and Edwards F M (eds) 1996 *Health effects of exposure to low levels of ionizing radiation* (Bristol: Institute of Physics Publishing)

[17] Jablon S and Kato H 1972 *Studies of the mortality of A-bomb survivors. Part 5. Radiation dose and mortality 1950–1970* Radiation Research **50** 649–698

[18] Mould R F 1983 *Cancer Statistics* (Bristol: Adam Hilger)

[19] Hammond E C 1975 in: *Persons at high risk of cancer* (ed) Fraumeni J F (New York: Academic Press)

[20] Royal College of Physicians 1977 *Smoking or Health* (Tunbridge Wells: Pitman Medical)

Chapter 23

[1] Last J M 1995 *A Dictionary of Epidemiology* 3rd edn (New York: Oxford University Press)

[2] Altman D G 1991 *Practical Statistics for Medical Research* (London:Chapman & Hall)

[3] Gordis L 1996 *Epidemiology* (Philadelphia: W B Saunders)

[4] Beaglehole R, Bonita R and Kjellström T 1993 *Basic Epidemiology* (Geneva: World Health Organisation)

[5] MacMahon B and Pugh T F 1970 *Epidemiology Principles and Methods* (Boston: Little Brown)

[6] Clayton D and Hills M 1993 *Statistical Models in Epidemiology* (Oxford: Oxford University Press)

[7] Breslow N E and Day N E 1980 *Statistical Methods in Cancer Research* **1** *The Analysis of Case-Control Studies* IARC Scientific Publication No. 32 (Lyon: International Agency for Research on Cancer)

[8] Pearl R 1929 *Cancer and tuberculosis* Amer. J. Hygiene **9** 97–159

[9] Carlson H A and Bell E T 1929 *Statistical study of occurrence of cancer and tuberculosis in 11,195 post-mortem examinations* J. Cancer Research **13** 126–135

[10] Mellin G W and Katzenstein M 1962 *The saga of thalidomide: neuropathy to embryopathy, with case reports of congenital anomalies* New England J. Medicine **267** 1184–1193 and **267** 1238–1244

[11] Millar J S, Smellie S and Coldman A J 1985 *Meat consumption as a risk factor in enteritis necrotans* Int. J. Epidemiology **14** 318–321

[12] Kannel W B 1990 *Coronary heart disease risk factors: a Framingham study update* Hospital Practice **25** 93–104

[13] Dawber T R, Kannel W B and Lyell L P 1993 *An approach to longitudinal studies in a community: the Framingham study* Annals New York Academy of Science **107** 539–556

[14] Norell S E 1995 *Workbook of Epidemiology* (Oxford: Oxford University Press)

[15] Freedman L S and Rubin S G 1979 *Lung cancer in areas in Liverpool defined by election ward boundaries* Unpublished report

[16] Mould R F, Wrighton K and Pickup D S 1977 *Mortality in a lodging house* Lancet **2** 1503 *and* Mould R F, Wrighton K and Pickup D S 1978 *Down and out in London and Liverpool* New Scientist 9 March 642–643 *and* Tucker A 1977 *Down and out in a unique hostel* Guardian newspaper 12 December

[17] Mould R F 1978 *The Iron Church: a History of Everton and the World's First Cast Iron Church Built 1812* (Liverpool: ESG Press)

[18] Leavitt J W 1992 *Typhoid Mary strikes back: bacteriological theory and practice in early 20th century public health* Isis **83** 608–621

[19] Editorial 1969 *Strange fever* MD Medical News Magazine **13** 220–226

Glossary

[1] Bailar J C and Gornik H L 1997 *Cancer undefeated* New England J. Medicine **336** 1569–1574

[2] Office of Population Censuses and Surveys 1979 *Cancer Statistics Registrations 1971 England and Wales* Series MB1 No. 1 (London: HMSO)

[3] Office of Population Censuses and Surveys 1982 *Cancer Statistics Registrations 1977 England and Wales* Series MB1 No. 8 (London: HMSO)

4 Office of Population Censuses and Surveys 1983 *Cancer Statistics Registrations 1978 England and Wales* Series MB1 No. 10 (London: HMSO)

5 Office of Population Censuses and Surveys 1993 *Cancer Statistics Registrations 1987 England and Wales* Series MB1 No. 20 (London: HMSO)

6 Office for National Statistics 1997 *Cancer Statistics Registrations 1990 England and Wales* Series MB1 No. 23 (London: The Stationery Office)

7 World Health Organisation 1997 *The world health report 1997, conquering suffering and enriching humanity* (Geneva: WHO)

8 Kentucky Cancer Registry 1996 *Cancer incidence report for 1995: age-adjusted cancer incidence rates by county* (Lexington: Kentucky Cancer Registry, University of Kentucky)

9 Tucker T C, Friedell G H and Ross F 1996 *The relationship between smoking and lung cancer in Kentucky* Unpublished report

10 Mould R F 1983 *Cancer Statistics* (Bristol: Adam Hilger)

Index